James Russell Soley

Report on Foreign Systems of Naval Education

James Russell Soley

Report on Foreign Systems of Naval Education

ISBN/EAN: 9783337035648

Printed in Europe, USA, Canada, Australia, Japan

Cover: Foto ©Andreas Hilbeck / pixelio.de

More available books at **www.hansebooks.com**

REPORT

ON

FOREIGN SYSTEMS

OF

NAVAL EDUCATION,

BY

PROFESSOR JAMES RUSSELL SOLEY, U. S. N.

————◆————

WASHINGTON:
GOVERNMENT PRINTING OFFICE.
1880.

LETTER

THE SECRETARY OF THE NAVY,

TRANSMITTING,

*In compliance with resolution of the Senate of January 19, the report of Prof.
James R. Soley on the foreign systems of educating naval officers.*

JANUARY 21, 1880.—Referred to the Committee on Naval Affairs and ordered to be
printed.

NAVY DEPARTMENT,
Washington, January 20, 1880.

SIR: In compliance with a resolution of the Senate, of the 19th
instant, I have the honor to communicate to the Senate herewith " the
report of Prof. James R. Soley on the foreign systems of education of
naval officers."

I am, sir, very respectfully,

R. W. THOMPSON,
Secretary of the Navy.

Hon. WILLIAM A. WHEELER,
Vice-President of the United States.

LETTER OF TRANSMITTAL.

UNITED STATES NAVAL ACADEMY,
Annapolis, Md., November 20, 1879.

SIR: I have the honor to transmit herewith a report on foreign systems of naval education, prepared in compliance with the order of the Navy Department of March 11, 1878, while on duty at the Paris Exposition. In addition to the material obtained at the Exposition, which was not large, I have utilized such official documents and papers as came into my possession from other sources. I also made a personal inspection of the naval schools and colleges of England and France, and of the German practice-ship for cadets.

In my visits to foreign establishments, I was received by the official authorities with the greatest kindness and attention, and given every facility in pursuing investigations. I have to express my obligations to the ministers of the United States in England, France, Germany, and Italy, for obtaining the necessary authorizations and documents to enable me to study the systems of naval education in those countries. I have also to acknowledge particularly the assistance given me by Admiral Sir C. F. A. Shadwell, K. C. B., President of the Royal Naval College, and Dr. Hirst, Director of Studies, at Greenwich; Captain Fairfax, Commander Lord Ramsay, and Mr. Aldous, of the Britannia; Captain Herbert and Commander Acland, of the Excellent; Vice-Admiral Bourgois, Préfet Maritime of Brest, and Captain Guépratte, commanding the Borda; M. A. Lebeau, Chef du bureau des Equipages de la Flotte, at the ministry of marine, in Paris; Lieutenant Collet, of the Polytechnic School; and Captain Zirzow, commanding the German practice-ship Niobe.

I have also to report that the documents, programmes of study, lectures, and other works relating to naval and military education, which I collected at Paris and elsewhere, to the number of about three hundred volumes, have been deposited in the library of the United States Naval Academy.

I have the honor to be, sir, very respectfully, your obedient servant,

J. RUSSELL SOLEY,
Professor, United States Navy.

Hon. R. W. THOMPSON,
Secretary of the Navy, Washington, D. C.

CONTENTS.

PART I.—GREAT BRITAIN.

8 CONTENTS.

PART I.
GREAT BRITAIN.

CHAPTER I.

GENERAL STEPS IN THE CAREER OF LINE OFFICERS.

The regular grades of executive or line officers in the Royal Navy are as follows:

Admiral of the fleet.
Admiral.
Vice-Admiral.
Rear-Admiral.
Captain.
Commander.
Lieutenant.
Sub-lieutenant.
Midshipman.
Naval cadet.

The career of an officer in the Navy begins with his nomination as a naval cadet. These nominations appear to be made at the will of the Admiralty, as far as the selected persons are concerned, and they number about 40 for each class or half-year. After passing the physical and mental examinations at Greenwich, the candidates receive appointments as naval cadets, and they join the Britannia, the training ship, at Dartmouth, in January or July following their examinations, as the case may be. After two years of study on board the Britannia they go to sea, not in any particular ship, but in any of the cruising ships to which they may be appointed. After one year's sea-service they are rated as midshipmen, though the degree of proficiency shown in the Britannia may reduce this time; in fact, in certain cases, extinguish it altogether; in which last case they are rated midshipmen as soon as they graduate. These cases, however, are rare. In all cases except the last, cadets are required to pass an examination for the rating of midshipman.

After five years' service in the Navy, including the time allowed on leaving the Britannia, and after having attained the age of nineteen, a midshipman comes up for promotion. Before promotion, he must, however, pass three examinations. The first is in seamanship, and is conducted on the spot, at sea, wherever the midshipman may be; and, on passing, he receives from the senior officer present an order as acting sub-lieutenant. As soon as he returns to England, he goes to the Royal Naval College at Greenwich for six months, at the end of which he is examined in navigation and in general subjects. After a brief interval, he goes to the Excellent for a course of 65 days,* followed by an exam-

---

* Course of instruction in gunnery ships, 1875, p. 14.

11

ination in gunnery. This is the last compulsory examination in his career as a naval officer.*

Having passed these three examinations, he becomes a sub-lieutenant, that is, one of a body of naval officers who are qualified and ready for promotion to be lieutenants, but are simply awaiting vacancies. In this promotion (from sub-lieutenant to lieutenant) seniority seems to govern largely, though there are a few cases where upper men remain in a lower grade, while their juniors are promoted; and a few junior sub-lieutenants are promoted, while their seniors remain sub-lieutenants; but the mass of promotions consists of men at the head of the list.

The promotions from lieutenant to commander and from commander to captain are by selection, some of them in accordance with special rules fixed by the Admiralty, as in the case of some gunnery lieutenants; but in the case of both promotions, they involve most extensive changes in the seniority of the officers concerned. For example, of the commanders promoted to the rank of captain since January 1, 1874, the one who now (January, 1879) stands seventh on that list was in 1873 No. 103 on the list of commanders; and of the 102 commanders who were then his seniors, only 6 are still above him; 40 are captains junior to him in rank, and 33 have not yet been promoted, the remaining 23 having disappeared from the active list. The changes in the seniority of officers in the promotions from lieutenant to commander are equally marked.

From the grade of captain upward, promotions are strictly by seniority, though, as the Admiralty instructions say, "reserving Her Majesty's undoubted right of selection." Since January, 1874, however, there has been no case in which the seniority of officers in these grades has been permanently altered, except that of the Duke of Edinburgh.

---

* In Fleet Circular No. 41, C, dated May 27, 1879, the Admiralty has announced its intention of instituting a course and examination for all officers in pilotage, subsequent to the gunnery examination. The course will last two months, and will be given on board the Duke of Wellington, flag-ship, at Portsmouth. The examination will be held at the Hydrographic Office, Whitehall. As this provision only takes effect on officers entering the service subsequent to the year 1875, the course will not go into operation before 1883.

I

CHAPTER II.

EXECUTIVE OFFICERS ASSIGNED TO SPECIAL DUTIES.

Besides the ordinary duties incident to the career of an executive officer* in the Navy, officers who are specially qualified, or who have special tastes for one branch of their profession, are given an opportunity and are even encouraged to devote themselves specially to it. At present there are four of these branches, and the number seems likely to increase. They are as follows:

1. Navigating officer.
2. Gunnery lieutenant.
3. Torpedo lieutenant.
4. Interpreter.

Officers choosing one of these branches have to go through a special course, or at least to pass a special examination. They are not exempted from their ordinary duties as executive officers, but they remain in the line of promotion, and they are given certain emoluments and privileges, mainly in the shape of extra pay† and special service; in some cases, also, of more rapid promotion. A distinguishing mark is affixed to their names in the Navy List, and, in general, their extra effort receives full and substantial recognition.

The question of encouraging specialties in naval education and administration is one which has been much discussed, and about which opinions differ; but the full trial which the system has received in England, and its marked success, make it worthy of the most careful consideration. In a profession as varied and as many-sided as that of a naval officer, there is room for the indulgence of every taste, and use for every form of talent. Of course, the first object is to produce officers who can manage ships and fight ships. Seamanship, navigation, gunnery, and steam-engineering, in their broadest sense, form, if not the foundation, at least the essential superstructure, of naval education. Every officer must receive a certain training in these branches to fit him for his ordinary duties and enable him to meet extraordinary emergencies. But there are other duties incident to the naval profession, and useful alike in peace and in war, which call for a high order of special talent and a high degree of special acquisition. For some one or other of these duties or branches, all officers have sufficient time, while tastes and aptitudes vary; and it is to employ this spare time, and to utilize these particular aptitudes, that an opportunity is given in the English service to cultivate specialties. It has not been found that this system

---

* The term "executive officer," as used in the English Navy, and as generally used in this report, refers to officers of the executive branch, or "line officers," as they are called in our own Navy.

† The table given in the Appendix, Note A, shows the rate of extra pay in each case.

injures the general efficiency of individuals who take advantage of it, while it adds incalculably to the strength and usefulness of the service as a whole. It gives the Admiralty men of high scientific training for every kind of work which needs such training; while the presence in the service of officers so trained tends to leaven the whole body and to raise the standard of professional knowledge.

It is upon this principle that the general system of education in the English Navy is largely based. The course of study and practice is uniform for all officers up to the final examination of sub-lieutenants. After that, every facility is given for the development of individual inclinations. Not only the four special branches mentioned above, in following which officers to some extent separate themselves from the line, but the great variety of voluntary elective courses for higher officers, at the College and on board of the Excellent, afford abundant opportunities. And the government, knowing well that special opportunities without special rewards form an imperfect incentive to most men's ambition, has held out many forms of direct, material compensation, in the shape of extra pay, prizes, medals, more rapid advancement, attractive lines of duty, and, by no means least, distinguishing marks on the Navy List.

1.—NAVIGATING OFFICERS.

Until recently the duties of navigation and pilotage were performed by a separate corps of officers, called navigating officers. Of these there still remain a large number in the several grades of staff-captain, staff-commander, navigating lieutenant, and navigating sub-lieutenant. Two of the old grades, navigating midshipman and navigating cadet, have ceased to exist, and the others will gradually pass away.

Formerly the duties of this corps were performed by officers called masters, of whom there were several grades. In 1863, and again in 1867, the corps of masters was reorganized, and its older members became the newly-established staff-captains and staff-commanders. Higher relative rank was given to the navigating officers, and their position was in every way improved; the cadets of the corps being educated with naval cadets in the Britannia, and passing subsequently through a special course of training.*

In 1872 the Admiralty decided to abolish gradually the separate corps of navigating officers, after its short-lived career, and to intrust their duties to line officers, as in the American service; except that certain officers are selected, who are to devote themselves during a certain period to this branch of the profession. Accordingly, the appointment of navigating cadets ceased, the last having been made in January, 1872. The new regulations which, somewhat modified, are now in force, provide that lieutenants under four years' standing, and sub-lieutenants,

---

* Navy List, January, 1871, p. 375. Circ. No. 3, C, January 6, 1870, § 30.

may, on their own application, be appointed to navigating and pilotage duties. They pass the regular sub-lieutenants' examination at Greenwich, and a special examination at the Hydrographic Office on pilotage in general, and the navigation of the English Channel in particular. Before entering on their duties, they must have had one year's sea-service as watch officers. Within five years from their appointment to navigating duties they must pass through a short course in gunnery in the Excellent. They are called on to perform navigating duties while in the grades of lieutenant and commander. But, contrary to the practice in the case of the old navigating corps, they may also be required to act as watch and divisional officers, and they are eligible for promotion to the highest rank in the service, their seniority not being affected by their choice of navigating duties. An arrangement is also made by which the present navigating sub-lieutenants can be transferred, if they prefer, to the executive branch; and the Admiralty is also authorized to transfer to the executive branch a small number of navigating officers of the grades of staff-commander and navigating lieutenant, who have distinguished themselves by some special service.†

It has very recently (May 27, 1879) been decided to establish a course of instruction for navigating officers on board the flag-ship at Portsmouth. This course lasts two months, and is preliminary to the pilotage examination at Whitehall.

2.—Gunnery and Torpedo Officers.

Lieutenants who desire to devote themselves to these branches of the service receive permission to qualify, on their own application, when recommended by the captain under whom they are serving. Candidates who have not previously served one year at sea as lieutenants are appointed to a sea-going ship, to complete that period as watch officers. The special school for gunnery officers is the Excellent, and for torpedo officers, the Vernon, both stationed at Portsmouth.†

The total period of instruction for gunnery lieutenants, including a vacation of a month in July and September, is about twenty months, distributed as follows:

Theoretical course at Greenwich, 9 months; vacation, 1 month; torpedo instruction at Portsmouth, 2 months; practical gunnery course in Excellent, 6 months; attendance at Royal gun factories, carriage department, and laboratory, at Woolwich, 3 weeks; review and examination, 3 weeks; leave, 1 week; total, 19 months, 3 weeks.

If it is found during any part of the course that a lieutenant is not likely to prove efficient as a gunnery officer, his name is submitted to the Admiralty, with a view to his removal from the books of the Excellent.

---

*Regulations of 1879, § 239.
† Course of instruction in gunnery ships, p. 17.

At the final examination in each course (theoretical and practical), two certificates are given, as follows:

THEORETICAL COURSE.

First class.—About 60 per cent. of the maximum.
Second class.—About 30 per cent.

PRACTICAL COURSE.

First class.—About 80 per cent.
Second class.—About 60 per cent.

FINAL CERTIFICATES.

First class.—First class in both theoretical and practical courses.
Second class.—First class in either course and second in the other.
Third class.—Second class in both courses.

The duties of gunnery officers are largely instructional, and they may perform watch and divisional duty with the other lieutenants of the ship in which they are serving. They receive extra pay according to the certificate given them at their passing examination. They are picked and selected men; and they may be said to be to the Navy what the engineers are to the Army—the scientific corps of the service.

3.—INTERPRETERS.

The examination for interpreters is held at the Royal Naval College, and is open to all officers below the rank of commander, on their own application.

The application states the language or languages chosen by the candidate, the choice being open between French, Spanish, German, Portuguese, Italian, and such others as may be designated by the Admiralty. The examination is oral and written, and of a searching character. The branches of the examination and the marks given for each are as follows:

> *a.* Oral:
>
> 1. Pronunciation and accent 50
> 2. Facility of understanding the language 75
> 3. Accuracy of expression 100
> 4. Fluency 75
> 5. Extempore translation:
>   (*a*) From the language into English 75
>   (*b*) From English into the language 100
>
> > *b.* Written:
>
> 6. Writing from dictation 75
> 7. Idiom:
>   (*a*) General 75
>   (*b*) Maritime 100
> 8. Composition 100
> 9. Grammar 125
> 10. Translation:
>   (*a*) From the language into English 100
>   (*b*) From English into the language 150
>
>   Total 1,200

Candidates are not qualified unless they obtain 50 per cent. in each subject, and 58⅓ per cent. of the aggregate; 75 per cent. gives a first class, 58⅓ per cent., a second-class certificate. Rejected candidates are re-examined six months later, but the second examination is always final. The names of officers who pass successfully are noted for employment as interpreters, in flag or senior officers' ships, each of which, on foreign stations, is allowed one officer who has qualified in the languages spoken within the limits of the command. Officers so employed act as interpreters in addition to their regular duties, and receive extra pay according to their certificate.

CHAPTER III.

ORGANIZATION AND EDUCATION OF THE STAFF AND CIVIL CORPS.

1.—ENGINEER SERVICE.

The grades in the Corps of Engineers are :
Chief inspector of machinery,
Inspector of machinery,
Chief engineer,
Engineer,
Assistant engineer.

To these may be added the grade of engineer student, which corresponds in some measure to that of naval cadet.

Vacancies for appointment as engineer student are open to public competition. The candidates must not be more than 16 nor less than 14 years of age, and must be the children of British subjects. They must also satisfy the Admiralty with regard to "respectability, good character, and physical fitness," the last being tested by a medical examination. The mental examination is held in May of each year by the Civil Service Commissioners. To avoid the expense incurred by candidates in traveling to any one place, simultaneous examinations are held in London, Liverpool, Portsmouth, Devonport, Bristol, Leeds, Newcastle-on-Tyne, Edinburgh, Glasgow, Aberdeen, Dublin, Belfast, and Cork. This provision is specially important in an examination which is open to unlimited competition, where a candidate can form no possible estimate of his chance of success ; and its absence would undoubtedly, as in the case of the examination for cadet engineers in the United States Navy, prevent many excellent candidates from presenting themselves.

The subjects for examination, and the corresponding marks, are as follows :

Arithmetic	300
Dictation	100
Composition	100
Grammar	150
Translation from French	100
French grammar	50
Geography	100
Algebra (including quadratics)	300
Geometry (Euclid, first six books)	300
Total	1,500

Candidates who fail to pass in the first three subjects, or in reading aloud, are disqualified, and their other papers are not examined. Those who show a competent knowledge of all the subjects, and who obtain an aggregate of not less than 50 per cent. of the maximum, are classed in one general list in order of merit, and are eligible for appointment as engineer students in one of the dockyards, according to the number of appointments which it is decided to make in that year. The successful candidates are entered as engineer students before July 1 of each year, and are required to join with their parents in a bond for £300 to enter, if required, into the naval service, as assistant engineers, at the end of their period of training. Parents of engineer students are required to pay £25 a year for each student during the first three years of his training, payable each year in advance. In case of failure of payment, the student is discharged. Board and lodging are, however, provided by the government, and students are required to reside in the dockyard.

The course in the dockyards covers six years, and is mainly devoted to practical training in the workshops and to instruction in iron ship-building. The students also attend the dockyard schools, where they have an extensive course in mathematics, and spend a portion of their time in the drawing office. Means are afforded them of acquiring the groundwork of the knowledge required by a naval engineer in regard to the working of marine engines and boilers, including those repairs that can be carried out afloat; the practical use of the instruments used in the engine-room, including the indicator; and, in general, they become acquainted with the duties of a naval engineer. While undergoing this course the students are under the supervision of a captain of the Steam Reserve and a staff of officers; and instruction in ship-building is under the direction of the chief constructor of the dockyard.

Engineer students are examined yearly under the direction of the President of the Naval College. They are also examined at the end of their fourth, fifth, and sixth years of service, by the engineer officers of the Admiralty, as to their practical knowledge of steam machinery. Two prizes are given annually in each dockyard to the students who show the greatest skill as workmen. Practical engineering is an essential subject of examinations, and students have to get 50 per cent. of the maximum in this branch in order to pass; students who fail to get a passing mark are allowed another year, and a re-examination; but those who fail a second time are dropped. Those who pass successfully at the end of the six-year course in the dockyards are admitted to the Naval College at Greenwich, for a theoretical course of one term, as acting assistant engineers. On the completion of this course, they are commissioned as assistant engineers, and are sent to sea.

Two assistant engineers are chosen annually from those who complete the course at Greenwich, to remain for two terms longer, taking a far higher course of scientific instruction. After graduation, they serve for one year at sea, and then become eligible for positions as constructing engineers at the Admiralty and in the dockyards. By this ingenious but very simple device the great body of engineer officers get a sufficient training, while the government has at its disposal in the same corps a few men of the highest scientific attainments, to fill those positions in the service where such attainments are needed.

No assistant engineer who has passed three terms at Greenwich is allowed to leave the service within seven years of the completion of his course, except upon payment of £500 to defray the cost of his education.

Test examinations are held for promotion to the grades of engineer and chief engineer.

2.—CONSTRUCTORS.

Though there is no organized corps of constructors in the English Navy, with a line of promotion and relative rank, like the engineers, yet the civil officers employed at the Admiralty and in the dock-yards to superintend the work of construction resemble in some respects such a corps, and perhaps will sometime be organized in such a way. At the head is the Director of Naval Construction, one of the most important officers in this department of the government; with him is associated at the Admiralty a staff of chief constructors, assistant constructors, examiners of dockyard and contract work, engineer inspectors, and draughtsmen. Each dockyard has also its chief constructor and its constructors. These officers, who have hitherto been drawn mainly from graduates of the School of Naval Architecture, will now be supplied by a small body of students, who, beginning their career as dockyard apprentices, are admitted, to the number of three a year, to a three-years' course at Greenwich; this course and the long course for assistant engineers are the highest pursued at the college, and they are of the very first importance to the naval service. To show the simple and admirable method by which these students are obtained, it will be necessary to go with some detail into the appointment and training of dockyard apprentices.

As in the case of engineer students, vacancies for appointments as apprentices at any of the five dockyards are open to public competition. Applications must be sent by May of each year to the superintendents of the dockyards, by whom lists of candidates are kept. Candidates must not be under 14 nor over 15 years of age; must give proof of age, character, &c.; and their physical fitness must be determined by a board of medical officers. Examinations are held simultaneously in June of each year, by the Civil Service Commissioners, at London and at the dockyards. The subjects and marks for the examination are as follows:

Arithmetic	350
Orthography	100
Handwriting	100
Grammar	100
English composition	100
Geography	100
Euclid (first three books)	150
Algebra, including quadratics	150
Total	1150

Candidates have a preliminary examination in the first four subjects, and, if they fail in any one of them, they are at once rejected. Those who pass the preliminary test then undergo a competitive examination, and, if they show a competent knowledge of all the subjects, they become eligible for appointment as apprentices in the various trades, according to their position on the examination lists, at the various dockyards. They are bound by indenture to serve in this capacity for seven years, at the end of which they receive a certificate of their character and conduct, and the progress they have made in their trade and in the knowledge of the prescribed subjects.

Three from among those who have passed five years at the dockyards are selected annually by competitive examination for study at the Naval College at Greenwich. They are entered as students in naval architecture, and they remain at college three terms, passing the vacation (July 1 to September 30) at one of the dockyards. The course at Greenwich is similar to the long course for selected engineers, and is of a very high professional and scientific character.

On passing the examination at the end of the course, the construction students may be sent to sea for a year,* after which they are appointed to some post for which they are fitted, at first, perhaps, as assistants to the foremen at the dockyards, and later to positions as constructors or draughtsmen, at the yards or the Admiralty.

On their entry into the college, they are required to give a bond for £250 to serve under the Admiralty for seven years after completing their apprenticeship.

Both dockyard apprentices and engineer students have instruction in the dockyard schools, comprising algebra, descriptive geometry, analytical geometry, calculus, and mechanics.

Half-yearly examinations are held, the papers for which are sent down by Dr. Hirst, the Director of Studies at Greenwich, and the examinations are supervised by the dockyard chaplains. The dockyard schools are also personally inspected from time to time by Dr. Hirst.

3.—CHAPLAINS AND NAVAL INSTRUCTORS.

There are, properly speaking, no grades in either of these corps; nor is there any specific examination for an appointment as chaplain. Naval instructors, however, are subjected to strict examinations before appointment; and chaplains frequently—in fact, in the majority of cases—qualify for this examination, and hold the two positions jointly, and perform the duties of both, throughout their whole career. There are now on the Navy List 96 chaplains and 71 naval instructors, of whom 46 persons hold both positions. Of the 71 naval instructors, 1 is the Director of Education, 30 are attached to the ships bearing midshipmen and naval cadets, 5 are in the training ships for boys, 9 instruct in the Britannia, 6 at the Naval College, 13 are performing duty as chaplains, and the remainder, 7, are unattached.

---

*This rule is not generally observed, however. Evidence before the Greenwich committee, Q. 1852.

Candidates for appointment as naval instructors must not be under 20 nor over 35 years of age. They must pass a preliminary examination in arithmetic, algebra, geometry, trigonometry, elementary mechanics, Latin, and French; but university graduates, who have passed with a certain distinction, may dispense with all the subjects but French. On passing the examination, candidates are. admitted to the Royal Naval College, and remain there one session of nine months, passing an examination at its close.

4.—MEDICAL SERVICE.

Although the education of medical officers in the Royal Navy forms no part of the scheme laid down for other officers, yet as it forms a branch of education in the naval service, it may be proper to give an outline of it here. The grades of medical officers are—

Inspector-general of hospitals and fleets.

Deputy inspector-general of hospitals and fleets.

Fleet-surgeon.

Staff-surgeon.

Surgeon.

Candidates for the medical service must not be under 20 nor over 28 years of age, and must have been licensed to practice under the Medical Act. Their physical fitness is tested in a preliminary examination. In the professional examination the principle of limited competition is introduced. The candidate is first examined in the required subjects, and he may then undergo a voluntary examination in elective subjects, the marks for which are added to those given in the obligatory examination, and assist materially in determining the successful competitor. The required subjects are—

Anatomy and physiology.

Surgery.

Medicine, including therapeutics and diseases of women and children.

Chemistry and pharmacy, and a practical knowledge of drugs.

The required examination is partly practical, and includes operations on the dead body, the application of surgical apparatus, and the examination of medical and surgical patients at the bedside. .

The voluntary examination includes comparative anatomy, zoölogy, natural philosophy, physical geography, botany with special reference to materia medica, French, and German; and any of the subjects may be selected by the candidate.

After passing this examination, candidates are required to attend the course of practical instruction in the medical school at Netley on—

1. Hygiene.

2. Clinical and naval and military medicine.

3. Clinical and naval and military surgery.

4. Pathology of diseases and injuries incident to the naval and military service.

On passing the examination at the close of the Netley course, and not till then, candidates are eligible for a commission as surgeon in the Navy.

After completing three years' full-pay service, surgeons may be examined for the rank of staff-surgeon, but they cannot be promoted until they have served five years, two of which must be in a ship actually employed at sea. Certain peculiar points are to be noted in this examination. A number of written questions, framed by the professors at Netley, and approved by the Director-General of the Medical Department of the Navy, are forwarded, sealed, twice a year, in January and July, to Haslar and Plymouth Hospitals, and to each of the foreign hospitals. The necessary arrangements are made and a day fixed by the commander-in-chief on the station and the chief medical officer, and notice is given to surgeons who are eligible. On the day fixed the candidates are assembled, and the papers are opened in their presence by the chief medical officer. At the close of the examination the work-papers, signed and sealed by the writers, are delivered to the senior medical officer, who transmits them, unopened, to the Director-General of the department, under whose directions they are finally examined.*

5.—MARINE ARTILLERY AND LIGHT INFANTRY.†

Admission to the Marine Artillery as probationary lieutenants is offered to the successful candidates in order of merit, according to the number of vacancies, at the open competitive examination for admission to the Royal Military Academy at Woolwich. This examination is held twice a year, in July and December, by the Civil Service Commissioners. The limits of age are from 16 to 18, and of height 5 feet 5 inches. The successful candidates are appointed lieutenants on probation, and join the Naval College at Greenwich for a course of two years.

On passing out of the College the probationary lieutenants are sent to the Excellent for a course in gunnery and torpedo instruction, at the conclusion of which they join headquarters and are instructed in their drill and military duties, for service ashore and afloat.

Admission to the Marine Light Infantry is obtained in the same way as that to the Marine Artillery, by open competition, except that the examination is held at Sandhurst instead of Woolwich. The limits of age are from 17 to 20, but they are varied for university graduates and lieutenants in the militia who are eligible for Army commissions. The successful candidates are appointed lieutenants without any further trial, seniority being determined as usual by the order of merit at the examination. On passing they join their divisions at once, and are instructed in their drill and military duties, going through a course of garrison instruction.

---

* Full information in regard to medical education in the English Navy will be found in the admirable report of Dr. Richard C. Dean, Medical Inspector, U. S. N., published by the Bureau of Medicine and Surgery.

† Regulations, May 14, 1877, Navy List, p. 501.

6.—PAYMASTERS, SECRETARIES, AND CLERKS.

The duties pertaining to these civil branches of the Navy are performed by what is essentially a single corps, though with a somewhat loose organization, having apparently a regular line of promotion through the four grades of paymaster, assistant paymaster, clerk, and assistant clerk. In general, the course of this promotion seems to be tolerably regular, though officers are sometimes appointed in other grades than the lowest. The clerks and assistant clerks act not infrequently as assistant paymasters, and they perform duties at sea similar to those of captain's and paymaster's clerks in the United States Navy. Both paymasters and assistant paymasters may also be detailed as secretaries to flag officers. A special examination enables them to qualify as secretaries, and they may also qualify as interpreters in the same manner as line officers.*

Two examinations for assistant clerkships are held semi-annually at the Naval College at Greenwich, in June and November. The limits of age for candidates are from 15 to 17 years, and they must produce certificates of birth, good conduct, and good health. They must also pass a medical examination.

The mental examination is competitive, the number of competitors for each examination being fixed by the Admiralty. It consists of two parts; a test examination and a voluntary examination for competition. The subjects and relative weight are as follows:

I.—TEST EXAMINATION.
Marks.
1. Writing from dictation in a legible hand 100
2. Writing a letter on a given subject 75
3. Writing the substance of a chapter or portion of a chapter read out, taking into consideration the time in which this exercise is performed............. 75
4. French; reading and translation from French into English, and from English into French, and grammar.. 150
5. Addition, simple and compound, with reference to time 50
6. Arithmetic generally 250
7. Modern geography and English history... 150
8. Scripture 100

II.—VOLUNTARY EXAMINATION.

9. Elementary mathematics, viz., algebra, including quadratic equations and problems producing them, and the first three books of Euclid 200
10. Latin; translation of passages from books usually read at schools, translation of English into Latin, and grammatical questions 200
11. The German, Spanish, or Italian languages, as in French 100
12. Elementary physics, viz., chemistry, heat, and properties of solids and fluids, electricity and magnetism.. 150
13. Drawing; free hand and from models 100

In the test examination, 40 per cent. is required in each subject in order to pass. Of the voluntary subjects not more than three may be taken, unless drawing be one, in which case four may be taken.

---

* Regulations of September, 1874, Navy List, p. 501.

Not less than 20 per cent. must be obtained in any one of the voluntary subjects in order that it may be reckoned towards the total. If a candidate fails to pass the test, he cannot appear at any subsequent examination; but if he passes the test, and yet is unsuccessful in the competition, he can compete once again at the following examination, even though he may be over age.

Test examinations are held for promotion to the grades of clerk and assistant paymaster.

CHAPTER IV.

The way in which the present complicated system has grown up can only be fully understood by reference to former regulations.

The Naval Academy was first established at Portsmouth dockyard in 1729, for the education of 40 students. The age at admission was between 13 and 16. In 1806 the name of the school was changed to the "Royal Naval College," and in 1816 the age was fixed at from 12½ to 14 years. The course lasted two years, and comprised various branches of elementary mathematics and English studies, somewhat similar to the present Britannia course. After leaving the college, the students served for a year as "volunteers of the first class," on board cruisers, and were then rated as midshipmen. After six years' service as midshipmen, and after passing an examination in seamanship and navigation, they became mates (the present sub-lieutenants), and were eligible for promotion to lieutenants. During the term of service at sea, some little instruction in navigation was given by the chaplains, or naval instructors, if there happened to be any on board.

Only a part of the young officers of the Navy went through the course at the Naval College, and those who did had no incentive to continue their studies after they left it. Accordingly, in 1837, the college was abolished, and the efforts of the Admiralty were directed towards the improvement of the corps of naval instructors.

In 1839 the Royal Naval College was again opened, but on an entirely different basis; in fact, it was practically another establishment. It was to provide "further means of scientific education" for a certain number of officers and mates, the latter of whom studied at the college for a year. At the same time the instruction given on shipboard was improved and broadened. By subsequent orders the college was extended so as to take in, in a certain measure, students in the higher ranks of the Navy and marines, officers qualifying for the marine artillery, masters, naval instructors, and engineers. Its intention was to teach advanced pupils, and it corresponded to the present college at Greenwich, as its predecessor had corresponded to the Britannia.

But the Admiralty, which, in 1837, discovered the want of higher education in the Navy, and to that end abolished the old college, in 1857 discovered the want of elementary training, and again opened a junior school, this time, however, without abolishing the other. The new school was the beginning of the present Britannia system, though much has since been changed in details. It comprised a stationary training ship, an easy entrance examination, and a course of fifteen months, afterward lengthened to two years. The limits of age at admission were fixed at 13 and 15 years, which were changed in 1859 to 12 and 14, then to 12 and 13, and lastly to 12 and 13½. In 1868 a special sea-going training ship was attached to the school, but this has since been discontinued, and cadets are now sent to sea in every variety of large cruiser. The course in the sea-going ship lasted a year. The examination for admission to the school was competitive, only half the number of candidates examined receiving appointments. The number nominated varied from 40 to 80, and the number appointed was always one-half; but competition was entirely done away with in 1875.

After leaving the special training ship, cadets were rated midshipmen and began their regular duties in ships of the fleet. Here they had still some limited instruction from naval instructors, or navigating officers, or officers specially detailed for the duty. A half-yearly examination of a somewhat crude character was held by the captain, and at the end of two years and a half (later eighteen months) midshipmen passed the thorough intermediate examination in navigation, chart-drawing, surveying, steam, French, and seamanship. In 1873 both the half-yearly and the intermediate examinations were discontinued, and in their stead full examinations were held in January of each year. These new annual examinations differed from the intermediate examinations in several points, but chiefly in the addition to the required subjects, of arithmetic, algebra, geometry, trigonometry, mechanics, and hydrostatics. The change was made on account of the general complaint that junior officers forgot or neglected the elementary mathematics they had already learned. In 1875 the annual examinations were placed in July, and the half-yearly examinations were revived in December, a regulation still in force. Meanwhile a more important change had been accomplished in the final abolition of the Naval College at Portsmouth, which had been in existence since 1839, and the opening of the new college at Greenwich, with a vastly improved organization, on the 1st of February, 1873.

It will be well to notice in this connection the School of Naval Architecture, first opened in 1811 at Portsmouth, and closed, for no particular reason, in 1832. It was reorganized, with considerable changes, as the Central School of Mathematics and Naval Construction, and closed, with as little reason as before, about 1853. In 1864 a third school was opened at South Kensington, which, in 1873, was united with the Naval College at Greenwich; and this last organization bids fair to be permanent.

It will be seen from the above sketch of the history of naval education in England that, while there has been undoubted progress, it has been after a long series of changes, experiments, renewed experiments, and expedients of all kinds, from which even now it cannot be said that a harmonious or satisfactory system has been evolved. In fact, it is rather a combination of makeshifts, resulting from a series of tentative and spasmodic efforts in almost every form which naval education is capable of taking. The naval administration never seems to have looked at the subject as a whole, from the beginning in the entering examination of cadets to the final stage at the promotion to sub-lieutenants, and to have worked out a systematic plan which should have both consistency and coherence. It appears rather to have adopted from time to time such partial views as were presented to it by advocates of a particular theory, by officers who leaned one way or another, a process which has sometimes resulted in its going back upon its own tracks, and making experiments which had been already proved failures. This is partly due to the want of attention hitherto given to the subject, a want which is now in a fair way to be met. Every year more is to be heard in the way of discussion of naval education, and every year more comprehensive and reasonable views seem to gain ground. That the government is likely to stop at its present stage in reforming the education of officers is very improbable; and as the Naval College at Greenwich is now firmly established, it will hardly be many years before further, and perhaps more radical, changes take place in the English system.

CHAPTER V.

The training school for naval cadets is at Dartmouth, a picturesque old town in South Devonshire, on the river Dart, about two miles from its mouth. Two old ships of the line, the Britannia and Hindostan, are moored head and stern in the stream, and on board of these two ships the cadets sleep, study, and live. They go ashore only for amusement, or in case of serious illness, the only parts of the establishment which are on land being the hospital, gymnasium, bowling-alley, park, and cricket-field. One-half of the cadets sleep in the Hindostan, and all have meals and musters in the Britannia, the two ships being connected by a bridge. The Britannia has six studies and one large lecture-room under the poop. The main deck is the sleeping place for the two upper forms, with baths in the bow, and the captain's cabin is aft on this deck. The middle deck is used for muster and inspection, and has the wardroom and cabins in the after part. The lower deck aft is devoted to the cadet's mess. The orlop contains a model-room and officers' cabins, and is also the men's berth-deck.

In the Hindostan, which is a two-decker, there are, as in the Britannia, six studies under the poop. On the main deck are officers' cabins and one study and place for muster. The lower deck is the berth-deck for the two lower forms.

All the masts and spars are removed from the two ships, except the foremast and bowsprit of the Britannia, which are set up and fully rigged.

Attached to the Britannia as a tender is the Dapper, a screw gun-boat, with engines of 262 horse-power, bark-rigged, and used for exercises in seamanship. There are also two launches, schooner-rigged, a schooner-yacht of 50 tons, six launches, and thirty gigs and dingeys; the last, for amusement.

The officers of the establishment consist of a captain, commander, two staff-commanders, and three lieutenants; a chaplain, and the requisite number of surgeons and paymasters; one chief naval instructor, eight naval instructors, two French masters, two drawing masters, and one Latin master; and warrant officers, comprising gunner and carpenter, and four boatswains. There are also three or four officers attached to the Dapper.

1.—EXAMINATION FOR ADMISSION.

Cadets are nominated by the Admiralty. The number seems not to be prescribed by law, but averages about 43 at each half-yearly examination for admission. The examinations for admission are held, as a matter of convenience, at the Royal Naval College at Greenwich, on the third Wednesday in June and the last Wednesday in November; but

the appointments date from the 15th July and January following. The
limits of age are fixed at not less than 12 nor more than 13½ years, at
the date of appointment.

A medical examination of the usual kind is first held, at which it
must appear that the candidate is free from any physical defect of body,
impediment of speech, defect of sight or hearing, and predisposition to
constitutional disease; and he must be generally active and well-devel-
oped for his age. Candidates rejected at the medical examination are,
upon approval by the Admiralty, finally excluded from the Navy.

The mental examination covers three days, and is conducted by the
Admiralty examiners, under the Director of Studies. It includes the
following subjects:

Hours.
1. Writing English from dictation.... ....
2. Reading and oral parsing...... ....
3. Arithmetic............... .... 2¼
4. Elementary algebra.... .... 1¼
5. Elementary geometry... 1
6. Latin.... 2
7. French.... .... 1½
8. Scripture history...... .... 1½

Candidates are required to make 40 per cent. on each subject; and
those who fail are allowed to come up a second time at the next examination, six months later. A third trial is never allowed, nor a second
trial if the candidate is over thirteen at the date when he should, if successful, have entered.

The character of the examination is simple, and the standard may
fairly be called high for boys of this age. The third subject, arithmetic,
includes proportion and vulgar and decimal fractions, and there are no
puzzling or difficult questions, nor any involving long calculations. No.
4, algebra, includes simple equations with one unknown quantity; and
the questions are chiefly simple examples in the four elementary processes,
with one or two very easy equations. No. 5, geometry, includes definitions, axioms, postulates, and demonstrations of the first twelve propositions in the first book of Euclid's Elements. The paper in Latin, No.
6, consists of translation of a passage from Cæsar or Nepos, the explanation of some of the more common constructions, and a few simple
sentences in Latin composition, and is a thorough test as far as it goes.
In French, No. 7, the paper is also elementary, and omits translation into
French, but includes grammar. The use of dictionaries is allowed in
both these examinations.

As has already been stated, the examination for admission, until the
year 1875, was competitive, the number of candidates designated being
double the number of vacancies. The system was changed, owing to the
severe effects of such a competition upon the nervous system of boys
of that age, and the excessive cramming that it fostered.

It should be added that seven nominations are given annually in the
colonies. In these cases, candidates pass their examination on board

the flagships abroad, and are then sent to England to join the Britannia, either in a returning man-of-war or in a mail steamer.

2.—COURSE OF STUDY.

The length of the course is two years, or four terms of five months each. The terms are from about the 1st of February to the 15th of July, and from the end of August to the 20th of December. There are three vacations: six weeks at midsummer, five weeks at Christmas, and two at Easter.

There are four forms,—or classes, as they would be called in America,—corresponding to the four terms; the first form being composed of cadets admitted last. The cadets are also divided into two watches, and each form is subdivided into two classes, half of each class being in one watch and half in the other. The eight naval instructors have charge of the eight classes in their allotted branches, and each instructs his own class during the whole time it remains in the ship, taking the two watches of the class alternately. The natural objections, in regard to unequal marking, &c., that present themselves to such an arrangement, are met by the fact that the real test of a cadet's work is the final examination, which is conducted and marked by the examiners sent down annually from the Royal Naval College for the purpose.

There are no recitations, in the American sense of the word, on board the Britannia, but the time passed with instructors is devoted to study, oral instruction, oral questioning, and practice, in an informal manner, according to the discretion of the instructor and the needs of his class. This time occupies twenty-eight hours a week—three hours every morning, and two and a half hours every afternoon, except Wednesday and Saturday. One hour on every day, except Saturday, is devoted to evening study. There is also a period of early morning study, for half an hour before breakfast, for the two upper forms.

The various branches of study are arranged in two groups for convenience of organization.* The first group comprises—

 Arithmetic.
 Algebra.
 Geometry.
 Trigonometry, plane and spherical.
 Navigation and nautical astronomy.
 Dictation.
 Essay writing.

The second comprises—

 Instruments, chart drawing.
 French.
 Drawing.
 Seamanship.
 Latin.
 Astronomy and physical geography.
 Natural philosophy.

---

* The tabular programme of study is given in the Appendix, Note B.

Of the twenty-eight hours of instruction, fourteen are given to each group—that is, three mornings and two afternoons; the watches alternating in the different periods between the two groups of study.

The following tables will show the distribution of time for the twenty-eight hours of mental work:

I.—WATCH IN STUDY.

Subjects.	Number of hours in each form.			
	First form.	Second form.	Third form.	Fourth form.
	Hours.	*Hours.*	*Hours.*	*Hours.*
Arithmetic	3	2½	....	....
Algebra	3	3	3	3
Euclid	3	2½	2	2
Plane trigonometry	3	3	3	2½
Spherical trigonometry	....	....	2	3
Dictation	1	....	....	....
Navigation	....	2	} 3	2½
Nautical astronomy	....	....		
Essay	1	1	1	1
Total	14	14	14	14

II.—WATCH OUT OF STUDY.

Instruments	....	1½	2¾	2¾
French	3½	3½	3½	3½
Drawing	2¼	2	2	2
Seamanship	4¾	3½	3	3
Latin	2¼	2¼	2	2
Physical geography	1¼	....	....	....
Astronomy and dictation	....	1¼	....	....
Physics	....	....	¾	¾
Total	14	14	14	14

NOTE.—For no very apparent reason the cadets engaged with the studies of Group I are said, in the official language of the school, to be the "watch in study"; while those engaged with Group II are said to be "out of study."

The instruction of the watch in study, or, in other words, all the instruction given in the first group of studies, is given by the eight naval instructors. The hour for evening study is also devoted to these subjects, under the direction of the naval instructors, each instructor directing his own pupils. The "early morning" study is given on three days in the week to seamanship, on one day to drill, and on the other two it is occupied in the same way as the evening study. This gives the naval instructors an aggregate of thirty-four hours a week of work with their students, an amount of work for which men of the same attainments could not be obtained in America.

The principal naval instructor is charged with the supervision of all instruction given by naval instructors and masters. Except in the lectures in physics, he gives no direct personal instruction. It is a part of his duty to visit frequently all the class rooms, and he regulates their police and discipline, under the captain.

No studying is done except at the prescribed times, and cadets are even obliged to have permission to take and use their text-books out of these hours. No marks are given in recitations, if that name can be applied to the questions asked by instructors during the period. The instructors merely keep a journal containing memoranda of the work done, with remarks as to the ability, progress, and conduct of the cadets under their charge. A daily report is made to the principal naval instructor of the attention of each cadet, and this report is inspected by the captain. Monthly reports are also made of the progress of each cadet. Cases of serious inattention are punished by one or two hours of extra drill. For occasional neglect, or for trifling offenses in the class-room, the principal naval instructor may stop the leave of cadets, except for one hour of exercise. On such occasions the delinquents are assembled to write impositions under the surveillance of the cadet sergeant-major.

The courses in Group I would not be considered difficult for ordinary students; but they must put to a severe test boys of from 13 to 15 years of age, whose only preliminary training is that indicated by the examination for admission. This is especially true of the subjects of plane and spherical trigonometry, theoretical and practical navigation, and nautical astronomy. The theoretical navigation includes astronomical geography, plane and middle latitude sailing, great circle sailing, and the Nautical Almanac. Practical navigation includes doing a day's work, and finding ship's position. The other subjects are algebra, arithmetic, and elementary plane geometry.

Of the subjects of Group II, instruction in charts and instruments is given by the staff-commander. It is confined to the three upper forms, and comprises the construction and use of charts, of the sextant, azimuth compass, theodolite, barometer, and thermometer. The cadets take and work out their own observations for latitude and longitude, error and rate of chronometer, &c. During winter the staff-commander is allowed to take cadets of the two upper forms from other instruction, to take observations, when the weather is particularly favorable. Each cadet is required to have his own sextant.

Instruction in seamanship is in charge of the senior lieutenant, assisted by the other executive officers, and by the warrant and petty officers. The officer of the day visits all branches of seamanship instruction frequently during study hours; and, assisted by the signalman, gives instruction in signals. An extra hour, two afternoons in the week (making thirty hours a week of instruction), is given to practice in signals in the third and fourth forms. This takes place between 5 and 6 p. m., on the middle deck of the Britannia.

About one-eighth of the whole working time of the Britannia is given to what is generally called theoretical seamanship. It is chiefly a course of book-and-model work, if we except knotting and splicing, boat-sailing, and some exercises with spars and sails in the Dapper, towards the end of the course. Of practical seamanship, meaning thereby the manage-

S. Ex. 51——3

ment of a ship under sail or steam, there is nothing in the Britannia course.

The text book is Nares's Seamanship. The first term course comprises the naming and identification of the parts of the ship, spars. and sails, and standing rigging; the points of the compass, signal-pendants, knotting and splicing, pulling boats, and steering boats under oars. The second term is devoted to the fitment of rigging on lower masts and yards, and bowsprit; names and uses of parts of the running-rigging; tackles, blocks, seizings, log, and lead; a little further elementary knowledge of the compass and of signals, and questions on sailing in the launch. In the third term, instruction is given in rigging spars generally; setting up rigging on the models; lead of running-rigging; the fitment of sails; working anchors (model); boat-sailing in general, and the rule of the road. Instruction in furling sails, and in working masts and yards, is given on board the Dapper. The fourth term is occupied with a general review.

Lectures in elementary physics are given in the "science room" by the chaplain and principal naval instructor to the third and fourth forms, one lecture a week to each. These lectures, which last one and one-fourth hours, take place during the regular hours of instruction, the cadets assembling at convenient times from the other class-rooms. The subjects of instruction are—

THIRD FORM.—Mechanics, Hydromechanics, Pneumatics, Acoustics.
FOURTH FORM.—Heat, Light, Magnetism, Electricity.

Instruction in Latin, French, and drawing is given by the masters in those branches. In Latin, the examination is not one to give any difficulty even to negligent students. It consists of a passage from the text that the class has read, and another easy passage, new to the pupils, for which they are allowed a dictionary. There are some questions in grammar, and a few easy English sentences, on the model of those they have studied, to be turned into Latin. The French examinations, conducted by the professor of that language at the Naval College, are more searching.

3.—EXAMINATIONS.

Examinations are held twice a year, at the end of each term. The papers are set and the examinations are conducted by the examiners sent down from the Naval College at Greenwich, under the supervision of Dr. Hirst, the Director of Studies, who also conducts personally part of the *viva voce* examination. The examinations are of great importance in the Britannia course, as they constitute the ultimate test of proficiency, the only exercise at which marks are given; and they alone determine the seniority of the cadets and their ability to remain at the school and continue the course. The final examination also fixes the amount of sea-service time allowed to each cadet, and consequently the date of his promotion to midshipman.

Cadets are required to obtain a certain percentage of the maximum; but the passing mark is so low in all branches except seamanship that very few can possibly fail to get it. In the latter branch, the passing limit is 60 per cent. for each form. In the other professional subjects, and in all branches of mathematics, students are required to get 30 per cent. in the lower forms and 40 per cent. in the fourth form. Deficient cadets of the lower forms are warned by the Admiralty, and a failure on a second occasion causes their dismissal from the service. Cadets deficient at the final examination are turned back for a term; or, at least, they cannot go out into the service until they have passed. With these rules there is no reason why any lad, unless guilty of the grossest negligence, should fail to pass the Britannia course.

Certificates of proficiency are given, at the final examination, of three different classes, the first, second, and third. The certificates have a substantial value beyond the honor they confer, in lessening the time of sea-service required before promotion. As has been stated, graduates of the Britannia must serve one year at sea as naval cadets before being rated as midshipmen. But cadets who take a first or second class certificate at their final examination are allowed to count a certain number of months of this service as performed, and accordingly reach their promotion earlier. The certificates are given in conduct, and in three groups of study—mathematics, seamanship, and the "extra" or non-professional subjects. The percentage required for each class is as follows:

I. Seamanship: Per cent.

First class 90

Second class 80

Third class... 60

II. Mathematics, and III. Extra subjects:

First class 70

Second class 50

Third class 40

The time allowed in months for each certificate is shown by the following schedule:

	First class.	Second class.	Third class.
Seamanship	3	1	0
Mathematics	4	2	0
Extra subjects	2	1	0
Conduct	3	0	0
	12	4	0

Cadets who obtain first class in four departments are thus allowed twelve months, and are rated midshipmen immediately on passing out of the Britannia. The certificate of conduct to which a cadet is entitled is determined by the captain, commander, and principal naval instructor, reference being had to the record of conduct and to the cadet's general behavior in study.

In all examinations, cadets are distinguished by numbers, the key being in the hands of the captain of the ship.

The following table shows the relative weight of studies in making up the marks of the cadets.

RELATIVE WEIGHT OF STUDIES: MAXIMA.

Subjects.	First form.	Second form.	Third form.	Fourth form.
Arithmetic	150	150	} 200	150
Algebra	175	175		150
Geometry	200	200	200	150
Trigonometry	175	175	350	300
Physics			150	175
Navigation, practical and theoretical		125	200	
Practical navigation				175
Theoretical navigation and nautical astronomy				175
Essay	100	100	100	125
Drawing	75	75	75	100
French	125	125	150	200
Latin	125	125	150	200
Charts			100	100
Instruments		50	75	150
Astronomy		150		
Physical geography	125			
Dictation	50	50		
Total	1,300	1,500	1,750	2,000
Seamanship	160	160	180	750

4.—DISCIPLINE.

The regulations of the Britannia contain few prohibitions, and, as might be expected from the age of the cadets, serious offenses are of rare occurrence. Coming as most of them do from the higher classes, and having been subjected to no other influences than those of a well-ordered home or school, bad habits and vicious tendencies have had no opportunity to form or develop, and offenses are all of that minor character which may be easily dealt with. Graver offenses when they occur are punished in a way suitable to the age of the cadets; and if, as sometimes happens, an offender is incorrigible, he is simply removed and restored to his friends. While there is the closest and most careful supervision, there is no very close restraint, because there is no need of it. On the contrary, the life is made attractive and easy by utilizing every means to this end that can be obtained from the surroundings.

The discipline is, in form, that of a ship-of-war, and is in the hands of the captain, the commander, and other executive officers. The severer punishments, of a somewhat varied and elaborate character, are ordered only by the captain. In the gradation of good and bad conduct, cadets are considered as being in one of three classes, and all are held to belong to the first class, unless they incur the penalty of being lowered to the second or third, according to the gravity of their offenses, for a limited number of days. During this time they are subject to special regulations.*

---

* Stated in detail in the Appendix, Note C.

In the cadet organization, the chief captains and captains assist the officer of the day in carrying out the daily discipline. The chief captain of the day, and the captain of the mess-room, taken in rotation, are especially charged with this duty; but the cadet officers at all times assist in keeping order at formations, in class-rooms, at mess, and on the berth-deck, when cadets are turning out or turning in. In general, they are to "do their utmost to uphold the regulations of the ship." The captains are especially charged with the protection of junior cadets; and certain privileges are allowed them, such as the exclusive use of a part of the mess-room.

5.—FEES AND ACCOUNTS.

Instead of receiving pay during the period of pupilage, as is the case in America, naval cadets on board the Britannia are required to pay the government for the benefits they receive. The fee is £70 per annum. In the case of sons of officers making application the fee is reduced to £40; but the number of those received at the reduced rate is limited to ten a term, forty in all at the school, and it is understood to be allowed only to those who need it. In addition to the regular fee, £40 or £70, as the case may be, cadets pay for their outfit, clothes, traveling expenses, text-books, instruments, and stationery, and even for the repair of their chests and locks.

On the other hand, the government supplies the mess-table, and gives the cadets a weekly allowance for pocket-money; and the expenses for all amusements are paid by the ship. The allowance is 1s. a week for each cadet, 2s. for captains, and 2s. 6d. for chief captains. Cadets are forbidden to open an account with tradesmen. Such money as may be required by a cadet for any special circumstances is advanced by the paymaster, under the authority of the captain, and charged to the cadet's account.

A supply of small articles of clothing, stationery, &c., is kept in store by the paymaster, and may be furnished to the cadets, at cost, with a slight percentage to cover expenses.

Cadets are forbidden to buy, sell, or exchange any clothes or other articles among each other. Even such articles as cricketing shoes and bats can only be purchased after a formal written consent of the parent has been given to the captain. Pocket-money is stopped for injuring government property.

6.—MODE OF LIFE, HEALTH, AMUSEMENTS.

SUMMER ROUTINE.

A. M.

5.25. Turn out cadets for No. 2 punishment.

5.30. Punishment No. 2 fall in.

6.30. Bugle. Cadets turn out. Bath. (If weather permits, cadets bathe from the shore).

6.35. Punishment No. 2 dismissed.

A. M.
7.05. Warning bugle.
7.15. Muster and drill.
7.50. Dismiss drills.
8.00. Prayers. Breakfast.
8.40. Bugle for cadet defaulters.
8.50. Muster for studies.
10.20. Interval, ten minutes.
11.55. Studies dismissed. Dress for dinner.
P. M.
12.10. Dinner.
        Cadets land (Wednesday and Saturday defaulters muster).
1.30. Return on board.
1.40. Muster for studies.
4.05. Dismiss studies. Mess-room muster.
4.20. Cadets land. Defaulters drill. Bath, if weather permits.
6.45. Return on board. Shift clothing.
7.10. Muster.
7.15. Tea.
8.00. Evening study.
9.00. Dismiss evening study.
9.15. Prayers.
9.45. Cadets turn in. Officer of the day goes round.

The only material difference in the winter routine is that in winter the cadets have two hours (nearly) on shore just after dinner, and on their return have studies till tea; while in summer, they have shore liberty for two hours, between afternoon studies and tea, and only a short recess on shore after dinner. On Wednesday and Saturday, the whole afternoon is spent on shore.

On Sundays, the routine is as follows:

A. M.
6.45. Cadets turn out. Bath.
7.45. Breakfast.
9.30. Divisions.
10.00. Divine service.
P. M.
12.15. Dinner.
1.00. Bible class (compulsory).
2.00. Bible class dismissed. Cadets land.
5.00. Return on board.
6.00. Tea.
7.30. Prayers.
9.00. Cadets turn in.

The mess-table of the cadets is carefully regulated, abundant, and simple. The routine of diet will be found in the Appendix, Note D. Cadets are wisely forbidden to receive parcels containing eatables.

For sleeping, cadets are divided between the two ships, two forms being assigned to each ship. They sleep in hammocks. A large staff of servants attend the cadets, and have the whole care of their clothes, boots, and bedding. All the duties of this character, which at most military and naval schools devolve on the cadets themselves, are in

England performed by servants. The necessity of this kind of training for boys in general may be doubted, but it is certainly useful in forming a habit of self-reliance in regard to the minor personal details of daily life, and in enabling young officers to regulate with judgment the same details among the men who will afterwards be under their charge Cadets are not allowed to go to the sleeping-decks during the day, except when it is necessary to go to their chests.

Close attention is paid to dress and other externals on board the Britannia, as their importance in the training of young naval officers deserves. A full outfit is required when the cadet first joins the ship, and minute regulations prescribe even the changes of underclothing. Great care is taken that cadets shall not suffer from the effects of exposure on shore or in boats, by careful inspection of clothing when they return on board after having been out in the rain—a precaution that the prevailing habits of boys of this age renders specially necessary. All repairs of clothes and boots are made on board the ship, and the proper stowage, care, and inspection of all clothing are provided for by the most precise regulations. When a cadet requires new clothing, his parent or guardian is informed of the fact, and a request is sent that the articles should be supplied at once. Inspections of clothing and of chests are made once a month by the lieutenants of divisions, and a report made thereon to the captain.

The uniform is of dark blue cloth, and is always worn on ship-board. Working suits are of thick flannel or pilot cloth, and each cadet must have two. For recreation on shore, white flannel cricketing shirts and trousers are always worn, and special shirts for football.

Though a ship lying at anchor is not in every way the most healthful place for two hundred people to live in for months together, yet the precautions taken on board the Britannia, and the regular and active life of the cadets, seem to prevent any of the ill effects that might arise from it. The bilge of the Britannia is pumped out twice, that of the Hindostan four times, a day. The holds are ventilated through shafts, a fire being kept up in each hold from 8 a. m. to 4 p. m. The ventilation of all other parts of the ships is provided for by regulations so minute as to designate not only the ports which shall be opened at night, but even the number of inches of opening required in each case.

On the return of cadets from leave their clothing is disinfected. If any cadet has had a contagious disease he is at once separated from the others, and his clothing and bedding subjected to a heat of $220°$ in the drying machine; and the same precautions are taken with clothing of others who have been near him. Cadets who are sick enough to remain in bed are not detained on board more than forty-eight hours; and if the disease is of a dangerous character they are transferred at once to the hospital on shore.

There is a sick call, or its equivalent, three times a day on board the Britannia. At the appointed hours the medical officer is in the sick bay, ready to receive and treat complaints of illness. A medical officer is on board, however, at all times.

The gymnasium is on shore, directly opposite the ship, in the park above the landing pier. Exercises are so arranged by forms and watches that all cadets have a dumb-bell drill and exercise in the gymnasium twice a week, the only distinction being that one exercise in the gymnasium is voluntary for the two upper forms. Dumb-bell drill lasts fifteen minutes, and takes place in the open air in clear weather. The hours for gymnastic exercise are not those best suited to its enjoyment, or to getting the greatest good from it, being before breakfast, or in the evening.

The Britannia and Hindostan are fitted with salt-water baths and fresh-water shower-baths; and every cadet has a bath in the morning immediately after turning out; fifteen minutes are allowed for the purpose. The temperature is carefully regulated, and is never allowed to go below 54° nor above 60°. Cadets also have a fresh-water hot bath, temperature about 90°, once a month in the evening. From May 1 to September 1, all cadets are required to bathe daily in the river from the shore. Swimming is regularly taught to those cadets who are not good swimmers on their admission to the school.

The chief amusements of the cadets are cricket, foot-ball, boating, and other sports which are found at every English public school. The park belonging to the Britannia covers the steep hill-side opposite the ship, above the shore of the Dart. Half-way up are the gymnasium and bowling-alley. On a level plateau at the top is a well-kept and spacious cricket-field and play-ground. As liberty on shore is extended to two hours a day, and on Wednesday and Saturday to the whole afternoon, there is ample time for amusement and recreation. Instead of staying on shore, if they prefer, cadets are allowed in the hours of recreation to go out in the "blue boats," but each boat must be fully manned, and the boats are not allowed to go beyond certain limits. On half-holidays, in fine weather, when over ten cadets volunteer to go, a sailing launch takes them outside in the harbor, for sailing or fishing; but the launch always has an instructing boatswain and petty officer, and must return in time for evening muster. The three small sailing cutters may be taken by cadets of the third form, and the three large cutters by those of the fourth, every afternoon in good weather. The cutters must always have at least three cadets, and a seaman instructor is in attendance when they leave the buoys to see that the sails are properly set.

On shipboard and in wet weather cadets can amuse themselves with reading, chess, and checkers. The library of the school contains about 1,000 volumes, and cadets can take out books and keep them a fortnight Books and periodicals are also kept in the mess-room, but novels which have not the approval of the principal naval instructor are forbidden. During evening study cadets who do not choose to study are allowed to "skylark" on the middle deck.

Besides the daily leave there are three regular holidays, or "*fête* days," as they are called, during the year. The first of these is the Cadets' Regatta, on a Wednesday afternoon before the Easter holidays, when races take place in four-oared gigs, in ten-oared cutters, and in sailing cutters. The competing crews are made up by forms, by studies, by all cadets over five feet, by all cadets under five feet, and by scratch crews. Prizes in money, ranging from 5s. to £1, for each boat, are given to the winners. These prizes, amounting to £15 in all, are paid by the ship.

The second *fête* is the Queen's Birthday, when a whole day is given; an excursion is formed by train, steamer, or other conveyance, a lunch is supplied by the mess-steward, and expenses are paid by the paymaster under an order from the captain.

The third *fête* is the Athletic Sports, in October. It takes place in the cricket field, and consists of walking and running matches, hurdle races, throwing cricket ball, and jumping. The prizes, amounting to the value of from £10 to £15, are furnished by the paymaster.

During the session cadets are granted temporary leave under certain regulations. On Saturday and Sunday afternoons they may visit their friends in the neighborhood, returning by 9 p. m., and on Wednesday, returning an hour earlier. Leave beyond Dartmouth is only granted once a week, and no leave is granted during examinations or in the fortnight after vacations, or at any time to cadets under punishment. All cadets are obliged to land for recreation every day, unless kept on board by the medical officer.

Great care is taken that the fourth form cadets shall have their proper share in the boats and games, and shall not be hazed or fagged by the upper forms. An instructing warrant-officer is sent on shore during recreation hours, whose special duty it is to prevent bullying or hazing in any shape. When older cadets are found engaged in it, the offense is considered as of the most serious nature.

In considering the good and bad points of the Britannia system, the principal defect is to be found in the course of instruction. It cannot be denied that the course, as indicated by the examination papers, is far in advance of the mental powers of average boys of the age prescribed for cadets. The reason that more do not fail to complete the course is to be found in the low standard of passing, and in the system of cramming, carried out by clever tutors, who are masters in the art of coaching pupils for examinations. In seamanship alone the passing mark is relatively high, but the course of book and model work is one that presents no difficulties and exacts little concentration. For the other studies, no one among the persons acquainted with the system in England seems to pretend that the students come anywhere near the ostensible standard or carry away anything like real knowledge of the subjects embraced in the programme. The statements made by some of the officers and professors of the Royal Naval College, in their evidence before the commission of inquiry in 1876-'77, are distinct and emphatic on this point. This is the Britannia's defect. But in respect to training other than mental, in all especially that goes to make character, it would be hard to find a better system, or one more judiciously and carefully applied. It is doubtless in view of this that the Britannia has been selected for the training of the two princes, one of whom will eventually be the heir to the throne. The regulations for the government of the school and the daily life of the pupils strike a well-adjusted balance between oversight and independence, restraint and freedom; and the officers upon whom devolves the task of carrying out the discipline, from the captain to the lieutenants, are men qualified to continue to the young cadets the highest influences of the best-ordered homes, and to engraft upon them, by their example and companionship, the practice of a manly self-reliance and a manly self-restraint.

CHAPTER VI.

SERVICE AT SEA AS MIDSHIPMEN.

At each half-yearly examination on board the Britannia, forty boys complete the course, and with it the first stage in their career as naval officers. Soon after they are detached from the training ship, they are ordered to a sea-going ship, still as naval cadets; and it is here that they get their first practical acquaintance with the actual duties of their profession. The Britannia certificates indicate the length of time necessary as naval cadet before passing for midshipman. When this time is completed, an examination is held on board the ship in which the cadet is serving. The general examination is conducted by the captain of the ship, assisted by the next officer; and the naval instructor—two, if possible—examines in navigation, in presence of the captain or commander. The subjects of the examination, with the relative weight of each, are given in the following table:

EXAMINATION FOR RATING AS MIDSHIPMAN.

	Marks.
I. Knowledge of former instruction	200
II. Ability to work a "day's work" by tables as well as by projection; to find the latitude by observation of the meridian altitude of the sun, moon, and stars; longitude by chronometer; and to work an amplitude	200
III. Knowledge of the use of the sextant and azimuth compass, and observations with them	100
IV. State of sextant and other instruments	50
V. State of log-books	100
VI. Knowledge of steering and managing a boat under oars and sails; knotting and splicing; rigging lower masts and yards; use of the hand and deep-sea lead	250
VII. Knowledge of great gun, rifle, pistol, and cutlass exercises	100
	1,000

Certificates are granted in three classes for final marks of 900 or over, 750, and 600 respectively. Cadets failing to get 60 per cent. are rejected; those who pass are immediately rated midshipmen. In this grade they remain four years and a half, at the end of which they come up for promotion to sub-lieutenants. The intermediate time is spent in continuous sea-service. In January, 1879, out of 224 midshipmen, there were only 16 that were not attached to a cruising vessel.

During the four years and a half of sea-service as midshipmen the "periodical examinations" are held in July of each year, and others, known as the "half-yearly examinations," in December.* Of the two, the July examination is the more important and elaborate. The papers in extra-professional subjects are made out at the Admiralty and sent to the various ships, on board of which the examinations are to take place.

* In the new Admiralty Instructions (1879) both examinations are designated "half-yearly examinations."

The examination is in three parts :

(1) The professional examination in seamanship (1,000),* gunnery (600),* and steam (400).*

(2) The Admiralty papers, ten in number,† viz :

1. Arithmetic and algebra.
2. Geometry.
3. Trigonometry, plane and spherical, practical and theoretical.
4. Navigation; practical (I) and theoretical.
5. Navigation ; practical (II).
6. Chart drawing.
7. Mechanics and hydrostatics.
8. French.
9. Steam; theoretical.
10. Extra paper of advanced questions.

(3) Work done in the course of the previous year, viz :

	Marks.
A latitude by meridian altitude of sun	5
A latitude by meridian altitude of moon	6
A latitude by meridian altitude of star	6
A latitude by altitude near the meridian	8
A latitude by altitude of the pole star	8
A latitude by double altitude of sun	18
A latitude by double altitude of star	18
A longitude by sun chronometer	10
A longitude by moon chronometer	13
A longitude by star chronometer	10
A longitude by sun lunar chronometer	20
A longitude by star lunar chronometer	20
Error and rate of chronometer (artificial horizon) from two observations taken on different days	30·
A variation by amplitude	8
A variation by altitude azimuth	10
A variation by time azimuth	10
Total	200

The examinations in seamanship, gunnery, and steam are conducted by the officer in command, assisted by the commander or senior lieutenant, gunnery lieutenant, and chief engineer, respectively, or other competent officer. The other examinations are conducted by the naval instructor and navigating officer, or, if there is no naval instructor, by the navigating officer alone. The observations are to be certified, as having been taken and worked by the junior officer, by the naval instructor, or officer acting in that capacity; and they are to be revised and marked by him. Observations may be taken until correct results are obtained; by which means every officer under examination has the opportunity, if he exercises any sort of diligence, of obtaining full marks for his sights. In addition to the observations, log-books are to be written up, examined, and reported on.

---

* Relative weights.

† The time allowed is three hours for each paper, except 8 and 9, for which three hours are allowed in the aggregate.

The examination occupies about two weeks. At its close, the work-papers, numbered but not named, are examined on board the ship by the officers conducting the examination. They are next forwarded to the flagship, where they are examined anew by the naval instructor on board, or by a substitute designated by the commander-in-chief, and finally they go to the Admiralty. Here they are a third time examined by Dr. Hirst, the Director of Studies at Greenwich; and an exhaustive report is made to the Admiralty in regard to them. The two previous reports are expressly ordered to be published.

The half-yearly examinations, held in December, comprise papers in arithmetic, algebra, trigonometry, and navigation, questions in seamanship and gunnery, and practice in gunnery, and in musket and cutlass exercise. The captain is also required to report on the general ability of his junior officers, specifying their knowledge of steam, and their qualifications in knotting and splicing, in drawing, and in observing; and to say whether they keep a daily and seamanlike reckoning, what instruments they possess, whether they are careful of instruments, whether they can use the mercurial horizon on shore, and what foreign languages they can speak.

All these reports and examinations are repeated yearly from the time when a cadet leaves the Britannia until he is promoted to sub-lieutenant. It will be noticed, however, that the examinations are not progressive in their character; the papers are not graduated according to the length of the officer's services or training, but one set of papers is given for all. The object of the examinations is therefore not to test progress, but to serve as a simple check upon indolence, and an incentive to keep up knowledge already acquired. Indeed, it is confessed that midshipmen make no advance in mental acquisition during the long period of sea-service. On the contrary, it has been a very general complaint that the longer the time that had elapsed since they left the Britannia, the less they seemed to know of the subjects they had studied. It was to meet this very complaint that in 1873 the periodical examinations were introduced, with a list of elementary subjects resembling that at the final examination in the Britannia. It is, perhaps, too early to see the result of this experiment, but it is asserted that already acting sub-lieutenants come to Greenwich better prepared, and that this is directly traceable to the periodical examinations. Perhaps it is also due to the introduction of a more systematic arrangement of the detail and duties of naval instructors at sea. Formerly, instructors seem to have been attached to ships in a hap-hazard way, without sufficient reference to the midshipmen on board, so that it often happened that a young officer passed his five and a half years of sea-service without ever seeing one of them. A better arrangement is now in operation, and as much is done as possible to correct the inherent evils of the system of carrying out a course of theoretical training on board of regular cruising ships.

An examination of Table E of the Appendix, which has been carefully prepared from the Navy List of January, 1879, will show the class of ships in which midshipmen and naval cadets are serving, the number on board of each, and the distribution of the naval instructors. From this it appears that all the ships, having junior officers on board, carry an instructor, except three, the Blanche, Danae, and Spartan, where the number of pupils is so small as hardly to warrant the expense; and even then the naval instructor's duty can be done by one of the ship's officers. The distribution of teachers and pupils, to be sure, is somewhat disproportionate; ships like the Alexandra and Minotaur, for example, carrying seventeen or more midshipmen and cadets each, and four other ships less than three each ; while all have a single instructor. But, on the whole, this could hardly be better arranged without sacrificing something far more important than the mathematical instruction given at sea—the whole groundwork of professional training. For it is here, and here only, that this training is given, and it is this training which, however important scientific principles may be, is the *sine qua non* of a naval officer's education. The Britannia gives little of this, Greenwich gives nothing, and the Excellent treats only one phase. It is at sea alone, on board a seagoing man-of-war, either practice-ship or regular cruiser, that a young officer can learn his profession. Hence the midshipman's best energies at sea should be directed to this end, and his mathematical deficiencies, which can be made up later, or which, under a better system, would have been met earlier, become a secondary consideration.

Whether midshipmen get an adequate professional training under the English system is a question about which it is difficult to get the data necessary for a satisfactory answer. Much can undoubtedly be done for them if the captain is so disposed; but unless there is some prescribed system for him to follow he is not likely to give them the attention they need. The captains of those great ships have had far too serious work on hand for the past few years to sacrifice anything even to the professional training of junior officers. What duties, for example, are given to the sixteen or twenty midshipmen on board the Alexandra, the Minotaur, the Achilles, or the Iron Duke, which train them in practical seamanship ? Of practical navigation, as far as it consists in taking sights and laying out courses, they undoubtedly get much ; but how are they to acquire the needful skill in the management of a ship under sail or steam, the practiced eye for wind and weather, the ready resource, the rapid and unerring judgment, that distinguished the old seamen, and that are almost as necessary for the seamen of to-day ? These things, which alone can make a man an efficient watch-officer, are at present only to be learned by the unsatisfactory process of "picking-up"; and it is a question whether more would not be learnt in a year's practice-cruise in a real training-ship than in five years of midshipman's duty on board an ordinary man-of-war. Certainly, if the first of these five years were so occupied the midshipman would arrive at his seamanship examination better prepared in the fourth and most essential subject, " the practice of maneuvering ships, under all circumstances of wind and weather"; and his promotion would find him better qualified for performing the duties of officer of the deck.*

At the end of four years and a half of service in that grade, midshipmen come up for promotion to sub-lieutenants. The examination is in three parts, (1) seamanship, (2) navigation and other branches of mathematics, and (3) gunnery. Of these, the first, in seamanship, is conducted at sea; the second, in navigation, &c., at the Royal Naval College at Greenwich, after six months of instruction; and the third, in gunnery, on board the gunnery ship Excellent, at Portsmouth, after a course of about three months.

The examination in seamanship is ordered by the senior officer present, on the application in the midshipman, after he has reached the prescribed limit of service. Application *must* be made within six months after this time. The examination is conducted by three captains or commanders. The subjects of the examination comprise (1) the state of the candidate's log-books, in which track-charts and sketches of headlands must be entered; (2) stowage of ship's holds; (3) masting ships, fitting rigging and sails, rigging ship, mooring and unmooring, shifting masts and yards, laying out anchors, and the details of a seaman's duty in all its branches; (4) *practice of working and maneuvering ships, as officer of the watch, under all circumstances of wind and weather;* (5) flags and signals, fleet evolutions, and stations of ship's company. The last (5) seems to be considered of less importance than the others.

An officer failing to obtain a certificate in seamanship is re-examined at the end of three months; failing then, he is granted a second re-examination after three months more. If he fails at the third trial, he is dismissed the service.

On receiving a passing certificate,—first, second, or third class, as the case may be,—the midshipman is granted by the commander-in-chief an order as acting sub-lieutenant. He is borne upon the Navy List as an officer of that grade, and his seniority depends on the date of passing the seamanship examination. He remains, however, an acting sub-lieutenant until he has passed the two other branches of his examination for promotion, in navigation and in gunnery. These examinations and the courses of instruction that precede them form part of the organiza-

---

* This defect of the English training system is alluded to by Captain (now Rear-Admiral) Lord Gilford, one of the Lords Commissioners of the Admiralty, in his evidence before the Britannia commission, Q. 1703. It should be added that the *Queen's Regulations and Admiralty Instructions*, recently published (February, 1879), give a full and succinct programme (p. 176) for instruction of junior officers in seamanship, and certain general regulations for work under the naval instructors. In order, however, to ascertain the value of these regulations, it is necessary to know how far the seamanship instruction is theoretical and how far practical; whether the regulations are susceptible of daily and regular application on board a ship of war at sea, and how minutely captains are obliged to follow them.

tion of two distinct and important establishments, the Royal Naval College, and the gunnery-ship Excellent. Each of these institutions has
many peculiar features, and each has a variety of objects aside from the
training and examination of line or executive officers. It will, therefore,
be well to leave at this point the consecutive description of a line officer's education, which has been followed hitherto, and describe the general plan and organization of the Greenwich College and of the Excellent, taking up in connection with each the course and examination of
sub-lieutenants in general subjects and in gunnery.

CHAPTER VII.

THE ROYAL NAVAL COLLEGE AT GREENWICH.

The Royal Naval College was opened in February, 1873, and at the same time the college at Portsmouth was discontinued. The change, however, was more than a removal from one place to another; it was almost the foundation of a new institution. Its aim, as put forth in the circular of January 30, 1873, announcing the opening, was to provide for the education of naval officers of all ranks above that of midshipman in all branches of theoretical and scientific study bearing on their profession. With this view courses were established in great variety for several different branches of the service. The plan was conceived in the most liberal spirit, with an entire absence of the professional narrowness which is apt to characterize such an institution. It was distinctly stated that great advantages were expected to accrue from the connection established through the college between naval officers and distinguished students of the various branches of science. Still further; the college was not only to afford higher instruction, but was to become a nucleus of mathematical and mechanical science, especially devoted to those branches of scientific investigation which have most interest for the Navy; and all this was to be accomplished without interfering in any way with the practical training of officers in the active duties of their profession.

The college has been fortunate from the beginning in its officers. The successive presidents, Sir Astley Cooper Key, Admiral Fanshawe, and Sir Charles Shadwell are among the most eminent of the flag officers of the English Navy: a body of men distinguished throughout the world for breadth of view, for maturity of judgment, and for mental vigor. The Director of Studies, Dr. Hirst, who carries on his shoulders the academic burdens of the college, is one of those rare instances of the harmonious union of a profound student and an efficient organizer, and has thus been enabled to fill, with extraordinary success, the difficult duties of his position. The professors, chiefly Cambridge men of high university standing, are eminently fitted to carry out the purposes of the college, and the professional department is supplied by able constructors and engineers, whom the Admiralty has always at command in sufficient numbers.

The seat of the college is the ancient and magnificent group of buildings formerly devoted to Greenwich Hospital. The buildings, four in number, are of Portland stone, and stand on a high terrace close to the southern bank of the Thames, forming three sides of a large quadrangle open to the river. With the exception of the northwestern hall, which was built by Charles II as a royal residence, they date from the reign of William III, who founded Greenwich Hospital as an asylum for pen-

S. Ex. 51——4

sioned seamen. For the next hundred and fifty years the establishment
was kept up, having all the time from one to three thousand disabled
pensioners within its walls.

At length, in 1865, it was found that the class whom it was intended
to benefit would prefer to commute their allowances, and to live at their
own homes, and an act of Parliament was passed giving them that
option. The great majority of the pensioners accordingly left the hospital, and four years later it was closed as an asylum.

When the commission of 1870, on the higher education of officers,
was engaged in its investigations, its attention was called to the vacant
buildings of Greenwich Hospital as a possible substitute for the insufficient building at Portsmouth. The question as to the relative advantages of the two places was submitted to a large number of naval officers,
and the answers of more than three-fourths, including some of the ablest
of the older officers, were decidedly in favor of Portsmouth. It is curious
that the reason generally given for this view was the fact that Portsmouth was a dockyard, and that, as one of the answers puts it baldly,
"for self-evident reasons, the proper place for a naval college is a naval
arsenal"; while in the United States it was satisfactorily shown forty
years ago that a naval college could hardly reach its highest development as an adjunct of a navy-yard. The committee were equally
divided; the civil members urging the claims of Greenwich as stead-
fastly as the officers did those of Portsmouth. The chief objection to
Portsmouth, acknowledged by everybody, was the inadequate accom-
modation afforded by the ill-arranged and dilapidated building. It was
necessary either to move or rebuild. The college also needed reorganization badly, but, if Portsmouth was really the right place for it, it
could of course be reorganized as well there as elsewhere. It is difficult
to see exactly the force of the arguments advanced by the advocates of
Portsmouth. The reorganized college was to be mainly devoted to
scientific and mathematical training, for which the dockyard could offer
no facilities, and for which it seemed a positive disadvantage. It would
be impossible to obtain the best men, as instructors, at Portsmouth;
whereas a college at Greenwich, by its nearness to London, its independent organization, and its freedom from the trammels of professional
conservatism and professional exclusiveness, could offer fair inducements
to the highest order of scientific talent. For the students themselves,
instead of the incessant din and turmoil of a great depot of naval equipment and repair, there was the quiet seclusion of the old hospital, with
the best libraries, museums, and lectures, within easy distance; instead
of the unvaried association with officers engaged in the busy details of
daily work, there was a possibility of obtaining breadth of view and fresh-
ness of thought by contact with the most vigorous thinkers, in other but
still kindred lines of study. The only branches of study in which the dockyards would give advantages were engineering and ship construction;
but it is doubtful whether any of the college students except engineers

and constructors were inclined to profit by them, and these would be given ample dockyard experience wherever the college might be placed. In spite of the generally expressed opinion of the older officers, the Admiralty decided to move the college to Greenwich. Perhaps sentimental considerations had an influence with some of the advocates of the change; and, now that the change is made, the force of such co siderations is still more apparent. Apart from the fact that only slight alterations were needed to convert the deserted hospital into a well-appointed professional college, there were professional associations about it which seemed to fit it peculiarly for a training place of naval officers—the painted hall, with its noble gallery of naval pictures, the relics of Nelson, and the naval monuments and trophies. At all events, whatever may have been the reasons, the change was made, and the 1st of February, 1873, saw the opening of the new Naval College.

The changes in the buildings as they stood that have been found necessary are not very considerable, and relate only to interior arrangements. The northwestern, or King Charles's, building is the most important part of the establishment, containing the offices of the President and the Director of Studies, and the studies—i. e. class-rooms—of the main body of students. The opposite building on the river front, known as Queen Anne's building, is fitted up as a naval museum, and contains besides the relics a full collection of models of great historical as well as professional interest. The South Kensington collection of models has been placed here, comprising a series of ships of the line from the Great Harry, of Henry VII's time, to the armor-plated ship of the present day. In the east wing are models of docks and dockyard works, of spars and boats, and of engines. There are also collections illustrating the equipment of ships and the various kinds of projectiles. Probably these collections will be much improved by additions made from time to time, which can readily be supplied by the dockyards.

The southeast building, known as Queen Mary's, and containing the chapel, is used by the engineer officers and students, and by the construction students, and is arranged in dormitories and a mess-hall for these classes of officers, who have a separate mess.

In the southwest building—King William's—is the painted hall, and, in the room beneath, the spacious officers' mess, with smoking and billiard rooms. The lecture-rooms, laboratories, studies, and sleeping-rooms are in these two western buildings—King William's and King Charles's. The laboratories are especially well-appointed, having been furnished with everything required for the advanced study of physics and chemistry, with the greatest care and at very considerable expense. There is also an observatory, where the use of fixed instruments is taught, which formerly belonged to Greenwich Hospital School, and which has recently been refitted and attached to the college. Instruction in the use of fixed instruments is to be given in the course in nautical astronomy. For marine surveying, the Arrow, a double-screw iron gunboat, is attached to the college.

The staff of the college consists of a flag officer as president, assisted by a captain in matters of discipline and in the internal arrangements of the college unconnected with study. The whole system of instruction is, under the president, organized and superintended by the director of studies. The latter is not a naval officer, and has nothing to do with discipline, nor does he give instruction personally. Besides the president, the captain, and the director of studies, the staff consists of the following:

Two professors of mathematics; one for engineers and constructors, one for voluntary students and gunnery students.
Professor of applied mechanics.
Professor of physics.
Professor of chemistry.
Professor of fortification (captain in the Royal Marine Artillery).
Assistant professor of fortification (lieutenant in the Royal Marine Artillery).
Mathematical and naval instructor, and lecturer in meteorology and naval history (naval instructor).
Four instructors in navigation (naval instructors).
Three instructors in mathematics.
Two instructors in French.
Instructor in German.
Instructor in Spanish.
Instructor in nautical surveying (staff commander).
Instructor in steam (engineer officer).
Instructor in applied mechanics (engineer officer).
Assistant professor of physics (engineer officer).
Demonstrator in physics.
Demonstrator in chemistry.
Instructor in naval architecture.
Assistant instructor in naval architecture.
Instructor in marine engineering.
Instructor in marine engine drawing.
Instructor in freehand drawing.
Naval instructor for examining duties.

This gives a total of 34 officers of instruction and government. There are also two officers in the gunboat Arrow, a medical officer, a store-keeper, and five clerks; in all 43 persons. In addition to the above, lecturers to the number of perhaps half a dozen are appointed from year to year.

The courses of study proposed at the beginning, which have mostly been carried out, are as follows:

1. Pure mathematics, including co-ordinate and higher pure geometry, differential and integral calculus, finite differences, and the calculus of variations.
2. Applied mathematics, including kinematics, mechanics, optics, and the theories of sound, light, heat, electricity, and magnetism.
3. Applied mechanics, including the theory of structures, the principles of mechanism, and the theory of machines.
4. Nautical astronomy, surveying, hydrography, meteorology, and chart drawing.

5. Experimental science:
   a. Physics, viz, sound, heat, light, electricity, and magnetism.
   b. Chemistry.
   c. Metallurgy.
6. Marine engineering, in all its branches.
7. Naval architecture, in all its branches.
8. Fortification, military drawing, and naval artillery.
9. International and maritime law, law of evidence and naval courts-martial.
10. Naval history and tactics, including naval signals and steam evolutions.
11. Modern languages.
12. Drawing.
13. Hygiene, naval and climatic.

In this extensive programme, the only studies which have not yet, in one shape or another, been attempted, are naval artillery, the law of evidence and courts-martial, and hygiene; the first of which may be assumed to be adequately taught in the Excellent.

Of the courses in the list, some are required and some are voluntary; but the privilege of choice is only given to those classes of officers whose attendance at the college is voluntary, and even then under certain restrictions. The voluntary students are officers of certain grades who are admitted to the benefits of the college on their own application. They receive half-pay during their attendance. These half-pay officers comprise captains, commanders, lieutenants, staff commanders, navigating lieutenants, and naval instructors. There are also other officers who are admitted at their own request, and who have some liberty in the choice of studies, but who are not half-pay officers. These are the officers of the Marine Artillery and Light Infantry, and chief engineers, on full pay, but unattached to any ship; and engineers and assistant engineers, borne for full pay in some vessel in commission. Still a third class is composed of students admitted at their own request, but, once admitted, pursuing a defined course. These are persons qualifying for naval instructors (either chaplains or candidates from civil life); private students in naval architecture and engineering, and lieutenants qualifying for gunnery and torpedo officers. The latter receive full pay and are borne upon the books of the Excellent. Lastly come those officers whose attendance and course are both compulsory. These are acting sub-lieutenants, who are also borne in the Excellent, acting navigating sub-lieutenants, probationary lieutenants of the Royal Marine Artillery, acting assistant engineers (later assistant engineers), and shipwright apprentices, otherwise known as construction students, or students in naval architecture. The three classes last-named receive special rates of pay. It may be added in regard to the construction students and the assistant engineers, that their attendance at the school is not exactly compulsory, in the first instance, as they are selected by competition; but having once entered for the position their course of study is a necessary consequence.

The following table shows the classes of officers pursuing study at the college, the number of each, the duration of each course, and its character (whether compulsory or voluntary). The number of officers of each class is given as it stood at a fixed date, the 1st of January of each year, and therefore the aggregate does not represent the total number of students who have attended a college course. The table is taken in part from the report of the Committee on the Naval College, and in part bas on data from the Navy List.

STUDENTS AT THE ROYAL NAVAL COLLEGE, 1875–1879.

Compulsory or voluntary.	Duration of course.	Description of students.	Members, January 1,				
			1875.	1876.	1877.	1878.	1879.
Compulsory .	3 sessions or years.	1. Assistant engineers.....................	6	6	6	6	6
	Do.......	2. Shipwright apprentices (construction students).	11	9	9	9	9
Voluntary ...	3 sessions ...	3. Private students { English	1	1	......	(?)	(?)
		Foreign	7	8	6	(?)	(?)
Compulsory .	2 sessions ...	4. Probationary lieutenants, Royal Marine Artillery.	6	4	12	10	7
	Do.......	5. Probationary lieutenants, Royal Marines.	......	7	......	......	......
	1 session	6. Probationary lieutenants, Royal Marine Light Infantry.	10	(*)	(*)	(*)	(*)
	Do.......	7. Officers qualifying for gunnery lieutenants.	16	19	18	20	16
	Do.......	8. Officers qualifying for torpedo lieutenants.	......	......	......	4	5
	Do	9. Acting assistant engineers...........	25	18	26	18	37
	6 months....	10. Acting sub-lieutenants	74	75	46	59	44
	Do.......	11. Acting navigating sub-lieutenants ...	16	11	4	10	2
	Do.	12. Officers qualifying for naval instructors.	1	1	......	1	1
Voluntary ...	In general, for 1 session.	13. Captains...........................	3	2	2	6	2
	Do.......	14. Commanders....................	3	......	7	2	4
	Do	15. Lieutenants....................	43	56	39	26	32
	Do	16. Staff-commanders ...	......	......	1	......	......
	Do.......	17. Navigating lieutenants	9	4	3	1	2
	Do.......	18. Marine Artillery officers .... }	3	8	6	{ 1	1
	Do.......	19. Marine Light Infantry officers....... }				{ 1	2
	Do.......	20. Chief engineers....................	2	2	......	......	2
	Do.......	21. Engineers	9	7	3	7	8
	Do.......	22. Naval instructors..................	......	......	......	......	......
		Total, excluding private students...	237	229	182	183	180

* Discontinued.

The session, or academic year, begins on the 1st of October, and ends on the 30th of June, except in the case of officers pursuing a six months' course (see table), who join at periods a month apart, as they return from sea. From the 30th of June to the 1st of October the college has vacation, and leave is given for ten days at Christmas and a week at Easter. The vacations for acting sub-lieutenants are six weeks at midsummer and a month at Christmas. The arrangement of courses enables officers who desire a higher theoretical education to have two or three years, or even four years, of study at the college, in the course of their career. All are obliged to come in the first place as acting sub-lieutenants. On their promotion to lieutenant they may be ordered to Greenwich while waiting their turn, and may thus get a second course, for the whole session. After a little interval they may apply for permission to qualify for gunnery duties, and they will then pass a third term of study. There is nothing to prevent them from coming back again, after another interval, as captains or commanders, for a fourth course. The studies will not be consecutive. A considerable period must elapse between each attendance—a period of four years at least in the case of half-pay officers, two of which must be passed at sea. Nor is it always possible for an officer to arrange his duties so as to carry out such a plan; but the opportunity is given, so that, other things being equal, officers who have the inclination and talents may reach a higher point of scientific culture.

The discipline of the college is rather that of a university than of a naval establishment, as is necessarily the case where very few of the students are less than twenty years of age, and a number have reached the prime of life. Such restrictive regulations as exist bear chiefly upon the younger students, especially the acting sub-lieutenants and acting engineer officers and the probationary lieutenants of the Marine Artillery. Only officers on full pay are required to wear uniform. Captains, commanders, and chief engineers are allowed to reside outside the college if they prefer; but all other officers must occupy rooms and join the mess, towards which they receive an allowance of 1s. 6d per diem. Special permission may be given to live outside, however, and, in general, the operation of the rule is not so severe as would at first seem to be the case, on account of the relatively small proportion of married lieutenants.

The difficult subject of dealing with breaches of discipline and enforcing regularity is in charge of the President, assisted by the captain of the college. A certain surveillance is maintained by the college police. For grave offenses a court-martial is as applicable here as elsewhere in the service, and the negligence or irregularity of voluntary officers may be dealt with by removal from the college. Misconduct on the part of the younger officers pursuing compulsory courses is a more delicate matter, and has always constituted the chief difficulty in naval shore colleges, where the course followed a long period of sea service. The offenses, as a rule, are not serious enough to warrant a court-martial or removal from the service; removal from the college would defeat the object of the college itself; and the offenders are too old to be punished in the ordinary way. Fortunately, at Greenwich, the presence of the older officers as students gives a certain tone to the establishment, by which the younger cannot fail to be affected, and the compulsory courses are so exacting that they engross pretty thoroughly the time of the junior officers.

The general method of instruction, like that of the Britannia, is a system of informal exposition, study, and practice with the instructors. Recitations in the ordinary sense can hardly be said to exist, certainly no recitations for which marks are given. The only recorded test of general results, upon which certificates are given, is the final examination, which occurs at the end of the session, or, in the case of sub-lieutenants, at the end of the six months' course. Examinations are generally held at Christmas and Easter, and in some courses they occur with great frequency,—once a fortnight, or even once a week. These examinations may be marked or not as the professor chooses; but the marks only serve to test progress, and have no bearing on the final result. Text books serve rather the purpose of collateral reading, and in mathematics they are used chiefly as collections of examples. The student is never led to place his chief reliance on the book, in acquiring any subject, but on the instructor; in this way the bad influence of text-books is reduced to a minimum, and the vicious practice of slavishly following a text hardly exists. At the same time the objections to a lax lecture-system are removed by the practice of following up the lecture at once, often in the same room, by test exercises accompanied by full and particular explanation, mainly on the subject of the lecture itself. An hour of blackboard exposition, with from two to three hours of work in the class-room immediately after it, is the usual way of passing the morning at Greenwich, with all classes of students. The class-rooms, or studies, as they are called, are furnished with tables, at which the students work, while the instructor moves about among them, answering as well as asking questions, and endeavoring to supply what is wanted by each of his pupils. Mathematics forms the backbone of all the courses. In fact there have been general complaints that the college leaned too much in this direction, without affording sufficient scope in others. This objection only refers, however, to the courses for half-pay officers, as the special students who are to make naval architects, designing engineers, and gunnery officers, can hardly have too much mathematics, and the other important class, that of sub-lieutenants, is not likely to be in any such danger.

The teaching staff is by no means confined to the corps of naval instructors. On the contrary, the most important work is in the hands of professors otherwise unconnected with the Navy. The principal among these are the two professors of mathematics, the professor of applied mechanics, the professor of physics, and the professor of chemistry. There are practically three professors of mathematics, but one of them bears the title of mathematical and naval instructor, and belongs to the naval corps. These three take charge of all the higher instruction in this branch. Each of them has an assistant, also a university man, and not in the Navy. One professor with his assistant instructs the engineer and construction students. The assistant also takes the probationary lieutenants of the Marine Artillery. The second professor takes the gunnery lieutenants, the second-year Artillery lieutenants, and the best of the half-pay officers. The third has the half-pay officers who are less advanced. For the separate group of acting sub-lieutenants, whose knowledge is of the scantiest character, there are four naval instructors. The professors of physics and applied mechanics have engineer officers as assistants. Fortification is taught by a captain in the Marine Artillery, naval construction by officers in the construction department at the Admiralty, and steam by naval engineers.

The practice of separating the duties of instructor and examiner is rigorously carried out, and no papers are set at the final examinations by members of the regular teaching staff. The examination of the acting sub-lieutenants is conducted chiefly by a naval instructor borne especially for examinations, who has no duties of instruction. He sets and marks the papers in mathematics, navigation and nautical astronomy, and meteorology. The examinations in steam, physics, French, and marine surveying are given by external examiners or members of the staff who do not teach sub-lieutenants. The general examinations of all students, voluntary and compulsory, in June, which last three weeks and include everything, are chiefly conducted by outside examiners specially employed for the purpose. Even here, however, the Director of Studies has supervision, and all papers, marks, and reports go through his hands. Indeed the number of papers for which he is nominally responsible is so great that it would be simply impossible for him to give all of them even the most cursory inspection. They include not only voluntary and required examinations at Greenwich, but also the examinations for cadetships and clerkships, the semi-annual Britannia examinations, the semi-annual dockyard examinations, and the examinations of junior officers afloat. The total reaches the extraordinary figure of over 270 sets of questions and 11,500 work-papers, a year; of which 176 papers are set and 4,000 work-papers are marked by Mr. Goodwin alone, the examiner of the college.

It is only necessary, however, for the Director of Studies to give such attention to the work of the outside examiners as will insure unity of purpose in the questions and keep them within the limits of the course. The examiners themselves include a number of specialists, generally of large experience in this branch of educational work. Among them may be mentioned Prof. A. W. Rücker, examiner in natural sciences, Brasenose College, Oxford; Prof. C. Niven, examiner in mathematics, Trinity College, Cambridge; Mr. Barnaby, Director of Naval Construction, and Mr. Wright, Engineer-in-chief of the Admiralty; Professor Unwin, of Cooper's Hill College; Professor Kennedy, of University College, London; Professor Karcher, of Woolwich, and several others. These examiners do not constitute a mere board of visitors; they receive a syllabus of the work gone over by each class, and upon this they actually give questions and mark papers, thereby insuring to the Navy a rightly-directed course of studies and a high standard of scholarship.

Special courses of lectures form a peculiar and important part of the college teaching. They may or may not form some part of a required course for one class of students, but they are open to officers generally and all are alike permitted to benefit by them; that is, all the voluntary students. Notices are also sent to the naval clubs, and officers who are not students are invited; and the attendance at the lectures is usually large. The lecturers are sometimes members of the college staff and sometimes not. Among the courses may be mentioned those on international law (chiefly relating to the laws of war at sea) by Prof. Mountague Bernard; on the safety of ships, the behavior of ships at sea, the structural arrangement of modern ships of war, and the tonnage and propulsion of ships, by Mr. White, of the construction department of the Admiralty; the magnetism of iron ships, by the Astronomer Royal; the atmosphere as a vehicle for the transmission of signals, by Professor Tyndall; the strength of materials, by Professor Cotterill; naval operations on shore, and the effect of artillery fire upon armor; practical shipbuilding, marine engines and boilers, naval history, meteorology, navigation, and nautical astronomy. All of these courses have been given once at least since the foundation of the school, and some of them are repeated from year to year; but it is intended that an elastic quality shall be given to this extra-collegiate teaching, and the arrangement of special courses lies very much with the President and the Director of Studies. The importance of such lectures, given by specialists who are masters of the art or science they undertake to teach, cannot be overestimated. Even if no examinations are passed and no questions answered, the mere privilege of hearing a discussion or exposition of living professional subjects; from a thoroughly scientific point of view, is an immense stimulus to professional improvement and a powerful agent against professional ignorance and ultra-conservatism.

A certain amount of outdoor practical work is part of the regular programme of the college. Wednesday is the field-day at Greenwich; on that day the Marine Artillery lieutenants have their field surveying, and all the open-air exercises take place. These include nautical surveying, in which the whole day is passed afloat, open-air sketching, military surveying, and practical instruction at the engineering works under a naval engineer. Some of these exercises belong to special courses, but all are open to half-pay officers, and they form an important and valuable feature in the Greenwich course.

In the conventional grouping of studies at the college will be found the most convenient arrangement for considering separate classes and courses. According to this the groups are as follows:

Group A, subdivided into A₁, first year students, A₂, second and third year students:

1. Assistant engineers.
2. Construction students.
3. Private students.
9. Acting assistant engineers.
20. Chief engineers.
21. Engineers.

B₁:
7. Officers qualifying for gunnery lieutenants.
8. Officers qualifying for torpedo lieutenants.

B_2, commonly designated half-pay officers :

13. Captains.
14. Commanders.
15. Lieutenants.
16. Staff-commanders.
17. Navigating lieutenants.
18. Marine Artillery officers.
19. Marine Light Infantry officers.
22. Naval instructors.

C :

4. Probationary lieutenants, Marine Artillery.

Separate class:

10. Acting sub-lientenants.
11. Acting navigating sub-lieutenants.
12. Officers and others qualifying for naval instructors.

The groups will be taken up in the following order :
1. Gunnery and torpedo lieutenants and half-pay officers (B_1, B_2).
2. Probationary lieutenants, Royal Marine Artillery (C).
3. Acting sub-lieutenants, and candidates for naval instructors.
4. Engineer and construction students (A).

1.—GUNNERY AND TORPEDO LIEUTENANTS AND HALF-PAY OFFICERS
(GROUP B).

There are two distinct divisions of this group: B_1, lieutenants qualifying for gunnery and torpedo officers, and B_2, half-pay and marine
officers. The essential subject in the course of instruction of both
divisions is mathematics. For half-pay officers it is the only obligatory
subject, though even here a certain latitude is given to captains and commanders. In mathematics the more advanced of the half-pay officers
receive the same instruction as the gunnery and torpedo officers, attend
in the same class-room, and pass through the same course; while the
remainder have a course, a study, and an instructor by themselves. In
other subjects, the courses for B_1 and B_2 are nearly the same, with this
general reservation, that gunnery and torpedo lieutenants are expected
to give almost undivided attention to three subjects, mathematics,
physics, and fortification,* which alone count in determining their final
certificate; while half-pay officers take their choice, after mathematics,
in nearly the whole range of subjects taught at the college.
    It will be well to go more fully into the details of instruction in this
group of students, especially in mathematics. Of the two divisions,
the first, composed of gunnery lieutenants† and the best half-pay officers,
is taken by the junior professor of mathematics, Mr. Lambert, and his
assistant. This again is divided into two sub-classes, one taking a some-

---

* Chemistry and marine surveying are substituted for fortification in the course for
torpedo lieutenants.
† The expression "gunnery lieutenants" in this section is used for the sake of brevity
to mean officers qualifying for gunnery and torpedo lieutenants.

what higher course than the other. The first hour in the regular morning period of instruction, from 9 to 10, is occupied with a lecture to both
sub-classes. In the second hour, from 10 to 11, the higher sub-class has
a second lecture of an advanced character, and the lower remains "in
study" with the assistant. The third hour, from 11 to 12, is passed by
both sub-classes in study together, working under both instructors. At
the end of each lecture a task is given to be done before the next lecture,
which may be worked out in study, or taken by the students to their
rooms. Much work is done by the officers in this way by themselves;
and occasionally the professor, of his own accord, visits officers' rooms
to give them assistance. An examination is held every Saturday on the
work of the previous week or fortnight. Marks are given and the list
is published on Monday. A record is kept of these marks, but they do
not count in any way in determining final position. The examinations,
however, are found to consolidate knowledge and serve as an incentive
to exertion. The division, in its course of one session, goes over algebra,
trigonometry, analytical geometry, statics, dynamics, hydrostatics, the
differential and integral calculus, and the applications of integral calculus
to problems in physical subjects, as dynamics, hydrostatics, and astronomy. The practical requirements of the Navy are kept steadily in view,
and the standard, in general, is constantly advancing.

The second division of group B, composed of the lower half of the
half-pay officers, is in charge of the mathematical and naval instructor,
who also has an assistant and who arranges his instruction in much the
same way as that of the higher division. The two instructors lecture
to different halves of their class for one hour, and the rest of the morning is passed in study with both instructors together. Test examinations are held every fortnight or three weeks. Officers who intend to
qualify for gunnery lieutenants not infrequently go through the course
of this division as half-pay lieutenants, and after an interval come back
for the higher course with the other division. The course for the second
division is much less advanced than the other. Most of the officers,
according to the statement of the instructor in charge, know next to
nothing of mathematics when they come, and they leave with moderate
attainments. They begin with the earliest elementary algebra and go
as far in trigonometry as the abilities of the class will allow. In addition to a fair knowledge of trigonometry they generally go through some
elementary mechanics. Calculus does not appear in their examination.
As to the character of the mathematical work performed by this class,
it is stated by Prof. W. D. Niven, their examiner in 1876, that, as a
rule, it is not satisfactory. There is a large proportion of the class who
do very little, and there are very few who make a really good appearance, while the gunnery lieutenants have given the greatest satisfaction.
Perhaps this difference is due to the fact that many half-pay officers
neglect mathematics intentionally in order to give closer attention to
other subjects; and, in any case, it may be expected that the standard

will be improved when the influence of the half-yearly examinations and of the improved course for sub-lieutenants is more fully felt.

The instruction in physics and chemistry comprises two formal experimental lectures a week in each branch, which are also attended by other classes of students, and at least two afternoons a week in each laboratory, though the gunnery lieutenants do not take the practical chemistry. The subject of physics is not treated mathematically; the lectures are purely experimental; they include hydro-mechanics, acoustics, magnetism, electricity, heat, and light. Questions are given after each lecture to be worked out, and ample opportunity is given of obtaining explanations, the assistant and demonstrator being in attendance every day from 9 till 1. In the physical laboratory such operations are performed as determining the specific gravity of bodies, making and reading accurately the barometer, determining the resistance of conducting bodies, and so on. Examinations are held at Christmas and Easter, but the marks given have no weight in the final count.

A thorough course in field fortification, for two afternoons in each week, is given by the professor, a captain in the Marine Artillery, to gunnery officers, to half-pay officers who desire it, and to officers of the Marine Artillery. The latter have also permanent fortification on two afternoons, and together with the Marine Infantry officers, semi-weekly lectures in military history, and a weekly exercise in topographical drawing. In general, with the exception of chemistry and steam, and the other branches which have been mentioned, marine officers share in the instruction given to half-pay students and attend the same classes.

The number of half-pay officers allowed annually on their own application is fixed by regulation for each grade, but the limit has not yet been reached. Twenty-five captains and commanders may be appointed in this way, four staff-commanders, ten navigating lieutenants, two officers of the Marine Artillery, and six of the Marine Light Infantry. The number of lieutenants allowed is fixed from year to year, and though classed with half-pay officers, they receive an addition to their half-pay of 3s. a day during their first year, and full pay during the second; the allowance being offered as an inducement to young officers to perfect their scientific education.

The length of the course for half-pay and gunnery officers is a single session of nine months; but those of the former class who arrive late in the academic year are allowed to stay through the whole of the next session, unless required for active service. No appointments are made in any year after Easter. A limited number, not exceeding ten in the case of lieutenants, are allowed to remain a second session for study, if recommended by the President and examiners.

Half-pay officers can hardly be said to have an examination for admission; certainly not an examination, the results of which are brought forward and kept for reference. Its object is merely to show what each officer knows and can do when he comes, in order to place him. It is said that each officer must show that he knows a fair amount of elementary mathematics; but no one is rejected, and in the case of the higher officers, captains and commanders, the examination hardly amounts to more than a conversation with Dr. Hirst, the Director of Studies, as to what they want to do, and where they are ready to begin.

There is some irregularity in the entrance and departure of half-pay officers, as they may join at other times than the beginning of the academic year, and they may be detached according to the needs of the service, before their nine months' course is ended. This is especially the case with lieutenants. The prospect of appointment to a good ship is held out as an inducement to join. It is obviously the desire of the Admiralty—indeed, it is so set forth officially—that young lieutenants should avail themselves of the advantages of the college while waiting their turn for appointment to a ship; and that special consideration will be given to these cases, in the selection of lieutenants for sea-duty. This very fact leads to irregularity in the time of joining and leaving. It often happens in this way that a half-pay officer leaves the college before the close of the academic year, in which case he has no examination. In the session of 1875–'76, for example, there were 117 half-pay officers, of whom only 37 attended the whole course, and only 56 were examined.

The optional subjects for half-pay officers, besides the lecture-courses, are physics and practical physics, chemistry and practical chemistry, navigation and nautical astronomy, steam-engineering, marine surveying, military surveying and drawing, military history, permanent and field fortificatication, French, German, Spanish, and freehand drawing. Of these electives, they are expected to take at least two, but the option is generally left to themselves. Each one makes out a programme at the beginning, to which he is expected to adhere. The President is directed to ascertain whether they are making satisfactory progress, and to report accordingly to the Admiralty, and removal may be resorted to, if they neglect their work. This has occasionally been done, with beneficial results.

The hours appointed in Group B to organized study under the different instructors are 33 a week, of which 16 are given to mathematics, and the other 17 to elective subjects. These 33 hours of instruction consist neither in recitation, nor in lecture, nor in practical exercise, alone; but in a combination of all three. In addition to the regular class-room work, it is expected that the voluntary students will work about twelve hours a week by themselves, making forty-five hours in all.

According to the report of the committee appointed to inquire into the workings of the Naval College, there seems to have existed considerable irregularity in the attendance of voluntary students at the lectures and in class-rooms. Full statistics on the subject, if they could be got, would be both interesting and valuable, as showing the working of the system of voluntary attendance. The committee, taking the table of attendance for a certain month in 1876, found that there were 104 hours of study during the month. Out of these, the average absences of the class amounted to 30 hours, nearly one-third, and in the case of one lieutenant to 83 hours, or about four-fifths of the whole time. The gunnery lieutenants, as might be expected, were far more regular, the average absence being about four hours in the month and the highest individual case fifteen hours.

A general examination is held at the end of each session. It is, to a certain extent, voluntary, but all half-pay officers are expected to come up in mathematics, and in the electives which they have taken from the beginning. Each subject has a maximum number of marks, the scale of which, at the first examination in 1874, was as follows:

```
* Mathematics ............................................................ 3,000
* Physics ................................................................ 700
  Chemistry ............................................................. 700
  Steam ................................................................. 400
  Marine surveying ...................................................... 300
* Fortification ......................................................... 400
  Military surveying and drawing ........................................ 400
  International law ..................................................... 250
  French ................................................................ 400
  German ................................................................ 350
  Spanish ............................................................... 350
  Freehand drawing ...................................................... 100
                                                                       ------
      Aggregate ........................................................ 7,350
```

As has been already said, a half-pay officer may neglect mathematics to a certain extent and make his numbers in other subjects. For example, in the examination referred to, the officer who came out at the head of his class got a mark of 3,245, of which 1,543 was for mathematics, and the remainder, 1,702, for his optional subjects, viz, physics, chemistry, steam, international law, French, and German, representing a possible aggregate of 2,800; in other words, he made 51 per cent. of the maximum in mathematics and 61 per cent. in his electives, his percentage on all that he studied being 56. This does not appear to be a very high standard for the head man; but it must be remembered that this was the first year of the college, and that not only are the courses now better organized, but the tendency will be each year for the voluntary students to come better prepared. Moreover, the officer in question might doubtless have done much better had he confined himself to a smaller range of subjects; but the general object of the half-pay students is not so much to reach perfection in any one branch of knowledge as to obtain such a familiarity with several branches as may be acquired in nine months. Viewed in this light, and in respect to the general improvement of the service, the 56 per cent. of the year's instruction obtained by this officer is a positive gain, though not, perhaps, as great a gain as might be expected, to himself and to the Navy at large.

---

* Starred subjects are the required subjects for gunnery lieutenants; and the maximum for this class may be said to be the aggregate of the maxima assigned to the three subjects, 4,100.

All officers that reach a certain standard, provisionally fixed in 1874 at 1,200, receive honorary certificates, and their names have the distinguishing mark of (G) on the Navy List. This simple and inexpensive distinction, awarded for academic attainments, has been found to have excellent results, and the Greenwich mark is valued by its possessors. It will be still more valued when it comes to represent, as it undoubtedly will in time, a higher standard of attainment. In 1876, 26 certificates were given among 56 officers who presented themselves for examination.

In addition to the honorary certificates, the three highest also obtain a substantial reward in the shape of an annual supplement to their pay for the next three years, the highest man receiving £100, the second £80, and the third £50.

Though the best half-pay officers have the same paper in mathematics as the gunnery lieutenants, there are certain special regulations governing the examination of the latter which require separate attention. The marks given in the required branches are as follows:

A. Pure mathematics:	Marks.
I. Arithmetic	100
II. Geometry (Euclid: first six books)	200
III. Algebra, including binomial and exponential theorems and logarithms	300
IV. Trigonometry	300
V. Co-ordinate geometry, including conic sections	400
VI. Calculus	500
	1,800
B. Applied mathematics:	
I. Statics	350
II. Hydrostatics	350
III. Kinematics and kinetics	400
	1,100
C. Physics	700
D. Fortification, including military surveying	500
Total	4,100

For a first-class certificate in this class a good knowledge of all the subjects is required and an aggregate of 2,500; for a second class, a fair knowledge of all the subjects except A. VI is required, and an aggregate of 1,200; that is to say, about 60 per cent. for a first and the very low limit of 30 per cent. for a second class. The final certificates are based upon those obtained in "theory" at Greenwich, combined with those in practice at Portsmouth; but the significance of a second or third class final certificate is very much reduced by the fact that one of its elements may be a degree of proficiency indicated by 30 per cent. of the maximum.

It will be noticed that mathematics, in one shape or another, constitutes about three-fourths of the marks at the final examination of gunnery lieutenants; hence their attention is chiefly given to these branches. Besides the three required studies, they may, if they like, take other courses that are given at the college; but their time is so occupied that, low as the passing standard is, they have enough to do without any extras. Moreover, the examination mark carries with it such direct and important consequences as to pay, &c., that all their efforts are concentrated on taking a first-class certificate.

The course of the gunnery lieutenants is very thorough, if lacking in breadth, and in general they form a body of earnest students. Dr. Hirst gives them the high praise of being the best among all the students he has known in colleges or universities; the most satisfactory and the hardest workers.*

2.—PROBATIONARY LIEUTENANTS OF THE ROYAL MARINE ARTILLERY.

The Marine Artillery lieutenants join the college either on October 1 or after the Easter holidays, according as they pass the examination for admission in July or December. The course lasts two years, and two examinations are held in each year. Those who pass satisfactorily at the end of the first session after entry remain at the college for another session, at the end of which there is another examination. On passing this, they receive their commissions, seniority being fixed by order of merit. Those who fail at the first examination are dropped from the service, while those who pass at the first, but fail at the second, are allowed to receive commissions in the Marine Light Infantry.

Being younger than any other class of students and having no professional preparation, the probationary lieutenants are placed in two classes by themselves, and receive instruction separately in some branches, and in others in connection with any class that may be pursuing a parallel course. Their course consists of mathematics, physics, fortification (field and permanent), military history, topographical drawing, freehand drawing, French, and German. In mathematics, the first-year men cannot be classified with any other students, and they therefore receive special instruction for eleven hours a week from the assistant instructor of Group A. In fortification also they are classed separately, and have a preliminary course of lectures and drawing by themselves. In all other subjects, except perhaps French, where the course is very limited, they join one or another of the numerous classes. Thus, in second-year mathematics and fortification, and in topographical and freehand drawing, they go in with gunnery lieutenants or half-pay officers; in military history with the marine officers, and in physics with the acting sub-lieutenants. Both classes have a daily drill of one hour under the professor of fortification—the only drill that takes place at the college.

3.—ACTING SUB-LIEUTENANTS AND CANDIDATES FOR NAVAL INSTRUCTORS.

This group comprises acting sub-lieutenants, acting navigating sub-lieutenants, and candidates for naval instructorships. The acting navi-

* The special regulations affecting officers preparing for torpedo lieutenants are given in Chapter IX.

gating sub-lieutenants do not call for much consideration, as the last two in the service are now completing their course at the college, prepara- tory to examination. The regulations governing their studies are simi- lar to those for acting sub-lieutenants.

The course for chaplains and others qualifying as naval instructors is intended to give the professional training that may be wanting in their previous education. The principal subjects are therefore navigation and nautical astronomy. There are five other required subjects, nautical surveying, the practical use of instruments, steam, meteorology, and physics; and three electives, chemistry, drawing, and advanced French. Instruction is given by the instructors who teach acting sub-lieutenants, and generally in the same classes, though, of course, the candidates for instructorships reach a much higher standard.

The third division of this group, that of the acting sub-lieutenants, is by far the most important, and their course requires as much attention as any at Greenwich. It represents a marked stage in the career of executive officers, and the examination at its close is one of the three great tests of fitness for a lieutenant's commission; the others being the seamanship examination, which precedes it, and the gunnery examina- tion, which follows the course in the Excellent.

As soon as practicable after passing in seamanship, acting sub-lieu- tenants are ordered home to pass in navigation at Greenwich.* Here they have a course of instruction lasting six months. As the dates of their arrival are uncertain, no one time in the year can be fixed upon for the beginning of this course. To have as little confusion as possible, fixed dates are announced in each month at which officers arriving at any time in the four weeks previous can join the college. Though this system of monthly commencements saves time for the officers, in not compelling them to wait for others to begin with them, it adds immensely to the work of the college by forming a number of classes pursuing the same course, but in six different stages of advancement. Papers, also, must of course be set every month. It is only adopted as a choice of evils, the needs of the service apparently preventing the return of a large number of officers at or about the same date, or their useful em- ployment on shore in case of early return, during a certain period of waiting.†

---

* Formerly the gunnery examination came first, but in June, 1878, a regulation was adopted reversing the order.

† Vice-Admiral Sir G. Hornby, recently Commander-in-Chief in the Mediterranean, not long ago remarked that the present strain upon the service, in giving all sub- lieutenants six months at Greenwich, was almost more than the Admiralty could bear, unless shortly they saw that they got some compensation from it in an educational point of view.

The course of six months' instruction and the examination at its close comprise thirteen subjects, of which the details are as follows:

1. ALGEBRA.—Fundamental operations, fractions, simple equations, involution and evolution, theory of indices, quadratic surds, quadratic equations, proportion, the three progressions, elements of the theory of logarithms, including the use of tables.
2. GEOMETRY.—Up to the standard of the sixth book of Euclid's elements.
3. TRIGONOMETRY.—Definitions and fundamental formulæ; solution of plane and spherical triangles.
4. MECHANICS.—Elements of statics, dynamics, and hydrostatics.
5. PHYSICS.—Mechanical properties of liquids and gases, optics, heat, and magnetism; including construction and use of the barometer, telescope, microscope, theodolite, sextant, dry and wet bulb thermometers, azimuth compass.
6. STEAM-ENGINE.—Heat, steam, boilers, engines, and propellers.
7. FRENCH.—Writing from dictation, pronunciation, grammar, translation from French to English, and from English to French.
8. WINDS AND CURRENTS.—Prevailing winds and currents, their geographical limits, and their changes at different seasons; cyclones, their characteristics, localities, seasons, rotation, and track; relation of barometric pressure to prevailing winds and storms; Buys-Ballot's law; leading theories of winds and currents; rains and rain seasons.
9. PRACTICAL NAVIGATION.—The sailings; dead reckoning; Mercator's chart; passage of bright stars over meridian; latitude by meridian altitudes, by altitude of the pole star, by altitudes near the meridian, and by double altitudes; longitude by chronometer, and by lunar distance; error and rate of chronometer by single altitude, and by equal altitudes; variation of compass by amplitude, by altitude azimuth, and by time azimuth; time of high water.
10. NAUTICAL ASTRONOMY.—Definitions and principles; investigation of all processes and formulæ; explanation of all corrections used; Sumner's method; problems.
11. NAUTICAL SURVEYING.—Use of charts; rating of chronometers; determination of meridian distance; selection and measurement of a base line; determination of latitude, longitude, and true bearings; triangulation; leveling; soundings; fixing position; tide-gauge; establishment of the port.
12. INSTRUMENTS.—Construction and use of the marine barometer, sextant, artificial horizon, azimuth compass, theodolite, and level.
13. OBSERVATIONS.—Of the sun, for determining latitude; error of chronometer and variation of compass.*

From the list of subjects and text-books it might be inferred that the course and examination were too severe to be completed in six months. Indeed no one familiar with education would undertake to get a class through these thirteen subjects, for the first time, in six months, however well prepared they might be. But as a matter of fact the acting sub-lieutenants do not go through these subjects at Greenwich for the first time; they have been through nearly all of them before in the Britannia. The real fact is that in the Britannia the students have been

_____

* LIST OF TEXT-BOOKS.—Hamblin Smith's Arithmetic, Algebra, Plane Trigonometry, Elementary Statics, and Hydrostatics; Todhunter's Spherical Trigonometry, Elements of Euclid, and Mechanics; Ganot's Natural Philosophy, or Balfour Stewart's Elementary Physics; Laughton's Nautical Surveying, and Physical Geography; Hull's Practical Nautical Surveying; Jeans's or Raper's Navigation; Evers's Steam and the Steam-engine.

crammed up to an extremely high and full course (for boys of their age), and that from that time till the final examination at Greenwich, they are trying to digest what they have studied, and to convert it into real knowledge. The second course hardly aims at more than an intelligent review of the studies—as far as there have been any studies—of the last seven years. As Sir Cooper Key said plainly in his testimony: "It must be remembered that all we (at Greenwich) teach, or nearly all we teach them, they ought to have known before they left the Britannia. The utmost that we go up to when they pass as sub-lieutenants afterwards is very little beyond what they are supposed to know when they leave the Britannia;" and he adds: "The advantage of our system is that we take care that they know it." Certainly, "the end proposed,'' as the committee's report says, "is a modest one." It is merely to go over, at the age of twenty, studies that were completed up to that point five years before.

The clearest proof of the practical identity of the two courses is to be found in the examination papers at the close of each. Comparing the final examination on board the Britannia with the final examination at Greenwich, we find that eight subjects are common to both, viz: Algebra, geometry, trigonometry, physics, theory of navigation, practical navigation, nautical surveying, and French. Besides these, the Britannia examination includes Latin, and that at Greenwich, steam, mechanics, and winds and currents—all three of an elementary character. In comparing sets of papers in the subjects common to both examinations, Dr. Hirst says:

The algebra paper is virtually the same; the difference is not worth speaking of. The geometry is the same. The trigonometry, the practical part of it at all events, is precisely the same. In spherical trigonometry there is no difference in character. In practical navigation there is the college sheet. In chart drawing [surveying] it is the same. In physics it is slightly simpler now than it was then. On the whole, I should say that the present examination is almost, if not quite, as simple as the one indicated by that paper.*

The paper referred to was one of the Britannia papers of 1873. Since then the examination has changed somewhat; but, taking the papers given at both places in July, 1878, Dr. Hirst's remarks are almost equally true of them. In these, the Greenwich questions in algebra, geometry, trigonometry, and nautical surveying cover a trifle more ground than those in the Britannia; a mere trifle, hardly perceptible. In physics they are quite as simple, though perhaps they involved reading another book. In French the college paper is perceptibly easier. The chief difference—and that is very slight—in favor of the college is in navigation, to which the students may be supposed to have been devoting themselves for the past five years, and which has double weight at the examinations. It should be added that the similarity in the two sets of papers is not accidental, because both were made out by the naval instructor who acts as special examiner at the college.

---

* Evidence taken before the Royal Naval College committee, Question 223.

There is an examination of sub-lieutenants for admission, but no one is rejected; in fact, as the candidates know so little at the start, a rejecting examination would tell very severely on them. It would defeat the object of the course, as the junior officers only come to Greenwich to prepare for the final examination.

The hours of instruction are from 9 to 1 in the morning, and from 2 to 4 in the afternoon; except on Saturday, when there is no afternoon study. In mathematics and navigation, the two important subjects, the four naval instructors have general charge of the classes. The first two months are occupied with algebra as far as the binomial (including it in the case of a few of the best men), and geometry (six books of Euclid); the third and fourth with trigonometry, mechanics, dynamics, and hydrostatics; and the last two with navigation and nautical astronomy, under the senior instructor. The course for the third and fourth months seems to be the most difficult for the students. All the pupils of any one instructor attend in the same room, though a part of them may be a month in advance of the others. The normal limit of a class is twenty-five men. The instructor's method is to have his class in his room for two or three consecutive hours, usually three in the morning and two in the afternoon; to spend the first part of this time in blackboard explanation, doing problems and explaining them; and then to give out similar problems to the class, who work them out during the rest of the time. While working on the problems they are constantly receiving assistance and explanation, as they may need it. In fact, the system is one of private tuition, applied to a class, instead of an individual. In theoretical navigation a set lecture is given, on which the students take notes. They then go to another room, where they work out the subjects of the lecture, under supervision, as before. These lectures, which number about thirty, are frequently attended by half-pay officers, especially towards the end of the course, when they have worked up their mathematics. The senior instructor also examines the sights taken by the sub-lieutenants for their final examination. The rest of the time is devoted to instruction in French and nautical surveying, and to lectures in steam (four a week), physics (one a week), and meteorology. Most of the lecture courses are supplemented by some personal instruction, and by questions or problems given out for answer and solution; but sub-lieutenants have no laboratory courses.

There are no marks given during the six months of instruction; the only test is the final examination. For this, the following scale of marks is adopted:

Marks.
1. Algebra 125
2. Geometry 125
3. Trigonometry 125
4. Mechanics 125
5. Physics 100
6. Steam-engine 100
7. French 100

8. Winds and currents... 100
9. Practical navigation 200
10. Nautical astronomy 200
11. Nautical surveying 100
12. Instruments 40
13. Observations 60
                                               —————

    Total 1500

Three certificates are given, requiring the following marks:

First class 1,250
Second class.. 1,000
Third class.. 750

Moreover, to receive a first or second class, $16\frac{2}{3}$ per cent. must be obtained in each subject; and no certificate is given if 50 per cent. is not obtained in No. 13, observations.

An officer failing to obtain a certificate is re-examined at the end of a month (in practice, more nearly two months), but if he fails on the second examination he is discharged from the service. The failures that resulted in rejection during the first four years after the foundation of the college reached the rather high figure of 30.* They were generally due, according to Admiral Fanshawe, to the candidates' knowing nothing to begin with, and "taking things very easily" in the first three months. Curiously enough, nearly every one of those rejected had passed his time at sea in a ship with a naval instructor. This was before the half-yearly examinations afloat were instituted, or, at least, before their effects began to be felt, and it is but just to say that they are expected to improve very much the preparation of sub-lieutenants.

It is not a little singular that, notwithstanding the low standard of the examination for acting sub-lieutenants, notwithstanding the previous study on precisely the same ground, and the almost individual instruction the pupils get, yet the practice of taking a private tutor during the whole six months prevails very extensively. About half the candidates are supposed to make use of such assistance, although the college instructors dislike and discourage it. Nor is it confined to the lower men; there are cases of officers who get a first class who have been coached for three hours a day during the whole six months. The private tutors, of whom there are several at Greenwich in constant employment, study closely the papers given each month, and cram their pupils accordingly. The examiner, on the other hand, labors to vary his questions in such a way as to defeat the object of the tutor and to probe the knowledge of the student, so that a constant struggle is kept up between crammer and examiner. It is only in the required course of sub-lieutenants that the practice of private tuition prevails.

A lieutenant's commission is given in June and December of each year to the acting sub-lieutenant who has passed the best examination during the preceding six months, provided he has obtained first-class cer-

_____
*A high figure, that is, for officers who have already been seven years in the service.

tificates in seamanship and gunnery, and has reached a certain standard in the college examination. A prize of books or instruments is given to all who obtain first-class certificates, and the number of each sub-lieutenant's certificate in seamanship and gunnery, as well as in the college examination, is noted in the Navy List. The *Beaufort Testimonial* is a prize, established in 1860, to be given annually to the officer who passes the best examination for the rank of lieutenant. Candidates are, however, required to state their intention of competing, upon joining the college, and they pass a special examination at the close of the course. The examination includes advanced papers in mathematics, nautical surveying, French, nautical astronomy, physics, and steam, and is far more difficult than the regular examination.

On the completion of the course at Greenwich, acting sub-lieutenants are granted ten days' leave, at the close of which they join the Excellent for a three months' course, preparatory to the examination in gunnery.

4.—ENGINEERS AND CONSTRUCTION STUDENTS (GROUP A).

This group includes—
Chief engineers.
Engineers.
Assistant engineers.
Acting assistant engineers.
Construction students.
Private students.

The class of chief engineers and engineers resembles in its formation the classes of voluntary or half-pay officers, in group B, but in its course of instruction, it belongs distinctly to group A. Three chief engineers and fourteen engineers (or assistant engineers that have not studied at South Kensington) may be admitted annually on their own application. They pass a test examination in elementary mathematics at the time of admission, and they remain one session at the college, passing an examination at its close. A limited number are allowed, on recommendation of the examiners, to remain a second year for study. The courses, which can best be described in connection with the junior students, are opened to the older officers chiefly because they lack the advantage of a systematic training, the South Kensington school having been opened only in 1864.

Acting assistant engineers have a required course at the college after their six years' course as engineer students in the dockyards. This course at the college, which lasts one year, corresponds to the required course of acting sub-lieutenants, and like that, it is closed by an examination, for confirmation of an acting appointment. On passing the examination, the students are commissioned as assistant engineers, and the date of their commission depends on the class of their examination certificate. Those who obtain a first class receive commissions dated on the same day as their acting appointment; those who take a second class have commissions dated six months later; while the commissions of the third class are dated on the day of their discharge from the college. It is at the final examination, as has been already stated, that the two best men in each year are selected to be retained at the college for further instruction.

The construction students are obtained from the dockyard apprentices, by the process already described—a competitive examination, at the end of a five-years' course of practical and theoretical training in the dockyards. As only three are selected annually from a large number of competitors, the classes consist of the very best material. The course at Greenwich lasts three years, in the first of which the construction students are classed with the large body of acting assistant engineers,* and in the second and third, with the small selected classes of assistant engineers pursuing an advanced course.

In the same group, and undergoing the same course, are the private students, who pass examinations at the beginning and end of their studies, and have the same certificates of proficiency as the Admiralty students. By an exceedingly liberal provision, similar to that which opens the doors of the United States Naval Academy to Japanese youths, foreigners as well as Englishmen, and even foreign officers, are admitted to the benefits of the college course. In this way there have been students from among the naval officers of Russia, Italy, Denmark, Sweden, and other states, who have gone through the course, and obtained their final certificate.†

The group of students comprising the foregoing classes are known collectively as "students in naval architecture and marine engineering." Leaving out the acting assistant engineers, who are pursuing their required course before promotion, it is a group small in numbers, but it is among the most important at the college, as it receives the highest professional and scientific instruction given there, and from it will be recruited, in future, all the designers and constructors of ships and engines in the Navy.

The courses for the two divisions of students (construction and engineering) are identical in theoretical subjects, but differ in the professional branches. The common course includes mathematics, physics, chemistry, and mechanics. All these courses are of a very high character, a fact which is made possible by the previous training that all the Admiralty students have received in the dockyard schools. The course in mathematics is equal to the highest university course in England. In

---

* That is to say, with the first division, comprising only the best men. The second division of acting assistant engineers takes a much less extensive course in mathematics than the higher.

† In view of the possibility—a possibility that is now, in two cases, at least, about to be realized—that it may be deemed wise to obtain the unequaled advantages of this course for American students, a copy of the Admiralty circular is given in the Appendix, Note F.

physics and chemistry, the course includes a long and thorough training, under the ablest professors, in laboratories that are among the best appointed in the world. The same is true of applied mechanics, the instructor in which, Professor Cotterill, is the highest authority on the subject that England has had since the death of Rankine. In the professional branches, nine hours a week are devoted to the design of ships and engines. Problems are proposed in the form in which they will actually arise when the pupil enters the service, and designs are made from which a ship or engine could be built. Some of these designs contain valuable original features, and they will always be of use to the authors during their future career. Finally, as the instructors are men holding positions in the construction branch of the Navy, they are enabled to give their students the very latest problems—those on which they themselves are engaged.

In general, the students in this group are separated from those of the executive branch, but they attend the general lectures in physics and chemistry, and some of their special courses are open to half-pay officers. The general arrangement of instruction is to give the morning to theoretical or scientific study, and the afternoon to practical or professional work. The first three morning hours are devoted to mathematics and applied mechanics, taken on alternate days. The fourth hour is divided for different days among chemistry, physics, and French. In the same way the afternoons are given on certain days to laboratory work, both in physics and chemistry, and on the other days to naval architecture and ship-building by the construction students, and to marine engineering and engine-drawing by the engineers. There are also additional hours in French, and a special course in descriptive geometry, for first-year students; and some of the lectures that are optional for all officers are pre-eminently of importance for this class. The most noticeable are those on the safety of ships, the behavior of ships at sea, the structural arrangements of modern war-ships, as illustrated by models in the museum, and the tonnage and propulsion of ships.

The course in pure mathematics is in charge of one of the professors, whose exclusive attention is given to this group of students. The method of instruction is similar to that prevailing in other classes and already described. That a point so much higher is reached, is due to various causes, but chiefly to the careful selection of students, and to their previous thorough course in the dockyard schools, a course including descriptive geometry, calculus, and mechanics. Moreover, the students are all of mature age. They can perceive the direct professional bearing of their theoretical studies, and they give every morning for three years to instruction in either mathematics or applied mechanics, to say nothing of the considerable time outside devoted to study. In the first year a division is made, so that the lower men may have studies suited to their capacity and attainments, and may not impede the progress of the best. The first division includes the construction students and the ablest of

the acting assistant engineers; the second, the rest of the acting assistants, who have not been diligent at the dock-yards, but who have just
passed the examination. It also includes the voluntary engineer officers,
who are older men, and who have had few opportunities of study. The
foreigners are arranged in one division or the other, according to their
advancement. The second division has a moderate course, intended
chiefly to broaden and deepen the knowledge acquired in the dockyard
schools. For example, this division does not read calculus. After the
first year, there is no necessity for a subdivision of classes. The chosen
students are nearly on a level of attainment, and, as there are only five
of them in each of the last two years, nothing could possibly be gained
by classification.

The course in construction deserves particular attention. On this
subject, the construction students have two instructors: an instructor
in ship-building, including ship-drawing and laying off, and an instructor
in naval architecture. Both these officers belong to the staff of the
construction department of the Admiralty; the instructor in naval
architecture being Mr. W. H. White, an assistant constructor, and the
author of a well-known treatise on the subject. The account of the
details of the course in his department is based on his own statements.

The instructor in ship-building begins his course when the students
enter the college, and employs them two or three afternoons in the
week, during the whole of the first year and about a third of the second. During that time the average amount of work performed includes the preparation of one or two sheer-drawings and midship-sec-
tions, in order that the students may acquire facility as draughtsmen,
and the laying off of the fore and after bodies for a wooden ship, of all
which the Admiralty students have already obtained considerable knowledge in the dockyards. Towards the close of each year the special
course of lectures on ship-building is given for the benefit of executive
and engineer officers, as well as construction students.

The course under the instructor in naval architecture is more extensive, and continues through the whole three years. In the first year attention is given exclusively to ship calculations during one or two afternoons of each week. Each student works independently, from the drawings of an actual ship, and is taught to perform, and in nearly all cases
does perform, the following calculations:

1. Proofs and applications of rules for finding the areas and the position of the center
   of gravity of plane curves, including Simpson's rules, the method of polar co-
   ordinates, and the geometric process.
2. Proofs and applications of rules for calculating the displacement of ships and position of the center of buoyancy, including the ordinary method by double application of Simpson's first rule, Dr. Woolley's rule, and the geometric process.
3. Construction and use of curves of displacement, tons per inch immersion, and
   area of immersed midship section.
4. Proofs and applications of rules for calculating the position of the metacenters for
   transverse and longitudinal inclinations.

5. Construction and use of metacentric diagrams.
6. Explanations and applications of rules for tonnage measurements, including build-
er's old measurement, register tonnage, freight tonnage.
7. Graphic representations of the longitudinal distribution of the weight and buoyancy in ships floating in still water, together with the construction of curves of
loads, shearing forces, and bending moments.

The work of the first session may be said, therefore, to comprehend
all the ordinary calculations made for a ship, with some special methods
that are useful as checks and aids in ship designing. As a rule, the
students are capable, in their subsequent work, not merely of applying
the methods learnt in the first year, but of selecting the process best
adapted to economize labor, and to secure accuracy in practice.

In the first three months of the second year, while the students are
partly occupied under the instructor in ship-building, they continue their
work on ship-calculations, and are taught how to perform the following
operations:

, 1. Calculations for statical stability and construction of curves of
stability.

2. Calculations for dynamical stability by Moseley's formula.

3. Calculations for and construction of the loci of the center of buoyancy, metacenter, and center of flotation of ships for all transverse inclinations, under certain assumed conditions.

The student who has learned to perform these three calculations
ought to have no further difficulty in dealing with any calculation connected with the buoyancy and stability of ships.

At the close of this course of ship-calculation the work of ship-design
is begun, and continued through the rest of the second and the whole of
the third year, occupying two or three afternoons a week. In general,
arrangement is made for the actual performance of the work by each
student, but in the case of ship-design the amount of work is generally
too great for single students to work out a design, and it is customary
for them to be grouped by twos and threes, care being taken that each
has a share in all the important parts of the work. This includes the
preparation of the drawings and calculations connected with the design
of a ship of some selected type. The character and extent of the work
done may be illustrated by the following outline of what would ordinarily be done in designing an iron-clad ship:

1. Preliminary estimate of dimensions and displacement required to fulfill the desired
conditions as to draught, trim, armor, armament, speed, and equipment, &c.,
as well as to provide sufficient stability.
2. Preparation of sheer-drawing, midship section, outline specification for scantlings,
plans of decks, battery, &c., required for making an estimate of the weights of
hull, armor, &c.
3. Calculations of displacement, and positions of the center of buoyancy and metacenters.
4. Detailed calculations for the weight of hull, armored and fitted, together with the
position of the center of gravity of the same in height and length.
5. Detailed estimates of weights of equipment and position of their center of
gravity.

6. Combination of the results of 4 and 5 to determine the weight and position of the center of gravity of the completed ship. The comparison of these results with those obtained in 3 shows whether the new design is successful as regards draught, trim, and stiffness, or whether any corrections are necessary in the form or stowage. Having made these corrections, the design passes out of the preliminary stage, and only its details require to be worked out.

7. Preparation of outline profile and plans of decks and hold, showing the main features of the stowage throughout the ship.

8. Preparation of sail draught, with calculations for area and center of effort of plain sail, and power to carry sail.

9. Calculations for and construction of curves of displacement, tons per inch, and immersed midship section; also for the metacentric diagram.

10. Calculations for statical stability, construction of curve of stability, estimate of i dynamical stability.

11. Calculations for speed under steam.

12. Calculations for longitudinal bending moments in still water and among waves, including the construction of curves of weight, buoyancy, loads, shearing forces, and bending moments.

13. Construction of equivalent girders, and estimates of longitudinal strains on the structure.

A student who has passed through this course ought to be thoroughly competent to undertake the design of any class of ship for which the fundamental conditions are furnished.

In the third year students also attend a special course of about 35 lectures on the stability and oscillations of ships, delivered by Mr. White. By that time they have advanced far enough in mathematics to take up the general theorems of stability, and the theory of the oscillations of ships; and one morning in each week is devoted to these subjects. The extent of this course is shown by the following abstract:

SYLLABUS OF LECTURES ON THE STABILITY AND OSCILLATIONS OF SHIPS.

I.—STATICAL STABILITY OF SHIPS.

1. General theorems of Dupin, with extensions and particular applications to ships.

2. Metacentric methods of estimating initial stability, "metacentric heights" for various classes of ships, with results of experience as to the amount of stiffness desirable.

3. Applications of the metacentric method, including: Inclining experiments to determine the vertical position of the center of gravity of ships. Variations in stability due to addition or removal of weights. Estimates for changes of trim of ships produced by moving weights, or adding or removing weights. Efficiency of various methods of water-tight subdivision in preserving the stability of ships when compartments are bilged. Stability of ships aground and partially water-borne. Estimates of power to carry sail.

4. Stability at finite angles of inclination, with details of the various methods of calculation that have been proposed by Atwood, Reade, and Barnes. Calculations for and construction of curves of stability. Examples of the forms of such curves for typical ships. Influence of freeboard, beam, draught, and vertical position of the center of gravity upon the forms of curves of stability.

II.—DYNAMICAL STABILITY.

1. Moseley's formula and other expressions for dynamical stability. Estimates of dynamical stability by direct calculation, or from curves of stability. Connection of dynamical stability with the safety of ships acted upon by suddenly-applied forces, such as gusts or squalls of wind. Reserve of dynamical stability.

III.—OSCILLATIONS OF SHIPS.

1. Unresisted rolling in still water.
2. Rolling in still water, with the effect of resistance included.
3. Still-water rolling experiments, their conduct, and the construction and analysis of curves of extinction.
4. Dipping oscillations in still water.
5. Rolling among waves—treated mainly in accordance with the investigations of Mr. Froude, with the effect of resistance neglected and included.　·
6. Principal deductions from the theory of rolling, illustrated by recorded observations of the behavior of ships.
7. Pitching oscillations.
8. Heaving and yawing.

It may be added that the professor of applied mechanics, Mr. Cotterill, gives a course of lectures each session on the trochoidal-wave theory and the propulsion of ships, including the stream-line theory of resistance.

It is not supposed that the college will replace the necessity of experience in designing ships or supply the facility and information only gained by long and extensive practice; but it is intended that it shall give a good grounding in the principles of design, and enable the student to proceed intelligently in whatever direction his after work may lead him. Evidence of the thoroughness of the training may be found in the fact that many graduates of the course are at once employed in the construction department of the Admiralty; and both Mr. E. J. Reed, the late chief constructor, and Mr. Barnaby, who at present occupies that position, have borne testimony to the useful work that they have done.

During the summer, which the other students spend in vacation, the young constructors and constructing engineers go for three months to the dockyards, or, in the case of the latter, to the Steam Reserves, where they put in practice their theoretical knowledge, renew their skill in practical work, attend trials of new and repaired machinery, and, in general, gain additional practical experience, to be turned to account in the rest of their course, and ultimately in their profession.

At the close of the course certificates of three classes are given for proficiency, as shown at the final examination; and it has been said by one of those best capable of judging in England, that the first-class certificate in naval architecture or marine engineering, given at Greenwich, is the highest class of certificate of the kind given in England, and represents the highest scientific attainments, and that a man's fortune may be said to be made who takes it.

CHAPTER VIII.

THE GUNNERY-SHIP EXCELLENT.

The establishment to which the name of the Excellent is applied is the third great institution for educational purposes in the English Navy, and is, in its way, quite as important as either of the others. But it differs from them in one respect, that it is devoted as much to the training of seamen and petty and warrant officers as to that of commissioned officers. When it is considered that it has resources equal to the training of at least 100 officers and over 1,000 men, at a time; that there are usually more than this number of men actually there, and that it is the only place where systematic instruction is given to officers at least, in this branch,* it will be seen of what great consequence it is in the naval system.

The great work of the Excellent is instruction in theoretical and practical gunnery. There is also a school course in mathematics for gunners and gunnery instructors (petty officers or seamen), but the main work of the establishment is with gunnery.

In this branch there are courses carefully arranged for five different classes of officers. The officers qualifying for gunnery-lieutenants take the longest and fullest course, lasting about six months. The other courses, each of three months, but differently divided, are taken by officers of the Marine Artillery, sub-lieutenants completing their course and examination for promotion, and voluntary, or, as they are commonly called, "short-course" lieutenants. The fifth and last course, only organized last year, is for captains and commanders. It extends over twenty days, and is purely voluntary, like that of the short-course lieutenants, and like the courses for the same grades of officers at Greenwich. The division of time is briefly shown in the following table, the subjects being taken up in the order named.

COURSES OF INSTRUCTION ON BOARD THE EXCELLENT.

Subjects.	Captains and commanders.	Gunnery lieutenants.	Short-course lieutenants.	Marine artillery.	Sub-lieutenants.
Heavy gun	5	20	20	20	15
Field exercise	5	18	18		14
Ammunition	4	10	10	10	10
Truck gun	1	5	5	5	5
Musketry	2	10			
Cutlass and pistol		8	3		3
Turret	2	4	4	4	4
Field gun	1	8	5		6
Diving		1			
Theoretical, including examination of guns		10		10	
Final		10		10	3
Examination		7		3	5
Total	20	111	65	62	65

* The Cambridge, stationed at Devonport, is also a gunnery ship, but is exclusively for the training of men.

To carry on these courses, as well as those for the instruction of gunners, gunnery instructors, seamen-gunners, and men qualifying for one of these positions, the Excellent has a staff of officers composed of a captain, commander, six gunnery lieutenants, a naval instructor, and sixteen gunners, besides the usual staff officers. There are also a number of gunnery instructors. Besides their duties with the men, the gunners instruct the officers qualifying for gunnery lieutenants, and the gunnery instructors instruct the sub-lieutenants and short-course lieutenants. The officers take theoretical instruction, battalion drill, and exercise at quarters. It will be noticed that by this arrangement much practical instruction is regularly given by petty or warrant officers to commissioned officers. This system, which might at first sight be thought open to objections, is found to work exceedingly well on board the Excellent and to occasion no difficulty. The gunners, gunnery instructors, and seamen-gunners of the English navy are highly trained men, of great intelligence, who have won their positions after one, two, three, or perhaps four courses of thorough training, with severe examinations; and with rare exceptions, they are fully equal to the work they are called upon to do in the instruction of their superiors.

The *matériel* of the establishment consists of two old ships-of-the-line, the Excellent and the Calcutta, lying in the stream off Portsmouth dockyard. There are also two screw gunboats attached to the Excellent, the Comet and the Skylark, and an old mortar-boat used in the Crimean war, which serves the purpose of a rolling-motion boat. The Glatton, one of the powerful armor-plated coast-defense turret ships, is attached as a tender to the Excellent, and is used for turret instruction. The Lord Clyde, one of the older 18-gun wooden armored ships, is now being fitted out to be used as a drill ship. As to guns, the Excellent has on her lower deck ten 100-pounder.smooth-bores, and on her upper, one 90 cwt. 7-inch revolving gun. The Calcutta's guns are two 9-inch, two 8-inch, three 7-inch 6½ ton, and one 7-inch 4½ ton. The Glatton carries two 25-ton guns in her turret. The Comet has one 18-ton gun, and the Skylark two 64s and one 40-pounder. The Lord Clyde has a broadside battery of eighteen 6½-ton guns. The rolling-motion boat has two 9-pounders, and there are also for practice a launch, armed with one 9-pounder, and a cutter, with a 7-pounder and a Gatling (.45 caliber). The pile-battery consists of two 9-pounders, and the battery on the island, used for field-gun exercise, of six 9-pounders and a Gatling (.65 caliber). It will be seen that this comprises nearly every description of gun in use in the English service, a fact of the greatest importance in estimating the efficiency of the institution.

Before going into the details of the various courses, it should be stated that they are largely pursued in common with the seamen and petty officers of various grades, reviewing or qualifying for higher ratings. In battalion drill the officer-students are only company or non-commissioned officers; but in squad and great gun drills they fall in and work with the men. There is no permanent battalion organization; in fact, owing to the frequency with which separate classes join, and the irregularity with which officers and men are attached and detached, it would be almost impossible to have such an organization, and it would not be of any great advantage. Battalion drill and landing usually take place on Thursdays; and field-gun exercise on Friday mornings, with quarters in the afternoon. Theoretical instruction is given in lectures on Saturdays, and in some branches by a lesson lasting a quarter of an hour before and after each drill. Moreover, each drill and exercise involve a certain amount of theoretical instruction. The rest must be done by the students themselves, with the help of their text-books, and such explanation as they may receive from time to time from the officers of the staff.

The text-books used in the Excellent are mostly official publications of very recent date, and are all works of the highest character, and specially adapted to the needs of the course. They comprise the Gunnery Manual; the Official Treatise on the Construction and Manufacture of Ordnance in the British service, 1877; Wood's Notes on Naval Guns; Motion of Rifled Projectiles; Britton's Review of the Rifle System ; Official Treatises on Ammunition and on Military Carriages; Manufacture of Gunpowder at Waltham Abbey; Rifle and Field Exercises and Musketry Instructions, 1877.

In general, there are two courses pursued on board the Excellent, known as the long course and the short course; the first of about six months, and the second of three months. These two courses may be taken as types of the work done in all the specific courses given to officers and men, all being modifications of one or the other of the two established systems. The long course answers pretty exactly for the gunnery lieutenants (i. e., officers qualifying as such), and the short course for the voluntary lieutenants and the sub-lieutenants, the distribution of time for each being that previously given in the table.* In these courses each day's work is marked out, the exercise taking the best part of each morning and afternoon except Saturday.

The details of the courses are given below :

1.—LONG COURSE.

1.—HEAVY GUN.

1st to 4th day.—Formerly a specific exercise was assigned to each of the first four days. For example, the first morning was wholly given to casting loose and sponging, and the afternoon to sponging and loading different guns; the second day was given to running in and out different guns, using high elevations, and so on. Now, however, instead of spending a whole day at one exercise, the exercises are varied through the whole preliminary drill in the Manual of Gunnery.

5th day.—Clearing for action, independent firing, and training for loading.

6th day.—Revolving gun.

7th day.—Electric firing.

8th to 11th day.—Remainder of the firings, viz, from platform, ship (100-pounder), rolling-motion boat, and gun-boat, according to the state of the tide.

12th and 13th days.—Transporting, dismounting, gear of carriage and slide.

14th day.—Lowering ports, chalking drums, and down ports.

15th day.—Shifting breechings, supply, and spare stores.

16th day.—Working guns in a seaway ; preparing for ramming.

17th day.—Signals.

18th day.—Diminished crews.

19th and 20th days.—Examination.

Each day's drill begins and ends with a lesson lasting a quarter of an hour. The lessons are arranged somewhat as follows: (1) Parts of the guns, carriages, and slides ; (2) sights, wood-scales, &c.; (3) weight of charges, bursters, &c.; (4) weight of projectiles ; (5) weight and dimensions of guns. These subjects are taken in the same order during each five days of the heavy gun course, the student advancing at each lesson. Instruction in the Manual is given for two hours, each morning and afternoon.

2.—FIELD EXERCISE.

1st to 3d day.—9–10.30. Manual and firing exercises. 10.40–11.45. Squad drill. Same in the afternoon.

4th to 6th day.—Same as first three days, except that squad skirmishing is substituted for squad drill in the afternoon.

7th to 17th day.—9¼–10. Drill each other. 10–10.30. Miscellaneous.* 10.40–11.45. Companies. 1.30–2.40. Company skirmishing.

18th day.—Examination.

3.— AMMUNITION.

1 day.—Equipment of boats manned and armed, and firing 9-pounder from the launch.

2 days.—General description of all fuses used in the naval service, method of fitting, and mode of supply.

¼ day.—General description of war and life-saving rockets, tubes, and fireworks, used in the service ; their use, supply, and stowage in ships.

¼ day.—Rocket-firing.

1 day.—General description of the ammunition in use for Woolwich guns, including filling shell.

1 day.—Firing 64-pounder from gunboat.

1 day.—Life-saving rocket firing.

¼ day.—General description of the ammunition in use for 64-pounder gun, and for boat and field-guns.

¼ day.—General description of the ammunition in use for B. L. R. guns.

1 day.—General description of magazines and shell-rooms, stowage, working, and ventilation.

1 day—Examination.

Total, 10 days.

4.—TRUCK GUN.

2 days.—Preliminary drill, as in the Gunnery Manual.

1 day.—Firing and shifting breechings.

1 day.—Transporting, dismounting, and exercise with diminished crews.

1 day.—Examination.

Total, 5 days.

Each drill begins with a lesson of a quarter of an hour.

---

* Miscellaneous subjects comprise manual and firing exercises, sword-bayonet exercise, exercise for receiving cavalry, funeral exercise, &c.

5.—MUSKETRY.

10 *days,* as laid down in the musketry instruction, including lessons, position drill, aiming drill, judging-distance drill, and practice; blank firing; volley, independent, and skirmishing firing, and moving-object practice.

6.—CUTLASS AND PISTOL.

1 *day.*—Cutting and guarding practice.
1 *day.*—Pointing and general practice.
1 *day.*—Attack and defense practice.
1 *day.*—Attack and defense practice, and pistol drill.
3 *days.*—Attack and defense practice, drilling each other, loose play, pistol drill and firing, and cutting lead.
1 *day.*—Examination.
Total, 8 days.

7.—TURRET.

2 *days.*—Preliminary drill.
1 *day.*—Practice.
1 *day.*—Practice and examination.
—
4 *days.*

Firing takes place as convenient. Steam is raised on the third day for each class to work the turrets.

8.—FIELD GUN.

1 *day.*—Preliminary drill, as in the Gunnery Manual, up to and including "Front limber up," and comprising such exercises as formation of gun's crew, stations, marching, inclining, taking ground, reversing, wheeling, &c.; front unlimbering, loading, firing.
1 *day.*—Remainder of preliminary drill, including unlimbering and limbering up, reversing in a narrow passage, ascending and descending inclines, and changing front in action.
1 *day.*—Gatling-gun drill. Embarking.
1 *day.*—Target practice with field guns.
1 *day.*—Target practice with Gatling.
2 *days.*—Drill in Part II, of the Manual; including action; retiring with the prolonge, shifting wheels, removing disabled carriages, &c.
1 *day.*—Examination.
—
8 *days.*

The drill is always carried out with limbers, &c., packed for firing, and practice takes place according to the state of the tide.

The following table shows the details of the practice in firing, and the number of rounds fired by each student in the different exercises:

TABLE OF FIRING-PRACTICE.

Instruction.	Number of rounds.	Where fired.	Gun.	Projectile.	Distance.	Remarks.
Heavy gun	4	Platform	9-pounder	Empty shell	1,200 yards	Every man to fill his shell and fit the fuse.
	2	Ship	100-pounder	Shot	1,000 to 2,000 yards	
	3	Rolling-motion boat	9-pounder	Empty shell	1,200 yards	
	2	Gunboat	64-pounder	do.	1,000 to 2,000 yards	
	1	Mortar boat	Mortar	Shell (blowing charge)	1,000 yards	
Manual	1	Rocket boat	Rocket		1,800 yards	
Ammunition	3	Gunboat	64-pounder	Shell, filled	Various	
	3	Launch	9-pounder	Shell	do.	
	4	Ship	64-pounder	Shell, empty	1,250 to 2,500 yards	
Truck gun	90	Rifle range	Rifle		Various	One round, 12-in. gun, for each class.
Musketry	10	do.	Pistol		25 and 50 yards	
Pistol	2	Turret ship	9-pounder	Empty shell	1,000 to 2,000 yards	
Turret	2	Range	do.		1,800 yards	One double shell for each class. Target on same bearing; time, 2′. Every man to fit his own fuse. One 64-pounder M. L. shrapnel and one case for each class.
Field gun	1	Cutter	7-pounder	Empty shell	Various	
	2	Gunboat	64-pounder	do.	About 1,500 yards	
	1	do.	do.	Shell, filled	do.	
	1	do.	40-pounder, R. L	Shot	do.	
	1	do.	10-inch	Common shell, or chilled shot.	Various	Three targets placed in a line and moored 600 yards apart; the gunboat steams at full speed parallel to them, bringing them on in succession.
	3	do.	64-pounder	Shell, empty	600 to 800 yards	
	2	do.	100-pounder	Shot	1,250 to 1,800 yards	

Total firings for each man:

```
Gun ..................................  35
Rifle ................................  90
Pistol ...............................  10
Rocket ...............................   1
       Total ......................... 136
```

2.—SHORT COURSE.

The short course is chiefly a modification of the other. Thus, allowing ten days for the exercise in heavy guns (the time allowed in the short course for men qualifying for gunnery ratings), the first three days cover the ground of the first four days in the long course; the fourth (short course) corresponds to the 12th and 13th; the fifth, to the 5th and 14th; the sixth, to the 6th and 15th; and so on. In field exercise a similar plan is pursued, but the course is even more curtailed. In ammunition, pistol drill, musketry, and other branches, the chief omissions are in the firings. The short course, varied to suit the requirements of the different classes of students, may be said to be the course pursued by acting sub-lieutenants, voluntary lieutenants, and officers of the Marine Artillery. A still shorter course is taken by the voluntary captains and commanders.

3.—ACTING SUB-LIEUTENANTS' EXAMINATION.

It remains to notice the important points in the examination of acting sub-lieutenants. This, as has been before stated, is the third and final examination of these officers before confirmation as sub-lieutenants, the other two being seamanship, conducted on shipboard, and navigation and kindred subjects, at the college at Greenwich.* It is therefore the last required examination in their career as officers. There are other examinations, it is true, which officers are required to pass who adopt certain specialties, as gunnery, navigation, &c., and there are examinations at the close of the lieutenants' course at Greenwich, but in all these cases the preliminary steps which involve an examination are voluntary with each individual. The examination in gunnery is practically, then, the last required examination for officers.

The time of the examination is five days, and the numbers given in each branch are as follows:

Relative weights for passing certificates.—Acting sub-lieutenants' gunnery examination.

Subjects.	Maximum.	Minimum for—		
		First class.	Second class.	Third class.
Heavy gun	100	90	80	70
Field exercise	100	90	80	70
Ammunition	100	90	80	70
Truck gun	50	45	40	35
Cutlass and pistol	50	45	40	35
Turret	50	45	40	35
Field gun and battery	50	45	40	35
Theoretical and general questions	100	90	80	70
Total	600	540	480	420

* Allusion has already been made to the new regulation, which goes into effect in 1883, establishing a fourth examination, in pilotage.

As to the requirements and character of this examination. In the branch relating to heavy guns, each sub-lieutenant is required to be able to take any number at a gun, and to drill in Part II of the Gunnery Manual, with detail. He is also to answer all questions connected with the instruction, parts of the gun, carriage and slide, dismounting the gear, adjusting and mounting the compressor, &c. He must be able to drill quarters efficiently, and to detect and correct mistakes.

In field exercise, candidates must put a squad through rifle exercises and squad drill, with detail, and be able to perform thoroughly the duties of officer and man in company drill, and to give general detail of any section. He must be able to give any of the selected sections of the battalion in Part III of the Gunnery Manual, and to answer other questions relating to instruction, and especially the details of equipment for landing.

In ammunition he is required to describe and state the use of the different fuses, and to explain the supply and stowage of projectiles, filling and using, and the proportion allowed; to give the names and describe the manufacture and use of the different powders in the service, and the stowage and working of magazines and shell-rooms, and to explain rocket and mortar boats, the manufacture of cartridges, the supply of boat's stores, and the working of boat's guns.

In truck-guns the examination is similar to that in heavy guns, but goes less deeply into the subject, on account of the shorter time devoted to it. Under the head of turret-instruction the examination covers the duties and position of the different numbers in loading, training, and running in and out turret-guns; the principle of turret-firing and the method of sighting, the supply of projectiles, and the construction and support of the turret.

In cutlass and pistol exercise the candidate is required to perform the exercise and put a squad through without detail.

In field-gun and battery he must perform any duty in the drill, with detail of the various manœuvers. In battery drill he must be prepared with any selected sections, and he must understand the method of embarking and disembarking field-guns from boats.

Finally, he must pass a paper on the theory of gunnery, and an oral examination composed of general questions on the Gunnery Manual.

On passing this examination the acting sub-lieutenants are confirmed in their rank, and become eligible for promotion in the order of their seniority. The number of each certificate is noted in the Navy List. The Goodenough gold medal is given to the acting sub-lieutenant who, among those who have taken a first class in seamanship, passes the best examination in gunnery. An officer failing to obtain a certificate in gunnery is allowed a month's further study, and at the end of it a re-examination. If he fail to pass this he is discharged from the service.*

---

* Specimens of examination-papers for sub-lieutenants and for gunnery lieutenants on board the Excellent are given in the Appendix, Note G.

4.—GUNNERY LIEUTENANTS' EXAMINATION.

The examination for gunnery lieutenants is of course much higher and more complete than that just described. It has already been stated what percentages are necessary to obtain certificates in the Greenwich examination. The passing limit is higher in the practical, or gunnery course, than in the theoretical; especially in the case of the second class certificate, the limit for which is so low (30 per cent.) at Greenwich, that it must require very little diligence or ability to pass. In the practical examination at the end of the Excellent course for gunnery lieutenants, the relative weights are as follows:

Maxima.

1. Heavy gun	220
2. Field-exercise	280
3. Ammunition, including laboratory course	280
4. Truck-gun	100
5. Cutlass and pistol	90
6. Field-gun and field-battery	120
7. Turret	90
8. Musketry	90
9. Drilling	80
10. General questions	100
11. Theory	200
12. Torpedo	400
Total	2,050

Practical:
First class, 1,650, about 80 per cent.
Second class, 1,250, about 60 per cent.

Theoretical (Greenwich):
First class, about 60 per cent.
Second class, about 30 per cent.

Final gunnery certificates:
First class.—First in both theory and practice.
Second class.—First in one and second in the other.
Third class.—Second in both.

Gunnery lieutenants may be re-appointed to the Excellent, on their own application, after an absence of three years. The course of instruction in such cases is of five months' duration.

CHAPTER IX.

TORPEDO INSTRUCTION.

Instruction in torpedoes is given on board the Vernon, at Portsmouth, which forms a sort of floating torpedo-school, as the Excellent is a floating gunnery-school. The Vesuvius, a double-screw iron torpedo-vessel, of 260 tons, is attached as a tender to the Vernon. The officers, for instruction and otherwise, are a captain, commander, three torpedo lieutenants, two engineers, and eleven gunners. There are also in the Vesuvius a torpedo lieutenant, a chief engineer, and a gunner. The lieutenants generally hold their position for two or three years.

Lieutenants desirous of qualifying as torpedo officers are, when employed, to be recommended to the Admiralty by their captains, through the commanders-in-chief. Appointments are made to the Vernon on October 1, of each year, and officers so appointed remain on the ship's books during the whole course. Candidates who have not served one year at sea as lieutenants are appointed to a sea-going ship to complete that period as officer of the watch. The total period of instruction, including a vacation of three weeks, is eighteen months, and is distributed as follows: Theoretical course at Greenwich, nine months; senior torpedo-course, two months; Whitehead instruction, three weeks; diving instruction in Excellent, two weeks; vacation, three weeks; and torpedo instruction in Vernon, five months.

The final examination is in March, after passing which the torpedo lieutenant, unless his services are absolutely required at sea, takes part in the torpedo course ending in June and assists in the instruction of officers. If not then required for sea-service he continues to assist in the senior course ending in September. Between these two courses a period of fourteen days' leave is allowed. If it is found during any period of the course that a lieutenant is not likely to prove efficient as a torpedo officer, his name is submitted to the Admiralty with a view to his removal from the books of the Vernon. Officers qualifying within the stipulated period are allowed to count the whole time they are borne in the Vernon as time on full pay. Officers who have taken more than the regular time to qualify count any surplus as half-pay time, and those who do not qualify at all count the whole time of their service in the Vernon as half-pay time.

The following is the list of subjects, with their relative weight, in the course for torpedo officers:

THEORETICAL COURSE—GREENWICH.

A. Pure mathematics:	Maximum marks.
I. Arithmetic and mensuration	100
II. Geometry, first six books of Euclid	200
III. Algebra	300
IV. Trigonometry	300
V. Co-ordinate geometry, including conic sections	400
	—— 1,300

B. Applied mathematics: Maximum marks.
I. Statics 300
II. Hydrostatics 300
III. Kinematics and kinetics 300
                                                                              —— 900

C. Physics, including heat, light, electricity, and magnetism:
I. Theoretical 500
II. Practical 300
                                                                              —— 800

D. Chemistry:
I. Theoretical 500
II. Practical 300
                                                                              —— 800

E. Marine surveying.. 300

Total 4,100

It will be noticed that the aggregate is the same as in the course for
gunnery lieutenants, though the course is somewhat different, and the
separate marks are differently distributed. The relative proportion of
marks for first and second class certificates is also the same as for gunnery lieutenants, 2,500 for a first class, and the very low limit of 1,200
for a second.

PRACTICAL COURSE—VERNON.

I. Electricity as applied to naval purposes 200
II. Submarine mines 100
III. Outrigger and Harvey torpedoes....................................... 100
IV. Attack of submarine mines, and defense against torpedoes......... 100
V. Whitehead torpedo... 200
VI. Ability as a lecturer 100

Total 800

For a certificate in the practical course 650 numbers must be obtained.
This is an absolute requirement for all officers to pass as torpedo-lieu-
tenants. Their further classification into 1st class and 2d class, for
which they receive different rates of extra pay,* depends wholly on their
certificate in the theoretical course at Greenwich.

Provision is made for a short course of instruction in torpedoes on
board the Vernon, similar to the short course in the Excellent; and
torpedo lieutenants, after a three years' absence, are allowed to return
to the school for a five months' course of instruction.

* See appendix, Note A.

CHAPTER X

Made up as the English system is of diverse elements, it has a certain unity throughout, which is due solely to the fact that the whole is practically under one head. Except for the gunnery training, for which the Excellent is responsible, and the seamanship, which is instilled by some process on board ship, the whole training of officers is under the direction of the Director of Studies at Greenwich. His control does not go so far as the devising of a general plan; that is a matter for the Admiralty. But the specific application of the plan in all its details rests with him; and it is safe to say that whatever may be the faults of the English system, they do not lie in the application of it. From the time the young lad of twelve or thirteen passes his examination for a cadetship down to his last voluntary course as a captain, through the Britannia, the course afloat, the sub-lieutenants' collegiate course, and the subsequent voluntary studies, his education is in the hands of Dr. Hirst, under the Admiralty rules, and it is managed with all the wisdom and judgment that the rules will permit. The importance of this single head for the whole system cannot be overestimated.

In the English service there seems to be a theory that a naval officer is a creature of a delicate and sensitive organization, whose regard for his profession and whose zeal for a high standard of professional attainment must be stimulated by surrounding him eternally with all its minor details, to an extent unknown in any other walk of life. To make a sailor, he must begin at twelve or thirteen, even though he does not go to sea for two years, to accustom him early to his duties. During these two years he must live on board a ship, and be able to climb the rigging, to familiarize himself with details; though the ship lies at anchor in a river, a few yards from the shore, and carries no spars but her foremast and head booms. He must sleep in a hammock to inure himself to hardship. In the opinion of the majority of officers, he must have his college for higher instruction in a naval port, or he will forget his duties; and he must pursue his scientific researches in a dockyard, because he will be surrounded by officers engaged in the work of the profession, with whom he can discuss articles in the professional magazines.

If the naval profession has become what many enlightened officers of the present day would have us believe, an occupation involving accurate scientific knowledge, the system of training in England has a tendency to grasp the shadow while losing the substance. The expedients adopted with reference to the higher education of voluntary students, and the admirable courses of instruction for officers who have taken up one branch of the service, notably in the Excellent and Vernon, do much to remedy the inherent defects of the system; and the promotion in two grades by selection excludes the most incompetent officers from positions of great responsibility. But it seems impossible that the injurious effects of the method of training pursued with young officers, during the first eight years of their professional life, should not be felt by the vast majority throughout their whole career. The peculiar features of this training have been already pointed out: the discouraging efforts in the Britannia to attain a point hopelessly beyond the young student's reach; the five years of desultory training on board the great cruising ships, passed in a struggle to retain and comprehend a mass of undigested facts and principles, crammed for the immediate purpose of passing an examination; and, finally, the review course, where the student first finds himself fairly on his feet, in his relations with his instructors. The fatal defect of the system has been aptly set forth in a remark of one of the Greenwich professors, in his evidence before the commission, where he says that the standard for sub-lieutenants is the same as that for cadets in the Britannia; but the essential difference lies in the fact that at Greenwich the students actually reach the standard, while at Dartmouth they do not.* No one who has had much experience in educational methods will deny that such a system must be productive of harmful results when applied rigorously to the training of a body of young men; and one is therefore led to the conclusion that the high scientific and professional attainments of many English naval officers are not in consequence, but in spite, of their early education.

---

* Question 626.

PART II.
FRANCE.

CHAPTER XI.

ORGANIZATION OF THE PERSONNEL.

I.—LINE OFFICERS (*officiers de vaisseau*).

The corps of line officers in the Navy is recruited from four different sources: (1) *Aspirants* or naval cadets, graduated from the Naval School at Brest; (2) aspirants chosen from among the graduates of the Polytechnic School at Paris; (3) first masters (*premiers-maîtres*), nominated for the grade of ensign after undergoing an examination; (4) auxiliary ensigns who have received the certificate of sea-captain (*capitaine au long cours*) in the merchant service, and who are admitted to the titular grade of ensign.

The grades in the corps of line officers are as follows:

Admiral.
Vice-admiral.
Rear-admiral.
Captain.
Commander (*capitaine de frégate*).
Lieutenant, first and second class.
Ensign.
Midshipman (*aspirant*, 1ière *classe*).
Cadet (*aspirant*, 2ième *classe*).

The vice-admirals and rear-admirals form the roll (*cadre*) of the general staff of the Navy. This roll is divided into two sections: the first comprising those in active service and those waiting orders (*en disponibilité*); the second comprising the reserve. Vice-admirals at the age of sixty-five and rear-admirals at sixty-two pass from the first to the second section, with certain exceptions. In time of peace employment is only given to officers of flag-rank of the first section; but during war officers of the reserve can be appointed to commands.

If at any time during war the full number of officers is inadequate to the needs of the service, the want is supplied by auxiliary officers chosen among the merchant-captains. These merchant-captains, whose grade in the Navy is that of ensign, correspond to the English Naval Reserve.

Promotions are made partly by selection and partly by seniority, from the lowest grade up to and including that of commander. Above this point they are wholly by seniority. In every case, however, a certain term of service must be passed in each grade before promotion to the next higher. In the case of midshipmen, ensigns, and lieutenants, the minimum period is fixed at two years of sea service, with two additional years of shore duty in the latter grade. Commanders must pass three years at sea, one of them in actual command; or four years in the grade, two of them at sea and two in command. Captains must have passed three years at sea in that grade, or two years at sea in command of a naval division. Rear-admirals must pass two years in that grade at sea in a squadron. The highest rank in the Navy, that of admiral, is an exceptional honor, conferred only upon a vice-admiral who has distinguished himself in the chief command of a naval force in time of war. At the present time there are no officers of this rank in the French Navy.

Officers are divided in equal numbers among the five naval *arrondissements*—Cherbourg, Brest, Lorient, Rochefort, and Toulon. Each officer is attached to the chief port of the district; and, if below the grade of commander, he is required to reside there. A roster of officers is kept at each port, upon which is based the detail for sea duty.

II.—ENGINEER MECHANICIANS (*mécaniciens-en-chef et mécaniciens principaux*).

The officers of this corps have supervision of the engines and machinery on board all the ships of the division to which they are attached. Their sea duties, therefore, correspond to those of fleet engineers in the United States-Navy. They also serve to some extent as engineers (not fleet officers) on board a few of the largest ships. On shore their special duty is in connection with the instruction of machinists (*mécaniciens*) and firemen. They are chosen from the master-machinists, and may therefore be considered, to a great extent, as occupying higher grades in the corps of machinists. They hold relative rank as commissioned officers, but they rank only with the lower grades of the line. The name engineer (*ingénieur*) is not applied to them, but is given only to the constructors and the hydrographers. The grades and numbers of the corps are as follows:

Mechanicians-in-chief, with relative rank of captain of corvette (lieutenant-commander) 2
Principal mechanicians, first class (rank of lieutenant).................................. 8
Principal mechanicians, second class (rank of ensign)................................... 40

III.—MISCELLANEOUS STAFF CORPS (*corps entretenus et agents divers*).

1. Engineer corps (*génie maritime*).

This is really the construction corps of the French Navy, and it includes the most accomplished men of science in the service. Their duties are the designing of ships and engines, the supervision of their construction and repair, either in dockyards or at private works, and the care of timber required for ship-building. The corps is composed of—

Inspector-general, taking rank with rear-admiral............................... 1
Directors of naval construction, after rear-admiral and before captain.......... 11
Engineers, first class, with captain.. 20
Engineers, second class, with commander... 20
Assistant engineers (*sous-ingénieurs*) of the first and second classes, with lieutenant. 52
Assistant engineers, third class, with ensign....................................... 14
Engineer pupils (*élèves du génie maritime*), with midshipman..................... 7

Officers of the engineer corps are selected from the graduates of the Polytechnic School at Paris. After admission to the corps, they pass through the school of application for the engineer corps at Cherbourg. At the close of the cruise they are examined for promotion to the grade of assistant engineer of the third class. Promotion to the grade of assistant engineer of the second class only follows after two years of service, but in all subsequent promotions three years of service in the lower grade are required as a qualification. Promotion to the three highest grades is by selection; to engineer of the second class and assistant engineer of the first class, half by selection and half by seniority; and to the lower grades, strictly by seniority.

One-sixth of the vacancies in the grade of assistant engineer of the third class are reserved for masters (*maîtres entretenus*), of the corps of naval construction, who have performed a year of service in this grade and passed the required examination.

2. Engineer hydrographers (*ingénieurs-hydrographes*).

The duties of engineer hydrographers consist in the preparation of charts and sailing directions; the summarizing of nautical and scientific documents collected by the *Dépôt des Cartes*; observations of tides; magnetic and meteorological observations; care of nautical instruments used on shipboard; and preparation for the press of scientific treatises undertaken by the Navy Department.

The hydrographic officers have the same relative rank in their respective grades as the officers of the *corps du génie*, the hydrographer-in-chief ranking with the director of naval construction. The grades and numbers are as follows:

Engineer hydrographer-in-chief	1
Engineers, first class	3
Engineers, second class	3
Assistant engineers, first and second classes	4
Assistant engineers, third class	2
Engineer pupils, according to the needs of the service.	

Pupils of the corps are taken from the graduates of the Polytechnic School, and have two years of sea service before their first promotion.

3. Medical corps (*corps de santé*).*

a. Surgeons' division (*service médical*):

Inspector-general, taking rank with rear-admirals		1
Medical directors, first class, } before captains {		3
Medical directors, second class, }		4
Surgeons-in-chief, with captains		16
Medical professors, } with the old grade of captain of corvette {		9
Chief surgeons, }		34
Surgeons, first class, with lieutenants		161
Surgeons, second class, with ensigns		163
Assistant surgeons, with midshipmen		150

b. Pharmacists' division (*service pharmaceutique*):

Pharmacist inspector, after rear-admirals		1
Pharmacists-in-chief, with captains		4
Professors of pharmacy, } with captain of corvette {		6
Chief pharmacists, }		2

---

* The organization and methods of the French medical schools have been fully described and discussed in the report of Medical Inspector Dean, U. S. N.

Pharmacists, first class, with lieutenants 12
Pharmacists, second class, with ensigns 16
Assistant pharmacists, with midshipmen 19

4. Examiners and professors of hydrography.

Examiners, with relative rank of captain 3
Professors, first class, with relative rank of commander 10
Professors, second class, with relative rank of lieutenant..................... 10
Professors, third class, with relative rank of ensign 3

The professors of this corps are employed in the schools of hydrography, of which the examiners have the general direction, in matters relating to instruction. The latter have also charge of the examination of persons in the merchant service who are seeking a license as sea-captain (*capitaine au long cours*) or master of coasting vessel (*maitre au cabotage*).

Professors of the third class are appointed by the President, after a competitive examination, from naval officers and sea-captains who apply, who are at least thirty years of age, and have seen two years' sea service. Promotions in the corps are made by the Minister of Marine, on recommendation of the hydrographic examiners. The examiners themselves are promoted by selection from professors of the first class.

5. Professors at the Naval School.

Professors of the first class 6
Professors of the second class 2
Professors of the third class.. 2
Professors of the fourth class 2

Professors at the Naval School are selected by the Minister from university graduates, and they have the same relative rank as professors of hydrography.

6. Pay corps (*officiers du commissariat*).

The administrative duties of this corps are of an exceedingly complex and extensive character, embracing not only matters relating to pay, provisions, clothing, purchases, and stores, but also to equipment, recruiting (*inscription maritime*), administration of hospitals and prisons, wrecks, sale of prizes, and other miscellaneous subjects.

The corps is composed of the following officers:

Commissaries-general, taking rank after rear-admirals.......................... 9
Commissaries, with captains 26
Commissaries' *adjoints*, with captain of corvette 50
Sub-commissaries, with lieutenants 180
Assistant commissaries (*aides commissaires*), with ensigns.................... 150
Commissary pupils (*élèves*) 30

Commissary pupils are appointed by the Minister of Marine. Candidates must have taken the degree of licentiate in law, or that of *bachelier-ès-lettres*. In the latter case, they must pass a competitive examination. Those who receive appointments are sent to one of the naval ports to pursue a course in naval administration under the direction of a superior officer of the corps. The extreme thoroughness of these courses may be seen in the able lectures of M. Fournier, on the laws of maritime police, delivered at the commissary school at Brest, and recently published. They are characterized, in the highest degree, by that systematic arrangement and lucidity of exposition which distinguish French administrative papers. The courses are two years for licentiates and three years for the others. At their close, the commissary pupils pass an examination for promotion. Eight vacancies for the grade of *aide commissaire* are reserved annually: two for ensigns, selected by the Minister, on their own application; two for graduates of the Polytechnic School; and four for commissary's clerks, between twenty-five and thirty-five years of age, bachelors of arts or of science, who pass successfully a competitive examination.

7. Commissariat clerks (*commis du commissariat*).

These are recruited from warrant and petty officers, seamen, and marines, and from non-commissioned officers of the Army who have completed their term of service.

8. Civil engineers of the Ministry of Public Works (*ingénieurs des ponts et chaussées*).

The Ministry of Marine obtains from the Ministry of Public Works a number of engineers to take charge of hydraulic works in the forts, and to superintend the construction and repair of public buildings belonging to naval arsenals and dockyards. They continue to form part of the corps of engineers of roads and bridges, but they are under the authority of the Navy Department, and they have relative rank with naval constructing engineers.

9. Inspectors of administrative service.

This corps is composed of thirty officers of high relative rank, the highest taking rank between rear-admirals and captains, and the lowest with lieutenant-commanders. Their duties consist in the supervision of the administration service at the principal naval stations. They are selected by competition among lieutenants, captains of Marine Artillery, assistant constructing engineers, and assistant commissaries.

10. Accounting officers and storekeepers (*agents comptables des matières*).

11. Victualing clerks (*commis aux vivres*).

12. Keepers of ships' stores (*magasiniers entretenus de la flotte*).

13. Overseers of public works (*personnel administratif des directions des travaux*).

14. Overseers of subsistence (*service des manutentions*).

15. Overseers of hydraulic works (*conducteurs des travaux hydrauliques*).

16. Chaplains.

17. Hospital attendants.

IV.—MARINE AND COLONIAL TROOPS.

1. Dockyard gendarmerie.
2. Colonial gendarmerie.
3. Marine Artillery (*Artillerie de la Marine et des Colonies*).

To this corps belong the very important duties of the supervision of the manufacture of naval ordnance, the armament and defense of sea-coast fortifications, the direction of ordnance work in the dockyards, and the colonial artillery service.

4. Marine Infantry.
5. Naval armorers.
6. Senegal sharpshooters.
7. Senegal Spahis.
8. Corps of Cipahis.
9. Corps of discipline (penal corps).
10. Corps of discipline (colonial penal corps).
11. Wardens of colonial prisons.
12. Colonial militia.*

V.—WARRANT OFFICERS, PETTY OFFICERS, AND SEAMEN.

MASTERS.—The word master (*maître*) is a generic term, used to desig-nate those persons of different grades who exercise direct authority, under the commissioned officers, over the crew at sea, and over work-men at shore stations. At sea, masters form a military corps called the *maistrance de la flotte*. In the dockyards and at shore stations, masters are charged with the direct supervision of public work of all kinds. The latter are quite distinct from the sea-going masters, and are known collectively as the *maistrance des arsenaux*.

The *maistrance des arsenaux* is composed of 262 persons, of the follow-ing grades:

1. Principal masters (*maîtres principaux*), first and second classes.
2. *Maîtres entretenus*, first, second, and third classes.

The lowest grade is recruited from the *contre-maîtres* and from the sea-going masters. Below the masters come the men employed in the dock-yard works, composing a body of artificers (*ouvriers*). These are also arranged in grades, and measures are taken to keep them per-manently in the service. The grades are—

Chief *contre-maîtres*.
Contre-maîtres.
Chief artificers.
Artificers.
Apprentices.
Chief day-laborers.
Day-laborers, of four classes.

Promotion takes place in these grades, at least from that of appren-tices, who may be appointed between fourteen and seventeen years of age. At seventeen or eighteen they may pass for artificers upon giving

---

* Fuller information in regard to the corps of officers in the French Navy may be obtained from a series of articles entitled "*La Marine Militaire de la France*," by M. J. Delarbre, auditor-general of the French Navy, in the *Revue Maritime et Coloniale*, vol. 52.

proof of sufficient aptitude. Pay is partly fixed for each grade, and partly graduated according to the merits of the workmen. At fifty years of age, and after twenty-five years of service, artificers are entitled to a pension.

The *maistrance de la flotte*, or sea-going masters, (also known as *officiers mariniers*) is composed of three grades, viz:

First masters.
Masters.
Second masters.

These are the warrant-officers of the Navy. They are divided into eight classes, according to the special branch of a seaman's occupation that they profess. The three grades run through all these classes. The classes or specialties are as follows:

Seamanship (*manœuvre*).
Gunnery.
Small-arms.
Helmsmen.
Machinists.
Carpenters.
Sailmakers.
Calkers.

Below the three grades mentioned, which form the *maistrance,** come the persons composing the crews of ships, known under the general name of *équipages de la flotte.* They are graded as follows:

Quartermasters (of the various specialties). These are the petty-officers.
Seamen, first, second, and third classes.
Seamen apprentices.
Novices.
Boys (*mousses*).

To these should be added, to complete the list, the special ratings of topmen (*gabiers*) and small-arm men (*marins fusiliers*).

It will be noticed that one of the eight specialties named above is that of machinists, whose principal duty, as might be supposed from the name, is that of directing the engines at sea. This branch includes, like the others, the grades of first master, master, second master, and quartermaster. Ranking with the quartermaster-machinist are the machinist pupils (*élèves mécaniciens*), chosen from the lower grades of their corps, from graduates of technical schools (*écoles d'arts et métiers*), and also from artisans (smiths, boilermakers, &c.) in civil life. Below them are firemen (*ouvriers chauffeurs*) of three classes corresponding to the three classes of seamen. All the above belong to the *équipages de la flotte*, the class including all sea-going persons in the Navy who do not hold a commission.

---

* The *maistrance* also includes sergeant-majors, quartermaster-sergeants, and captains and sergeants of arms.

The corps of commissioned officers in charge of fleet engine service has been already alluded to. To distinguish them from the warrant-. officers of the same branch they are designated in this report as mechanicians, while the others are spoken of as machinists, although the same word *mécanicien* is used in the French Navy to apply to both classes. They must, however, be considered in connection, as the mechanicians are appointed directly from the highest grade of machinists.

The whole number of grades of commissioned officers, warrant officers, petty officers, and men, attached to this branch of the naval service, is as follows:

Commissioned officers:

Mechanicians in chief.

Principal mechanicians, first class.

Principal mechanicians, second class.

Warrant-officers (*officiers mariniers, maistrance*):

First master machinists.

Master machinists.

Second master machinists.

Petty-officers and men:

Quartermaster-machinists.

Machinist pupils.

Firemen artificers (*ouvriers chauffeurs*), first, second, and third class.

Firemen (*chauffeurs*) and *agents inférieurs*, assimilated to the third or lowest class of ordinary seamen.

CHAPTER XII

The following establishments are included in the general system of education of officers of the French Navy.

1. POLYTECHNIC SCHOOL.—Although the Polytechnic School is designed for the preliminary training of candidates for all the scientific branches of the public service, or at least for all branches in which scientific knowledge is required, and has no direct connection with the Ministry of Marine, it cannot be omitted in any complete description of naval education in France. It furnishes a certain number of midshipmen (about four a year) to the line of the Navy, and of assistant commissaries to the pay corps ; and from it are derived two-thirds of the officers of the Marine Artillery, and all the pupils of the corps of engineers (*génie maritime*) and of hydrographers. In most of these cases its function is distinctly that of a preparatory school, and the general instruction given by it in science and mathematics is supplemented by special and professional training in the schools of the selected corps.

2. NAVAL SCHOOL.—Two schools for the Navy were founded in 1810, at Brest, and at Toulon. In 1816 they were united in one, which was placed at Angoulême. In 1827 the Naval School was removed to Brest, where it has since remained. The school has no buildings on shore; like the English cadets' school, it is placed on board an old ship of the line, the Borda, anchored in the roads of Brest. The Borda is commanded by a captain, and is under the immediate supervision of the Préfet of the second maritime arrondissement. It has a staff composed of a commander, 12 professors, and 8 lieutenants.

The examination for admission to the Naval School is competitive. The number of candidates admitted is about 45 a year, and the course lasts two years. At its close, and upon passing the required examinations, the students, who have up to this time been called simply pupils (*élèves*). become cadets (*aspirants de Qième classe*). They are then embarked in the practice-ship, which is, to all intents, a separate establishment. Here they are joined by the four *aspirants* who have been graduated in the same year from the Polytechnic School, and who pursue the studies of the practice cruise with the graduates of the Borda.

3. TRAINING SCHOOL FOR LINE OFFICERS (*École d'application des aspirants de marine*).—The training school into which the graduates of the Borda pass is on board the practice-ship Flore. The course lasts nearly a year, and consists in the study and practice of the professional branches.

4. ENGINEERS' TRAINING SCHOOL (*École d'application du génie maritime*).—The school of application for engineers or constructors is a very

ancient establishment, having been founded at Paris in 1765. It was removed in 1801 (*An X*) to Brest, and later to Lorient, re-established in 1854 at Paris, and finally established at Cherbourg in 1872. It is a school of ship and engine design and construction, and it corresponds in a general way to the advanced classes for constructors and designing engineers in the Royal Naval College at Greenwich. Its pupils number about four a year, and come to Cherbourg after graduation at the Poly- technic School. The course lasts two years, and at its close graduates are appointed assistant engineers of the third class.

5. MEDICAL SCHOOLS.—The medical schools are three in number, at Brest, Rochefort, and Toulon. The course of study is two years; and the students, after passing the examinations at the close, become assist- ant surgeons or assistant pharmacists.

6. TORPEDO SCHOOL (*Ecole des défenses sous-marines*).—The torpedo school was established at Boyardville, in the island of Oléron, in 1869, and reorganized in 1876. It is under the command of a captain, and under the general supervision of the Préfet Maritime of Rochefort. In- struction is given by officers detailed as professors, and also by warrant and petty officers. A small vessel, the Messager, is attached to the school.

7. MACHINISTS' SCHOOL (*Ecole théorique et pratique des mécaniciens*).— Formerly there were two schools for machinists, one at Brest and the other at Toulon. Originally founded in 1860, and reorganized in 1862, these schools were united early in the present year in a single estab- lishment at Toulon, with a greatly-improved course and organization. The object of the school is the training of the warrant and petty officers of this branch for higher grades, and the selection by a competitive examination of those who are worthy of promotion. Although no com- missioned officers attend the courses of the school, it must nevertheless be considered a part of the general system of education of officers; partly because the principal mechanicians are derived directly from the first master machinists who have passed through the school, and partly because those who, in the French Navy, are classed as part of the *Maistrance*, in the English or American service would hold a commission.

8. GUNNERY SCHOOL (*Ecole d'application de cannonage*).—This school, similar in purpose to the Excellent in the English Navy, is situated at Toulon, on board the Souverain, a screw steamer of the first class, mounting 25 guns. The brig Janus is attached as an "annexe" to the Souverain. The Arrogante, the floating battery which sunk not long ago off the Iles d'Hyères, was also a part of the establishment. The school is intended for the education of both junior officers and men in practical gunnery. The former are both instructors and students, and they numbered, in 1878, 22 lieutenants, 15 ensigns, and 16 midshipmen.

9. ARTILLERY SCHOOL (*Ecole d'artillerie*).—The artillery school is at Lorient, and is exclusively for the training of officers of the Marine Artillery.

The Polytechnic School has been so often and so thoroughly described that it is hardly necessary to go here into its details. It is, as has been already stated, the preparatory school for the scientific branches of the public service. It is under the control of the Ministry of War, and its organiz1tion is military. At its head is a general officer of the Army. With him are associated a colonel, as second in command, and a number of commissioned and non-commissioned officers of the Army in charge of the details of discipline. The functions of all these officers are purely administrative. They have nothing to do with instruction, but belong to the military organization of the school. This military organization is not, however, an essential feature of the establishment. It exists chiefly for the simplicity and ease which it gives to the machinery of school government, and for the benefits derived from military discipline in the training of youths for administrative service. The students wear a uniform, and have a battalion organization and occasional drills; but the military part of their training is entirely subordinate to the scientific.

The academic staff is composed of a director of studies, fifteen professors, twenty tutors (*répétiteurs*), and three drawing masters. The method of instruction is quite different from any that exists in America; and, it might almost be added, in the rest of the world, outside of France. The class meets the professor in the amphitheater, where, after a short interrogation of a few of the students on the subject of the previous lesson, a lecture is delivered, the students taking notes. The class is then broken up into small sections and sent to recitation rooms, where a second hour is passed with the tutor or *répétiteur* in going over the lecture, of which a lithographed summary (*feuille autographiée*) is delivered to the class. This officer's duty, as his name implies, is to *repeat* the instruction given by the professor. During the hour of study, notes on the lectures are carefully written out and explanations are given by the *répétiteur*, as may be necessary. This system gives a close personal character to the instruction, which would be entirely wanting in an ordinary lecture-system.

It is a well-known fact that the system of instruction by lectures to large classes, when not supplemented by searching tests, is only suitable for voluntary students. At a university such a system is possible, because the choice of studies and the degree of application rest largely with the student himself, and the object is not so much to compel all to reach a certain standard, as to afford to each one the means of reaching the standard fixed by his own capacity and endeavor. At a school like the Polytechnic, however, whose diploma carries with it an appointment in the public service, and which graduates a number just large enough to fill vacancies, *all* the students must be brought up to the standard of attainment that the future career exacts. Nothing can be left to individual choice or individual volition. A pupil who is dull or indolent cannot be passed over, nor should he, except in extreme cases, be discharged; he must be made to understand and made to study. The duty of whipping in the laggards cannot well be undertaken by the professor, engaged as he is in carrying a large class through his subject; and it is with this work that the *répétiteur* is really concerned, the work of supplementing and making personal the instruction given to the class, by the closest attention to individual wants and the most careful tests of individual attainment. Various methods are adopted to test the diligence and acquirements of the students. First, there is the recitation preceding the lecture, to which the professor is required to devote from fifteen to thirty minutes of his total time of an hour and a half. Another hour and a half is given to the study and practice with the *répétiteur*, following the lecture. At intervals, brief written examinations (*exercices d'application*) are held, the class being divided into sections in charge of the professor and *répétiteurs*. Each examination consists of two sessions; the first of two and a half hours, the second of one and a half hours. At the first, the problems or exercises are worked out; at the second, they are made the subject of explanation and recitation. In physics and chemistry these exercises consist of laboratory work. After every five or six lectures, short oral examinations (*interrogations particulières*) of five or six students at a time are held by the *répétiteurs*. These come irregularly, and the student who is to be called upon has only a brief notice; so that he must be ready for them at all times. Of course these examinations take up a considerable time, and the instructor must make short work of it; twenty minutes only are allowed to each student. At the end of each course a fuller oral examination (*interrogation générale*) is held on the subject of the course. In this, half an hour is allowed to each man. Lastly, there are the examinations at the end of the year.

It must be understood that the pupils of the Polytechnic School are carefully selected at the start, and that the examination for admission insures a high standard of preliminary attainment. It is an open competitive examination, conducted by boards at various cities designated as centers of examination. The candidates must be between sixteen and twenty years of age. The competition is exceedingly close, the number of candidates being usually about four times as great as the number of appointments. The programme of examination includes arithmetic, geometry, algebra, trigonometry, analytic geometry of two and three dimensions, descriptive geometry, physics, chemistry, French, German, and drawing. Even with this high standard and close competition for admission, it is found that the severity of the course at the school tells hardly upon the weaker men; and it is a matter of observation that some at least of the graduates, on being subsequently admitted to the special schools of application, show signs of mental exhaustion.

The following table gives an outline of the course of study at the Polytechnic School:

FIRST YEAR.

Calculus (*analyse*): 40 lessons, 4 written examinations (*compositions*), 4 reviews of examinations (*conférences*).

Mechanics: 40 lessons, 4 written examinations, 4 reviews.
Geometry: 32 lessons.
Stereotomy: 26 lessons.
Physics: 30 lessons.
Chemistry: 32 lessons.
Mechanical drawing: 20 lessons.
Freehand drawing: 42 lessons.
Literature: 24 lessons, 4 written examinations.
History: 25 lessons.
German: 25 lessons, 2 written examinations.

SECOND YEAR.

Calculus: 40 lessons, 4 written examinations, 4 reviews.
Mechanics: 40 lessons, 4 written examinations, 4 reviews.
Astronomy: 28 lessons.
Physics: 30 lessons.
Chemistry: 32 lessons.
Architecture: 36 lessons, 4 written examinations.
Topography: 5 lessons.
Military art: 19 lessons.
Mechanical drawing: 20 lessons.
Freehand drawing: 48 lessons.
Literature: 24 lessons, 4 written examinations.
History: 25 lessons.
German: 25 lessons, 2 written examinations.

To these must be added: (1) the hours of study with the *répétiteur*, following each lecture; (2) the particular oral examinations (*interrogations particulières*), at irregular intervals, of which there would be, for example, about eight in each year's course in calculus; (3) the general oral examinations at the end of each course; and (4) the annual examinations. The last are conducted by an outside board.

It will be noticed that the programme divides the lessons about equally between three groups of studies—mathematics, scientific subjects, and miscellaneous subjects, including history, literature, German, and freehand drawing. The best part of each working day is, however, given to the first two groups, and the school is pre-eminently a school of mathematics and science.

The following table shows the branches of the government service for which the Polytechnic School gives the preparation, and the number of annual appointments in each; though the latter must be taken as approximate, the numbers varying slightly from year to year according to the needs of the service.

Number of annual appointments.	Service.	Department.
	ARMY.	
50	Artillery	
25	Engineers	} Ministry of War.
3	Staff corps (*état-major*)	
	NAVY.	
4	Executive or line officers (*officiers de vaisseau*)	
10	Marine Artillery	
4	Naval architects (*génie maritime*)	} Ministry of Marine.
1–2	Engineer hydrographers (*ingénieurs hydrographes*)	
2	Pay corps (*officiers du commissariat*)	
	CIVIL.	
18	Civil engineers (*ingénieurs des ponts et chaussées*)	} Ministry of Public Works.
3	Mining engineers (*ingénieurs des mines*)	
1	Telegraph service	Ministry of the Interior.
1	Department of powder and saltpetre	} Ministry of Finance.
2	Tobacco department (*administration des tabacs*)	

In the graduating class, the student who stands at the head in the final classification has his choice of career among all the vacant appointments. The second man on the list has the choice of what is left by the first, and so on to the foot of the class. The choice of the highest men is almost invariably fixed on the departments of mines, and of roads and bridges, as they offer the highest inducements in the way of pay and emoluments. After these come naval construction, the engineer corps of the Army, and the tobacco department. The artillery and staff corps follow next, and are nearly on an equality, while the remaining branches come at intervals up and down the list, according to the inclination of individuals, and following no general rule.

A word should be said in regard to the causes of the success of the school, a success so remarkable that its influence on the public service of France has become a matter of history, and it has taken a place among the very first of schools of its class in the world. Founded in 1799 by the foremost scientific men of the day, such as La Place, La Grange, Monge, and Fourcroy, it has for eighty years supplied the government with its ablest civil and military engineers. Its system is that which was devised by Monge at its foundation, and which, with very slight modifications, has been retained ever since. The system is essentially one of lectures delivered by the ablest professors that can be obtained, supplemented by compulsory private tuition and private examination by energetic tutors. Whether the system is one susceptible of general application is a question, but its results at the Polytechnic School are undoubted. It has been very generally imitated in other French schools for higher education, and the *lycées* and communal colleges, which give the greater part of secondary instruction in France, have also their professorial lectures and interrogations by répétiteurs. It is a fact to be noticed that a very similar system has been introduced at the Royal Naval College at Greenwich, though there has hardly been time, as yet, to test its working.

One immense advantage possessed by the Polytechnic over other schools lies in the rewards which it offers to successful competitors. The stimulus given by competition is shown in the results of every year's work, and is acknowledged by those who know the school best to be one of the most powerful incentives to effort, if not the most powerful. It is not so much that one career holds out overwhelming inducements, as that a great variety of careers is presented, and students are willing to do their utmost to obtain the privilege of following their individual inclinations. In this, perhaps, lies the secret of the school's success, more than in the skill of the instructors or the method of instruction.

It is not [said a recent director of studies] our interrogations, our examinations, and other contrivances, which make the pupils work; it is not even the system of competition by itself; it is not mere personal ambition, nor the desire of taking the first place on the list. All these conditions exist at the schools of application, as for instance, at Metz,* and yet the results in regard to industry are widely different. It is the inequality in the value of the prizes offered to them, *the choice of careers open to them*, that is the great incentive to work at the Polytechnic; and no institution, though it may imitate the Polytechnic system, can secure similar results unless it holds out similar inducements to pupils.

Though competition produces good results at the school itself, it is of more doubtful benefit to some of the branches of service whose officers the school supplies. This is especially true of the artillery, which is generally the last choice of the graduates, and which is, therefore, recruited from the lower part of the list. Complaints have been made by military officers that the scientific corps of the Army get only the fag-ends of classes (*queues de promotion*), whose position indicates either a want of ability or of effort, or both. Of course, among these students there is little or no competition. The only remedy for this defect would be to graduate a larger number than the vacancies require, but this does not seem to be a part of the policy of the school.

The principal advantages derived by the Navy from the Polytechnic School consist in the high preparatory training given to the officers of the three corps of constructors, hydrographers, and marine artillery. All of these are closely connected with the Navy proper, the hydrographers being occupied with the coast survey and the preparation of charts, the artillery with the manufacture of naval ordnance and with sea-coast fortification, and the constructors with the design and construction of ships and engines. The value of the training for these officers cannot be overestimated, especially in the case of the constructors, as the graduates that select this corps are usually among the higher men of their class. With regard to the other corps, the line officers and com-

* The School of Application for the artillery and engineers, now at Fontainebleau.

missariat, each of which receives a mere fragment of the graduating class, the object of the system is not quite apparent. In the case of the line officers, it is chiefly to give the Navy a few men peculiarly fitted for certain forms of purely scientific work which arise from time to time in the service. For this purpose the course is better fitted than that of the Naval School at Brest; though it is doubtful, even with this difference, whether any great advantage is gained, especially in the case of those naval graduates of the Polytechnic who take a low number in their class. Certainly the difference is not so great as to make the lower men of the Polytechnic classes necessarily more available for scientific purposes, during their whole career, than the best men from the Borda.

At the close of the Polytechnic course, the graduates who have selected the Navy as their future career are embarked on board the practice ship Flore, to make their first cruise, in company with the graduates of the same year from the Borda. Having no professional training whatever, they start at a disadvantage; and they must perform double work to put themselves on a level of professional acquirement with their contemporaries. At the end of their cruise they pass their examination for promotion and become midshipmen (*aspirants de 1ère classe.*)

CHAPTER XIV.

THE NAVAL SCHOOL (*Ecole Navale*).

The school at which nearly all the cadets of the line of the French Navy receive their education is on board the old wooden line-of-battle ship Borda. The Borda is anchored in the roads of Brest, about a mile and a half or two miles from the town. The interior of the ship is cut up and rearranged to suit the needs of the school, as in the case of the Britannia, though the details of arrangement in the Borda are quite different. The poop extends to to the main-mast, and contains the cabins of the commanding officer, although the latter does not live on board. The spar-deck forward of the poop is used as a gymnasium. The comb-ings of the main hatchway are removed, and the deck flushed over; and there is the usual supply of rings, parallel and horizontal bars, &c., that form the outfit of a small gymnasium. On the upper gun-deck is the mess-hall, with pantries and offices forward. The students sleep on the lower gun-deck. On this deck are also the two study-rooms (*salles d'étude*), one for each division. All these rooms are forward of the main-mast. In the after part of the ship, on the two gun-decks, are the officers' quarters and wardroom, and also the lecture-rooms. These are two in number, and are built in the shape of amphitheaters, the floor being laid in steps rising towards the back, and extending from the lower to the upper deck. The furniture in all these rooms is of the simplest character, consisting of small stationary tables, desks, and benches. The library is small, inconveniently placed, and contains few recent books. The space between decks in the room devoted to this purpose is too small to admit of standing upright. The library seems to be little used. The battery of the Borda consists of B. L. R. guns of the most recent type, of 12, 14, and 16 c. m. Two corvettes are attached to the establishment, one a sailing-vessel, the other a screw-steamer.

1. – PERSONNEL.

The Naval School, like all other adjuncts of the station at Brest, is under the general authority and supervision of the Préfet Maritime of the II arrondissement. At its head is a captain; next in rank to the captain is a commander (*capitaine de frégate*), who has the same general duties as commanders on board sea-going vessels.

The authority of the Préfet Maritime is not confined to mere formalities. He makes all the administrative arrangements necessary for carrying on the working of the school, especially those in regard to police and interior service. He makes inspections at discretion, but he is required by regulation to inspect regularly three times a year—in January, April, and August. The two first inspections are preceded by the inspections of the material administration of the school, made by the commissary-general, and by the commissary in charge of equipments. Reports are made to the Ministry of Marine of the result of each of these inspections. In regard to various details of government the captain frequently advises with the Préfet; and, finally, the latter is the presiding member of the board of improvement (conseil de perfectionnement).

The captain of the Borda is the director of the studies of the school as well as of the discipline. The commander, or executive officer, has charge of the interior police and service, of the conduct of students, and of practical and professional instruction. He keeps a conduct book and a punishment book, and gives the students a mark for conduct every quarter. With instruction in the scientific and miscellaneous branches he has nothing to do, the instructors in these branches being wholly under the direction of the captain.

The instructors consist of eight lieutenants, twelve professors, and one principal mechanician. The lieutenants have charge of the courses in seamanship, naval architecture, gunnery, and practical navigation, there being two for each branch, one taking the upper and one the lower class. Four of the lieutenants act as chiefs of sections (chefs d'escouades) and keep a constant and careful oversight of the members of their sections. They transmit orders to their respective sections, and receive complaints or requests from them. It is their duty to regulate all those minor matters of detail, pertaining to the daily life of the pupils, that are not covered by general instructions. A close personal relation is thus established between the members of each section and their chief. He sees that all articles in their possession, such as clothing and books, are properly kept and cared for, and that they have no unauthorized objects. He keeps their weekly allowance books, and, in case of permission to incur extra expenses, for special instruction or what not, he is required to certify that the lessons have been properly given, or that the articles purchased have been duly received. Of course these duties involve frequent inspection and constant personal intercourse; and the chief of section is the person to whom the student naturally looks for advice and assistance, and who is to aid and stimulate his efforts to perfect himself as a naval officer. The other lieutenants perform the ordinary duties of officer of the day; but all of them, including the chiefs of sections, have a share in the regular detail of the ship's duties; and all have to note delinquencies on the part of the students and to enforce the discipline of the ship. They can only inflict reprimands; cases requiring severer punishment must be referred to the captain or commander. The senior lieutenant is in charge of the two auxiliary vessels; the instructor in practical navigation has the direction of the observatory on shore; while a third lieutenant is charged with the small-arm practice.

The twelve professors at the Naval School are divided as follows:

Professors of analysis and mechanics 2
Professors of astronomy and navigation 2
Professor of physics and chemistry 1
Professors of literature, history, and geography........................ 2
Professors of English 3
Professors of drawing 2

The professors do not reside on board the ship, but are brought off from shore every morning in a small steamer attached to the school, in time for the morning lecture.

The other officers of the school are, a principal mechanician, instructor in steam-engineering; a chaplain; an assistant commissary, who has charge of all administrative matters other than military or academic; three surgeons; and an accounting officer and storekeeper (*agent comptable économe*). The latter officer has the direction of a variety of matters, such as the care of public property, the preparation of estimates for stores and materials, receipts, disbursements, and purchases, and the correspondence of the captain of the school with the parents of pupils.

To the list of officers should perhaps be added the members of the two boards of examination, whose functions are subsidiary to the main purpose of the school, though the members are not attached to the Borda. One of these boards conducts the examination for admission, the other the annual and final examinations. The latter board is composed of naval officers of high rank, together with a member of the corps of hydrographic examiners. The pharmacist-in-chief at Brest usually examines in chemistry.

The crew of the Borda numbers about 150 men. The warrant and petty officers are carefully selected by the captain of the Borda, and six or eight of them, in addition to their regular duties, assist in the instruction of the students in practical exercises connected with the specialties to which they belong. These special branches include seamanship and gunnery, and the specialties of helmsmen, topmen, machinists, and captain of arms. A similar number of non-commissioned officers of the marine artillery (*adjudants*) perform the details of disciplinary service. At their head is the captain of arms, and the whole force comes directly under the executive officer. Their duties include the hourly oversight of the pupils, the frequent inspection of their desks, chests, and lockers, and of all their belongings, and duty as watchmen by day and by night. They are directed to enforce the regulations of discipline, and to report all infractions. In fact, in all matters of detail they perform the police of the ship.

There are four boards or committees that occupy an important place in the organization of the school. The first of these is the committee on improvements (*conseil de perfectionnement*.) It is composed of the Préfet Maritime, as president; the captains of the Borda and of the Flore, the sea-going practice-ship of cadets; and the members of the two examining boards, of admission and graduation. It meets annually, revises the programme of study, and considers and proposes other changes in the organization and methods of the school. These changes are sub· mitted for approval to the Board of Admiralty (*conseil d'amirauté*) at the Ministry of Marine.

The council of instruction or academic board (*conseil d'instruction*) is composed of the captain and commander of the Borda, the two examining boards, three professors or instructors, appointed for one year by the Préfet, of whom one is in the professional, one in the scientific, and one in the literary or "general" department, and the commissary.* The duties of the board are to consider and report upon measures proposed by the secondary council, or referred to it by the Minister. The latter include the distribution of scholarships (*bourses*), and of indemnities for outfit.

The secondary council of instruction (*conseil secondaire d'instruction*) is composed of the same members as the council of instruction, except the examining boards. It acts as an advisory board to the Préfet Maritime and the captain of the school, by whom various questions relating to academic organization are submitted to it. It has also the initiative in all propositions relating to the instruction and course of study, and it conducts the re-examination of· deficient pupils, making recommenda· tions as to the final disposition of doubtful cases. It considers propo· sals for the purchase of scientific works, periodicals, and apparatus. Its other duties include the preparation of the term and yearly class-lists, and, in general, it attends to those matters of detail which concern the academic interests of the school. The captain of the school is pres-ident of the council, and any of the instructors or professors who are not members may be required to attend its discussions, but only with a consulting voice.

The fourth of the governing boards is the council of administration (*conseil d'administration*). It is composed of the captain and commander, the commissary, and the two senior lieutenants, chiefs of sections. It keeps the running account with the Ministry of Finance, and with the pupils, and it has general charge of receipts, disbursements, and pur-chases. The "accounting officer and storekeeper" (*économe*) acts as its agent.

2.—EXAMINATION FOR ADMISSION.

The examination for admission to the Naval School is one of the most important parts of the French system of naval education, on account of its scope, its method, and its close relation to the system of public in-struction in the country. It is competitive in character. Its require-ments are high and extend over a considerable range of subjects. Fi-nally, it is based directly on the programmes of study in the *lycées*, the principal schools for secondary instruction in France. It has several

---

* It must constantly be borne in mind that the word commissary denotes a member of the administrative corps of the Navy, and has no connection with what in English is understood as the commissariat.

other noticeable features, but it is to these three that its important effects are chiefly due.

The method of organization is simple. The examining board consists of four examiners, chosen each year by the Minister of Marine, two in scientific, and two in literary or general subjects. A captain in the Navy is president of the board, but his duties are confined to administrative matters. Junior officers are assigned to take charge of the examination-room. A special examiner is appointed to mark the drawings handed in by candidates.

The examination is in two parts, written and oral. The written examination is held first, and candidates are required to obtain a certain mark—35 per cent. in mathematics, 25 per cent. in literary subjects, and 15 per cent. in drawing. Those who fail to reach this standard are excluded from further competition. To save candidates the expense of a long journey, the simple and excellent method is adopted of having different centers of examination, at any one of which candidates may present themselves. The centers are ten in number—Paris, Brest, Cherbourg, Lorient, Rochefort, Toulouse, Toulon, Bastia (Corsica), Algiers, and Lyons. At these places written examinations are held simultaneously on the 11th, 12th, and 13th of June in each year. As most of the places designated are naval stations, or stations where a number of naval officers are constantly on duty (all, in fact, except Lyons and Toulouse), the service of conducting the written examination is attended with no special expense to the government.

Useful as this system is found in France, its advantages would be even greater in the United States, on account of the immense distances to be passed over by candidates from all the Congressional districts, in reaching any given point. At present it happens, and under the existing system it must continue to happen, not infrequently, that young men whose means do not warrant the expense are obliged to take a journey of 1,500 or 3,000 miles to present themselves at an examination which they are totally unprepared to pass. The only way to avoid this is by holding examinations simultaneously at different centers. The principal navy-yards furnish convenient points, with all the materials ready at hand. In this way Boston and Portsmouth would be centers for New England; New York and Philadelphia for the Middle States; Annapolis, Norfolk, Port Royal, and Pensacola for the Southeast, and San Francisco for the extreme West. To these might be added Cincinnati, Chicago, Saint Louis, and New Orleans for the center. Such a system would present great advantages, and at the same time be easy of application and attended with little or no expense.

Persons desiring to compete in the examination for admission to the Naval School are obliged to enter their names as candidates at the prefecture of the department in which they reside, between the 1st and 25th of April, preceding the examination at which they intend to present themselves. They must be at least fourteen, and not more than seventeen, years of age on the 1st of January preceding their application. This condition is rigorously applied, and no dispensations are ever granted to candidates above or below the limits. The average age of candidates admitted has been found to be about 16 years. At the time of entering their names, candidates are required to present the following papers:

1. Certificate of birth.
2. Certificate of French nationality.
3. Physician's certificate of vaccination.
4. Choice of center of examination.
5. Bond of parent or guardian for payment of fee for board and tuition, amounting to 700 francs a year.
6. Bond of parent or guardian for payment of outfit, amounting to about 1,000 francs.

The regulations for conducting the written examinations are prescribed with considerable minuteness, particularly with a view to prevent irregularity or unfairness in marking. The questions are the same at all the centers of examination. They are sent in sealed envelopes from the Ministry of Marine to the prefects and subprefects in whose jurisdiction the examination is to be held, and by whom they are transmitted to the naval officers in charge of the examination.

The examinations are held with closed doors, at the day and hour prescribed. Not more than twenty candidates can be placed in one room, and warrant officers are detailed for the surveillance of these rooms; at least one to every ten candidates. At the beginning of each day's session the officer in charge opens the envelope in the presence of the candidates, and reads aloud the questions. At the close of the session he collects the papers and transmits them to the ministry of marine. Here the duties of this officer cease. The papers are sent to the president of the board of examiners, who detaches the headings containing the name of the writer, after having placed on both headings and papers a corresponding series of numbers. In this condition, numbered, but not named, the papers are turned over to the proper examiners to be marked. When the marking is finished, the board meets and draws up a list, still without the names of candidates, of the marks given to each numbered paper. The president then opens, in the presence of the board, the sealed envelope in which he had previously placed the headings, and the final report of the written examination is drawn up. This is published in the *Journal Officiel*, and is the only notification received by the candidates of their success or failure at the preliminary examination.

The oral examinations follow immediately upon the written. Like the first, they are held at various cities, the board of examiners making a tour for this purpose. The places of the examination are the same as before, except in the case of Bastia and Algiers, candidates from Corsica and Algeria presenting themselves at Toulon. The first examination is held at Paris, July 1, and at the other cities in succession. Each candidate must pass a medical examination before a board composed of the president of the examining board, a commander, and a naval surgeon. This always precedes the oral examination. Candidates who have failed at the written examination, or who have absented themselves from any of the tests, or who have made use of any improper means of assistance, are ruled out before the oral examination begins.

The final classification is prepared under the direction of the jury of examination, composed of the examining board, together with two naval officers of high rank, sitting at Paris. The marks, ranging, according to the usual scale in France, from 0 to 20, are multiplied by the prescribed coefficients, and the sum of the products gives the final mark. In case two candidates have the same mark, the oral examination decides their final position; and if the result is still the same, greater weight is given to the scientific branches. Thirty additional marks are given to candidates who have taken the degree of bachelor of letters. From the final list the minister appoints the members of the entering class at the naval school in the order of classification, and in accordance with the number required. A letter is accordingly sent to each successful candidate, which he is to present to the major-general (chief of naval staff) on his arrival at Brest, before the beginning of the session. After a second medical examination, and after making the necessary deposit, he is regularly entered at the school.

The scope of the examination for admission is defined in the programmes of certain classes in the *lycées*. These are the principal public schools in France of a general character, and they are so closely connected with the special schools for professional training that some further allusion to them will be necessary. The subjects of the examination, with the coefficients of each, are as follows:

1.—GENERAL SUBJECTS (*Partie Littéraire*).

I.—Written examination:

	Coefficients.
French composition	5
Latin translation, with dictionary	3
English (without dictionary)	3
	11

II.—Oral examination:

History	5
Geography	5
French	6
Latin	4
Greek*	2
English	7
	29

Total: General subjects	40

*After 1880, Greek is to be omitted from the programme, and Statics is to be added.

2.—MATHEMATICS (*Partie Scientifique.*)

I.—Written examination:

Arithmetic and Geometry	8
Algebra and Trigonometry	5
Descriptive Geometry	4
	17

II.—Oral examination:

Arithmetic	9
Algebra	8
Geometry	11
Trigonometry	5
Descriptive Geometry	6
	39

Total: Mathematics	56

3.—DRAWING.

Sketch of a head	4

Aggregate	100

Handwriting and spelling are taken into account in marking the papers.

The detailed programmes in each subject are prescribed by the min-. istry of public instruction for the schools. These schools are of two kinds, *lycées* and communal colleges, and it is at one or the other of these that candidates for the Naval School are expected to prepare for the examination. The recorded candidates preparing at the public schools were thus distributed in 1876:

LYCÉES.

Henri IV (Paris)	1
St. Louis (Paris)	2
Amiens	1
Bordeaux	8
Brest	106
Dijon	1
Grenoble	1
Lille	2
Lorient	21
Rouen	1
Toulon	7
	151

COMMUNAL COLLEGES.

Cherbourg	34
Dieppe	3
Rochefort	30
	67

Total in public schools	218

These figures show how great a part the public schools play in preparing candidates for the Naval School, especially those of the five naval ports, which furnish nine-tenths of the whole number. As far as instruction goes, both lycées and colleges may be classed together, though the latter are usually less completely organized and provided with a less numerous staff of instructors; but in general they follow the programme of the lycées as far as they can. This includes nine regular classes, with a division (*bifurcation*) in the programme, above the second class, for students who desire to confine their attention more particularly either to science or to literature and philosophy. The higher classes are known as the classes of elementary mathematics and the higher mathematics, on the one hand, and of rhetoric and philosophy, on the other. The highest classes reach a standard as high as that of the sophomore class at our highest universities, or of the senior class at many of our colleges. Leaving out the last two years, the French schools correspond nearly to our best Latin and high schools. They are admirably organized and carefully inspected; their methods of instruction and programmes of study are based on sound principles, and they produce the best results. Of course it is a great advantage to a professional school to be able to draw its pupils from such a source, and thus to be assured that they have a preliminary training at once broad, thorough, and complete. There is no greater drag to the efficiency of an institution for higher education than the want of good fitting schools—a want such as is felt at the present moment at our Military and Naval Academies, and nowhere felt more severely. This want in France is supplied to the fullest extent and in the most satisfactory way. The great national schools, including the Polytechnic, Saint-Cyr, the normal school, and the naval school at Brest, receive the great majority of their pupils from the best schools that exist in the country. They are thus enabled to make the qualifications for admission high, to do away with all those elementary branches of instruction that form no proper part of the work of a professional school, and to concentrate the energies of their pupils upon studies of a high character, connected more or less directly with the profession in view. The high standard of admission and the competitive examinations are not productive of cramming; on the contrary, they are far less so than a lower standard would be, even in a test examination, where the preparation is given in inferior schools, with a month or a fortnight of private tuition just before the examination. At the examination for admission at Annapolis, limited as it is in scope and trifling as are the subjects that compose it, a large number of the weaker candidates are prepared by this latter process. In France, where the programme of the examination is based directly on that of the schools—in fact, is identical with it in most subjects—there is no such difficulty. There is no need of private tuition; a candidate who wants to pass the examination cannot do better than attend the regular school course that leads up to it; and even in the case of a pupil who is weak in one of the required branches, the school system is sufficiently flexible to admit a slight modification in his favor.

The programmes of the lycées that define the requirements for admission to the Naval School are given below in the form in which they appear in the circular published for the information of candidates. Specimens of the questions given at the written and oral examination are given in the Appendix.*

FRENCH.

Programme of third class and of the grammar classes (4th, 5th, and 6th).

French grammar: Study of the French language and literature; explanation and recitation of French authors. Designated works: La Fontaine, Fables; Fénelon, Télémaque; Voltaire, Charles XII; Montesquieu, Grandeur et Décadence des Romains.

LATIN.

Third class and grammar classes.

Latin grammar: Explanation and recitation upon Latin authors. Authors designated: Cæsar, Gallic war; Virgil, Æneid, I, II, and III books.

GREEK.

Third class and grammar classes.

Greek grammar: Explanation and recitation upon Greek authors. Authors designated: Xenophon, Anabasis, or Expedition of Cyrus.

ENGLISH.

Third class.

Grammar: Revision of syntax; idioms; proverbs; general rules of prosody.
Exercises: Explanation and recitation; exercises of conversation; money, weights, and measures of England taught in English.
Authors designated: W. Irving, Christopher Columbus (abridged by the author); Macaulay, Lays of Ancient Rome.

HISTORY.

Third class.

History of Europe from the fifth to the end of the thirteenth century (395–1270).
Gaul under the Roman Empire: Invasion of the Barbarians; the Germans, their establishments in Italy, the States they founded.
The Kingdom of the Franks: Clovis, Brunehaut, Dagobert; conquests in Germany; government and institutions; the Salic law.
Justinian; his wars and legislative work.
Mahomet: Conquests of the Arabs; high character of the Arabic civilization.
Pepin d'Heristal; Charles Martel; Pepin the Short.
Charlemagne; his wars and his government. Re-establishment of the empire. Louis le Débonnaire. Treaty of Verdun. Charles the Bald. The Normans. Dismemberment of the empire into kingdoms and of France into great fiefs.
The feudal system. State of the church in the tenth century.
The Empire. Otto the Great. The quarrel of investitures. Gregory VII. Innocent III and Innocent IV. Frederick Barbarossa and Frederick II.

---

*Note H.

Norman conquest of England. Henry II. Magna Charta.
The Crusades. The Kingdom of Jerusalem and the Latin Empire of Constantinople.
Progress of royal power in France. Enfranchisement of the communes. Louis VI.
Philip Augustus. War of the Albigenses. Reign of Saint Louis.
Arts, letters, and schools in the twelfth and thirteenth centuries; commerce and industry.
Tabular summary of the different States of Europe in 1270.

NOTE.—Ancient history is studied in the lower classes of the Lycées; modern history (after 1270) in the higher classes. The historical course at the Naval School corresponds to the latter course.

GEOGRAPHY.

Third class.

Physical, political, and statistical geography of Europe (omitting France).
Configuration of Europe. Extreme longitude and latitude. Oceans that bound Europe; gulfs and straits; islands; peninsulas and capes; description of the coasts, Geological formation; mountain system; chains and peaks; principal summits, plateaux and plains. General direction of water-courses. Basins, rivers, streams, lakes. Isothermal lines, maritime and inland climates, winds and rains. Races, languages, religion.
British Isles: Rapid review of the physical geography. Territorial formation; surface, configuration, and limits; great divisions. Principal cities. Agriculture (cereals, industrial products, pasturage, cattle raising). Fisheries. Mines (coal and iron). Metallurgical (iron and steel) and textile industries. Canals and railways. Navigation. Commerce. Government and administration. Army and navy. Budget and debt. Population. Possessions in Europe. Colonies.
Netherlands.—Rapid review of the physical geography. Territorial formation; surface, configuration, and boundaries; provinces. Principal cities. Agriculture (dikes and polders). Fisheries. Canals and railways. Navigation. Commerce. Government and administration. Population. Colonies.
Grand-duchy of Luxembourg.—Mines and commerce. Government and administration.
Ger i any.—Review of physical geography. Territorial formation; surface, configuration, and boundaries. Kingdom of Prussia and its old and new provinces. Secondary states. Principal cities. Agriculture. Fisheries. Mines. Metallurgical and textile industries. Canals and railways. Navigation. Commerce. Government and administration. Germanic Confederation of 1815. Zollverein. North German Confederation of 1866. German Empire of 1871. Army, navy, and finances. Population.
Austria-Hungary.—Review of its physical geography. Territorial formation; surface, configuration, and boundaries; provinces and countries included in it. Principal cities. Agriculture. Mines. Industries. Railways. Coast and river navigation. Commerce. Government and administration. Army and finances. Population. Races and languages.
Switzerland.—Review of its physical geography. Territorial formation; surface, configuration, and boundaries. German, French, and Italian cantons. Principal cities. Agriculture. Commerce. Means of transit. Government and administration. Population.
Portugal.—Review of physical geography. Territorial formation; surface; configuration, and boundaries. Ancient division into provinces. Principal cities. Agriculture. Industry. Commerce. Government and administration. Population. Colonies.
Spain.—Review of physical geography. Territorial formation; surface, configuration,

and boundaries. Ancient division into kingdoms and provinces. Principal cities. Agriculture. Industry. Commerce. Mines. Means of transit. Government and administration. Population. Colonies. Republic of Andorra.

Italy.—Review of physical geography. Territorial formation. Ancient division into States. Surface, configuration, and boundaries. Countries and provinces. Principal cities. Agriculture. Mines. Industries. Railways. Navigation. Commerce. Government and administration. Army and finances. Population. Republic of San Marino.

Greece.—Review of physical geography. Territorial formation; surface, configuration, and boundaries. Mainland, Archipelago and Ionian Isles. Principal cities. Agriculture. Navigation. Commerce. Government and administration. Population.

Turkey in Europe.—Review of physical geography. Territorial formation, surface, and boundaries of the Ottoman Empire in Europe, in Asia, and in Africa. Great divisions of Turkey in Europe; Archipelago, and Candia. Principal cities. Agriculture. Navigation. Commerce. Government and administration. Population. Races and religions.

Principalities of Roumania, Servia, and Montenegro.—Review of physical geography. Principal cities. Agriculture. Navigation of the Lower Danube. Commerce. Government and administration.

Russia.—Review of physical geography. Territorial formation, surface, configuration, and boundaries of the Russian Empire in Europe and in Asia. Great divisions of Russia in Europe. Principal cities. Agriculture. Fisheries. Mines. Industry. Railways. Sea-coast and interior navigation. Commerce. Government and administration. Army, navy, and finances. Population.

Norway and Sweden.—Review of physical geography. Territorial formation; surface, configuration, and boundaries. Principal cities. Great divisions. Agriculture. Fisheries. Mines. Canals and railways. Navigation. Commerce, government and administration. Population.

Denmark.—Review of physical geography. Territorial formation; surface, configuration, and boundaries. Principal cities. Agriculture. Railways. Navigation. Commerce. Government and administration. Population. Iceland, Faroe Islands, and colonies.

General review.—Comparison of the extent and resources of different States. Weights, measures, and money. Density of population. Military forces.

(Demonstration and exercises at the black-board and wall-map. Map drawing.)

GEOGRAPHY OF FRANCE.

Fourth class.

Configuration and dimensions of France ; surface ; extreme latitude and longitude.

Seas and coasts ; gulfs, islands, peninsulas, capes, dunes, rocks and shoals, salt marshes, lagoons, principal ports, sea and inland frontiers, territorial losses of France in 1871.

Contour. Mountain chains, peaks and plateaux; altitude; perpetual snows, glaciers, torrents, passes, roads, tunnels, waters, waterfalls, and basins; rivers and their tributaries; canals; lakes, ponds, marshes.

Climate and principal productions.

Political geography. Ancient provinces, with their capitals. Departments by provinces, and departments by basins ; their principal cities ; other important cities. Principal railways. Population. Colonies.

(Demonstration and exercises at the black-board and wall-map. Map drawing. Study of the military map of France.)

ARITHMETIC.

Third class.

Decimal numeration.

The four processes with whole numbers.

Decimals. Processes and examples.

Characteristics of divisibility by 2, 3, 5, 9, and 11.
Definition of prime numbers and of numbers prime to each other. Resolution of a number into its prime factors (no theoretical developments). Greatest common divisor and least common multiple.
Common fractions. Reduction of fractions to lowest terms; reduction to fractions having a common denominator. Processes with fractions. Conversion of common fractions into decimals.
Square and square root of whole numbers and decimals.
Metric system.
Ratio and proportion. Ratio of two magnitudes. Proportional magnitudes. Problems upon proportional magnitudes. Questions of interest and discount; formulas for working them out.

The class of elementary mathematics reviews the third-class programme, completing it by a few lessons on the properties of prime numbers, repeating decimals, and the errors arising in the extraction of roots.

ALGEBRA.

Class of elementary mathematics.

Review and completion of the second-class programme (see below).
Discussion of the formulas that resolve a system of equations of the first degree with two unknown quantities. Examples.
Equations of second degree with one unknown quantity. Double solution. Imaginary values.
Properties of trinomials of the second degree.
Questions of maxima and minima which may be resolved by equations of the second degree.
Principal properties of arithmetical and geometrical progressions.
Theory of logarithms deduced from progressions. Logarithms to the base 10. Tables. Characteristics. Introduction of negative characteristics to extend logarithmic calculations to numbers less than 1.* Use of tables. Compound interest and annuities. Application of logarithms to these question.

Second class.

Algebraic operations (not including the division of polynomials).
Equations of the first degree. Examples. Equations of the second degree with one unknown quantity. Applications to some of the problems of arithmetic and geometry.

GEOMETRY.

Third class.

Right line and plane. Broken line. Curved line. Angle. Generation of angles by the rotation of a right line about one of its points. Right angle.

---

* To define the logarithms of numbers less than 1, it is sufficient to extend to these numbers the fundamental property of logarithms. Let a be a number less than 1, and let $P = a \times 10^n$, supposed to be greater than or at least equal to 1. P will have a logarithm, and if it is desired to extend to this product the fundamental property, we shall have—

$$\log a \text{ or } \log \frac{P}{10^n} = \log P - n.$$

Thus it is proper to call the logarithm of P, diminished by n, the logarithm of a. In reducing this to a single characteristic, it is evident that it will contain a number of negative units, equal to the position that the first significant figure of a occupies to the right of the decimal point.

Triangles. Simple cases of equality. Properties of the isosceles triangle. Cases of
equality of right-angled triangles.
Geometrical position of points equidistant from two given points. Geometrical position of points equidistant from two intersecting right lines.
Parallel lines. Sum of the angles of a triangle, of a polygon. Properties of parallelograms.
The circumference of the circle. Mutual relations of arcs and chords, of chords and
their distances from the center. Tangents. Intersection and contact of two circles. Measurement of angles. Inscribed angles.
Use of the rule and compasses in construction, on paper. Tracing of perpendiculars
and parallels; use of the triangle. Determination of angles in degrees, minutes, and
seconds. Protractor.
Elementary problems upon the construction of angles and triangles. To draw a tangent to a circle through a given point without. To draw a tangent parallel to a
given line. To draw a line tangent to two circles. To describe on a given line a
segment which shall contain a given angle.
Measurement of areas. Area of the rectangle, parallelogram, triangle, trapezoid; of
any polygon. Approximate area of a figure bounded by any curve whatever. Theorem of the square described on the hypotenuse of a right-angled triangle. Numerous numerical applications.
Elements of surveying. Use of the chain and the surveyor's square.

Second class.

Proportional lines.
Similar polygons. Conditions of similarity of triangles. Ratio of the perimeters of
similar polygons. Relations between the perpendicular let fall from the right angle
of a right-angled triangle upon the hypotenuse, the segments of the hypotenuse,
the hypotenuse itself, and the legs. Theorem relating to the square of the side of
a triangle, opposite to a right angle, an acute angle, or an obtuse angle. Theorem
relating to the secants of a circle passing through the same point.
Problems: to divide a line into equal parts, into parts proportional to given lengths ;
to find a fourth proportional to three given lines, a mean proportional between two
given lines ; to construct on any line a polygon similar to a given polygon.
Regular polygons; inscription of regular polygons in a circle; square, hexagon.
Method of determining the ratio of the circumference to the diameter; applications.
Area of a regular polygon ; of a circle, of a circular sector. Ratio of the areas of
two similar figures.
The plane and the right line in space. Perpendiculars and oblique lines drawn to a
plane. Parallelism of right lines and planes. Dihedral angles. Perpendicular
planes. Elementary notions of trihedral and polyhedral angles.
Construction of plans ; use of the metric scale, the semicircle, square, and plane table.
Topographical surveying ; sea-level, sight; method of describing the height of a
point ; lines of level ; interpretation of topographical charts.

Class of elementary mathematics.

The course begins with a rapid review of the course of the third and second classes,
amplifying certain points, especially in regard to the inscribing of regular polygons
(case of the decagon) and the determination of the ratio of the circumference to the
diameter by the isoperimetrical method. The review is finished by exercises and
problems upon the comparison of areas; the construction of a square equivalent to a
given polygon ; the construction of a square whose ratio to a given square is equal to the
ratio of two given lines ; the construction of a rectangle equivalent to a given square,
the sum or difference of whose adjacent sides is equal to a given sum or difference;
applications to the construction of roots of equations of the second degree with one
unknown quantity.

Geometry in space.

The plane and the right line. Conditions in order that a right line may be perpendicular to a plane. Properties of the perpendicular and the oblique lines drawn from the same point to a plane. Parallelism of right lines and of planes.
Dihedral angles. Generation of dihedral angles by the rotation of a plane about a right line. Dihedral right angle. Measurement of dihedral angles.
Properties of planes perpendicular to each other.
Trihedral angles. Cases of equality and of symmetry. Properties of the supplementary trihedral angle. Limit of the sum of the faces of a convex polyhedral angle. Limits of the sum of the dihedral angles of a trihedral angle. Analogies and differences between trihedral angles and rectilinear triangles.
Polyhedrons. The prism, parallelopiped, cube, pyramid. Plane and parallel sections of the prism and the pyramid
Measurement of volumes. Volume of the parallelopiped, the prism, the pyramid, the frustum of a pyramid, and the frustum of a triangular prism.
Symmetry in polyhedrons. Plane of symmetry. Center of symmetry.* Comparison of the faces, the dihedral angles, the homologous polyhedral angles of two symmetrical polyhedrons. Equivalence of their volumes.
Similar polyhedrons. Cases of similitude of two triangular pyramids. Ratio of the volumes of two similar polyhedrons, similarly placed.

The round bodies.

Right cylinder with circular base. Measure of the lateral surface and of the volume; extension to right cylinders with any base whatever.
Right cone with circular base. Sections parallel to the base. Convex surface and volume of the cone, and of the frustum of a cone.
Sphere. Plane sections; great circles; small circles. Poles of a circle. Given a sphere, to find its radius by a plane construction.
Tangent plane. Angle of two arcs of a great circle.
Spherical triangles. Analogy with trihedral angles. Measure of the surface generated by a line turning regularly about an axis drawn in its plane and through its center. Area of the zone of the entire sphere. Exercises.
Measure of the volume generated by a triangle turning about an axis drawn in its plane, and through one of its angles. Application to a regular polygonal sector turning about an axis drawn in its plane, and through its center. Volume of a spherical sector, of the entire sphere, of a spherical segment. Exercises. Approximate volume of a solid bounded by any surface whatever.

Properties and definitions of certain curves.

Definition of the ellipse by the properties of the foci. Drawing of the curve by determining points, and by a continuous motion. Axes. Vertices. Radius vector.
General definition of a tangent to a curve.
The radii drawn from the foci to any point of the ellipse make equal angles with the tangent at this point. To draw a tangent to the ellipse to a point taken on the curve, through an exterior point. Normal of the ellipse.
Definition of the parabola by means of the property of the foci, and by its directrix. Tracing of the curve through points, and with a continuous motion. Axis. Vertex. Radius vector.
The tangent makes equal angles with the line parallel to the axis, and the radius vector, drawn through the point of contact. To draw a tangent to a parabola

---

* The study of symmetry with reference to a point, reduces itself to that of symmetry with reference to a plane, by rotating one of the two figures through an angle of 180°, about an axis perpendicular to the plane and passing through the center of symmetry.

124 NAVAL EDUCATION—FRANCE.

through a point taken on its curve, and through an exterior point. Normal. Subnormal.

Relation between the square of an ordinate perpendicular to the axis, and the distance of this ordinate from the vertex.

Definition of the helix, considered as resulting from the unrolling of the plane of a right-angled triangle upon a right cylinder with circular base. Distance between the turns of the helix. The tangent to the helix makes a constant angle with the side of the cylinder. To construct the projection of the helix and of the tangent upon a plane perpendicular to the base of the cylinder.

PLANE TRIGONOMETRY.

Class of elementary mathematics.

Trigonometric lines. Relations between the trigonometric lines of the angle. Expressions of the sine and the cosine as a function of the tangent. Formulas relating to the sine, cosine, and tangents of the sum and the difference of two arcs. Expressions of sin $2a$, cos $2a$, and tang $2a$. Given cos a or sin a, to calculate sin $\frac{1}{2}a$ and cos $\frac{1}{2}a$.

To adapt the sum of two trigonometric lines to calculations by logarithms; e. g., the sine, cosine, and tangent. Construction of trigonometric tables. Use of tables.

Relation between the angles and the sides of a right triangle or of any triangle whatever. Resolution of right triangles. Resolution of any triangles whatever in the four cases that may present themselves. To determine the area of a triangle as a function of given parts.

Application of trigonometry to the different questions presented in surveying; distance of an inaccessible point. Measurement of heights. Three point problem.

DESCRIPTIVE GEOMETRY.

Class of elementary mathematics.

Inadequacy of ordinary drawing for the representation of solid bodies. Usefulness of a geometric method, which, by graphic operations executed on one and the same plane, determines exactly the form and the position of a figure of three dimensions.

Projection of a point, of a right line, of any line, on a plane. Plane of projection. Traces of a plane. True length of a line that joins two points given by their projections. Angles of a right line with the planes of projection. Representation of a plane by its traces. Angles made by a plane with the planes of projection.

Method of revolutions. Exercises.

Intersection of two planes. Intersection of a right line and a plane. Distance of a point from a plane. Distance of a point from a right line. Angle of two right lines; of a right line and a plane; of two planes.

Projections of a prism, a pyramid, a cylinder, a cone with circular base, executed from models.

Plane sections of polyhedrons.

Method of plans cotés.

STATICS.*

Class of elementary mathematics.

Forces. Condition of equality of two forces. Their numerical representation. Comparison of forces with weights by means of the dynamometer. Translation of the point of application of a force to any point in its direction, supposing it to be invariably connected with the first point.

Composition of two forces applied at the same point. Theorem of the moments with reference to a point taken in the plane of the forces. Composition of any number of forces applied at the same point. Conditions of equilibrium. Composition of

---

*After 1880.

two parallel forces. Couples. Composition of any number of parallel forces. Centers of parallel forces. Centers of gravity; their determination in certain simple cases, as the triangle and pyramid. Composition of a system of forces applied to a solid body; their reduction to two forces, one of which acts at any given point. General conditions of equilibrium.

Simple machines.

The lever: General condition of equilibrium; relation between the power and the weight. The balance: Ordinary balance, Roman balance, scales of commerce. The pulley: Equilibrium of the fixed pulley; of the movable pulley. Combinations of pulleys. The capstan or windlass: Equilibrium; relation between the power and the weight. The inclined plane: Equilibrium of a body placed upon an inclined plane.

Elements of kinematics and dynamics.

Uniform rectilinear motion; velocity. Varying rectilinear motion; mean velocity; velocity at any instant. Rectilinear motion uniformly varied; acceleration. Acceleration at any instant in a varying rectilinear motion. Composition of two simultaneous rectilinear motions, varying uniformly or otherwise. Uniform motion of rotation about a fixed axis; angular velocity.
Law of inertia.
Law of relative motion: The inference that a constant force, acting upon a material point which starts from rest, or which has an initial velocity in the same direction as the force, imparts to it a uniformly varied motion. Converse proposition. Two constant forces are proportional to the accelerations that they produce in acting separately upon the same material point which starts from rest, or which has an initial velocity in the same direction as the force.
Mass: Its measure by means of weight.

Work of forces.

The work of a constant force applied to a point whose displacement is rectilinear. Unit of work. To show that, in simple machines, in a state of uniform motion and acted upon only by a power and a resistance, the motive power is equal to the resistance. Influence of passive resistance. In practice, the motive power always exceeds the useful resistance.

3.—ACADEMIC YEAR.

The course at the Naval School is two years. The academic session on board the Borda begins on the 1st of October and lasts nearly ten months. There are three terms or trimesters, of three months each. The regular programme of studies extends only over the first eight months, June being the *mois de pioche*, which is occupied in studying for the coming examinations and in special exercises. The examinations take up nearly the whole of the tenth month, July.

After the close of the annual examinations, the students of the second class are embarked on board the Bougainville, a small vessel attached to the school. This a screw steamer, with engines of 120 horse-power, built specially for the school. In this they take a short practice cruise, during which they visit various points on the neighboring coast of France, including Cherbourg on the one hand and Lorient on the other. Sometimes the cruise extends as far as Ferrol, in Spain. During the cruise, the pupils are stationed with the men, and work the ship, receiving at the same time instruction of a practical character in seamanship, navigation, and steam-engineering, and performing exercises in drawing. For the first few weeks, the first class also takes part in the cruise, in the same or another vessel; and during this time, the practice ships manœuver in and about the roads of Brest; and when the second class goes to sea, the members of the first class, which has now been graduated, are sent home for a six weeks' vacation. The Bougainville returns to Brest early in September, and the second class men have leave for a month. On the 1st of October, they return for the second year course in the Borda, while the graduates, now for the first time admitted with a specific rank to the Navy, are embarked on board the Flore, for a more extended cruise. They are called *aspirants de deuxième classe,* a term nearly equivalent to cadets, and they are still undergoing instruction, though instruction of a more distinctly practical character.

During the course on board the Borda, the young men under instruction are known simply as pupils, *élèves.* There are two classes or divisions, the first and second, being those respectively in their second and first year. The average number in a class is between 40 and 50. In June, 1878, there were 53 of the *anciens,* or first class, and 44 of the second. Each class is divided into two sections (*escouades*) for purposes of administration, and each of the four sections is intrusted to the particular attention of a lieutenant. Each section is again divided into three subsections (*séries*) for the interrogations. Each subsection numbers about eight members.

4.—COURSE OF INSTRUCTION.

The methods of instruction adopted in the Naval School of Brest are similar, in a general way, to those of other French schools and colleges, and resemble nothing in America. Recitations in the ordinary sense do not exist, and text-books are almost unknown. A few books of reference are used, including the nautical almanac, a work on physics (by Almeida), English and French grammars, and the admirable series of manuals published under the authority of the Ministry of Marine.* These books are not, however, used as text-books. The main feature of the system of instruction is the *cours,* or lecture. This is delivered in a more or less formal manner in the amphitheater, and the students take notes, which are inspected from time to time by the instructors. They also receive, after the lecture, a full summary of the points treated; a summary so

---

* This series includes—

1. Manual of the seaman-gunner (*matelot canonnier*).
2. Manual of the topman (*gabier*).
3. Manual of the small-arm man (*marin fusilier*).
4. Manual of the coast-pilot (*pilote côtier*).
5. Manual of the helmsman (*matelot timonier*).
6. Manual of torpedoes (*défenses-sous-marines*).
7. Manual of fencing and gymnastics.
8. Manual of miscellaneous exercises (diving, &c.).

full that it amounts pretty nearly to the lecture itself. It is reproduced on lithographic sheets (*feuilles autographiées*) similar to those in use at the Polytechnic and elsewhere, and contains all the necessary drawings, diagrams, and tables. The time allowed for the *cours* is from an hour to an hour and a quarter. It is followed by a short recess, and by a period of study from one to two hours long, under the supervision of the same or a subordinate instructor, but always on the same branch of study. Thus a lecture in mechanics is followed by study in mechanics; a lecture in naval architecture by study in naval architecture, and so on. During the study period, assistance and explanations are given; various exercises are performed, in some cases an abstract of the previous lecture being made; and the *interrogations* are held. The latter are a peculiar feature of the system, and form the complement of the *cours*. They are conducted orally, and are something between a recitation and a monthly examination. They are of two kinds, particular and general; and each student has an interrogation of both kinds in each term. The particular interrogations take place at irregular intervals, during the study period, and are held without previous notice; the captain designating each day the students who are to be interrogated. They cover the ground gone over since the beginning of the term. The general interrogations are held at the end of each term on the studies of the whole term. They are longer and fuller than the others, and they have a greater weight in determining the term-mark, counting double in the two first terms, and three times as much in the third term. The professor may call upon students to answer questions at other times, during the hours of lecture, for example, but this is not done to any very great extent.

The general interrogation of the third term covers the course for the whole year, but it is independent of the final annual examination. This, of course, has the same scope as the last interrogation, and the latter seems to be conducted chiefly with a view to preparing the students to pass it. The examination is conducted by an outside board appointed for the purpose, of which the Préfet Maritime is president, and the principal working member in scientific subjects is a professor of hydrography. It is thorough and searching in character, and upon it chiefly depends the advancement or discharge of students of doubtful capacity. The essential point, however, to be noticed about it is that it is conducted independently of the school, and thus acts as check upon instructors.

The marking system is exceedingly simple. The maximum is, as usual, 20. The term marks, as has been stated, are ascertained by combining the marks for the two interrogations, giving the second double weight, and, in the case of third term, triple weight. The term-marks are combined with equal weight to determine the school-mark for the year; and the latter, combined with the mark for the annual examination, given by the examining board, fixes the mark in each subject for the year. The examination thus counts as much in the final standing as the whole year's course. To obtain the final mark of each student for the year, the mark in each branch is multiplied by its coefficient, and the sum of the products is taken, increased by the captain's marks for conduct and aptitude. The co-efficient of the former is 4, and of the latter 2. The two years have equal weight in determining the mark for the whole course.

Daily reports are made by instructors to the captain, of marks given at interrogations. On Sunday the marks for the past week are published. Those pupils whose mark is less than 10 in practical navigation, or less than 5 (25 per cent.) in any other branch, are deprived of the privilege of going ashore. The captain also sends a weekly report to the Préfet Maritime. At the end of each term merit-rolls are made out, in which the rank of students is determined by the method given above. Deficient students are reported to the secondary council of instruction, which re-examines them, and makes a report of the result, with recommendations as to the final action to be taken. In forwarding this report the captain is allowed, in the case of deficient students whose conduct or character is bad, to add a recommendation for their expulsion, in place of suspension or permission to withdraw.

After the annual examination, merit-rolls are made out for the year and for the course. Upon the latter depends the seniority of the graduates as cadets or *aspirants*. At this examination, also, students who show a want of capacity or of aptitude for the service, or whose marks fall below the standard, are reported. Those whose deficiency is due to illness are allowed to repeat the studies of the year, but the others are discharged and sent back to their friends (*rendus à leurs familles*).

An honorary distinction, but one accompanied by no authority or responsibility, is given at the end of each term to the students who take rank in the first quarter of their class on the merit-rolls for the term. The distinction consists in the title of *élève d'élite*, and the privilege of wearing an anchor on the collar of the jacket; and in the case of the first third of the *élèves d'élite*, two anchors. The honor is forfeited for any of the more serious offenses, and is not conferred upon pupils who have undergone imprisonment. The *élèves d'élite* are allowed to go ashore on all the liberty days. A prize of a sextant or a telescope is given to the three highest men on the roll of the graduating class.

The following table gives the branches of study, the co-efficients, and the number of hours per week of lecture, study, and exercises, the last being partly theoretical or instructional, and partly practical. The table applies equally to both classes, as both have the same branches of study on alternate days, and the same aggregate division of time.

Course of study and co-efficients.

	Co-efficient.	Number of hours per week.			
		Cours.	Study and interrogation.	Exercises.	Total.
General subjects:					
Literature, history, and geography	7	2	3½	......	5½
English	5	2	3½	......	5½
Drawing	3	......	......	3½	3½
Scientific subjects:					
Analysis and mechanics	10	2½	4	......	6½
Astronomy and navigation	10	2½	4	......	6½
Physics and chemistry	7	1¼	2	......	3¼
Practical navigation	7	......	......	2¼	2¼
Professional subjects:					
Seamanship*	8	......	......	15¼	15¼
Steam-engineering*	6	......	......	4¼	4¼
Gunnery*	6	......	......	2½	2½
Infantry } Small-arms* }	4	......	......	4¾	4¾
Naval architecture	4	1	1¾	......	2¾
Total	......	11¼	18¾	32½	62½
General study (*étude libre*)	......	......	10	......	10
	......	11¼	28¾	32½	72½

*Theory and practice. In the number of hours allowed to seamanship is included the time taken up in getting on board the corvette and returning to the ship; and in gunnery, the time consumed in going ashore and returning.

The total amount of work performed under supervision—over ten hours a day (including Sunday)—is reduced slightly by the infrequent privilege of an afternoon on shore; but even with this deduction it would be considered excessive according to American standards, especially in a school where the pupils had only a month or six weeks of vacation in the whole year. The principle of work under supervision is, however, one of the essential features of all French education.

In the daily arrangement of studies, the morning is devoted to scientific subjects; the afternoon, from one o'clock to four, to professional study and exercises (the latter including drawing), and the evening, after five o'clock, to miscellaneous subjects, including naval architecture. The working day begins with general study (*étude libre*) from a quarter before six to seven. At a quarter past eight the *cours* for the day begins, twice a week in astronomy and navigation, twice in analysis and mechanics, and once in physics and chemistry. The two subjects in each group are not pursued simultaneously—the second is taken up after the first is finished; but in the allotment of time they are considered as single groups or departments. The *cours* lasts one hour and a quarter and is followed by study, on the same subject, under supervision. During this period of study, lasting two hours, the interrogations are held. In the evening studies the same system prevails, the *cours* being from five to six o'clock, and the study from 6.15 to 7.45. In the afternoon the distinction of lectures, study, interrogation, and exercise is not preserved, but instruction is given during the allotted hours, lasting from 1 to 4.15, in the most convenient manner.

This is the programme for five days in the week; Thursday and Sunday are exceptional days. They are not holidays, by any means, for holidays can hardly be said to exist on board the Borda, except at rare. intervals; but they serve to vary the monotony of a dead level of routine. On these days the students begin with an hour of practical navigation at a quarter before six in the morning. Part of the forenoon is occupied on Thursday with infantry drill on shore, and on Sunday with inspection, mass, study, and fencing. The rest of the forenoon and the whole of the afternoon on both days are spent on board the corvette, performing manœuvers in the roads. In case of bad weather, the students remain on board the Borda, having study hours till noon, and various seaman-ship exercises (*école de matelotage*) in the afternoon.

During June, with the exception of three infantry drills, the forenoon and evening (from 4.30 p. m.) are given to study. One week is wholly taken up by the third term interrogation. The three hours in the after-noon are variously spent in exercises, visits to the arsenal and other places, and special study. On five days the students go on board the corvette and work ship. The two divisions spend three or four after-noons separately in the dockyard, one day in charge of the gunnery in-structor, another day of the engineer, and another of the instructor in naval architecture. The first class is taken to the gun-cotton factory by the gunnery instructor. The other exercises consist of drawing, infan-try, and great guns for both classes; observing, at the observatory on shore, and sabre-drill, for the first class; and boat-drills and signals for the second class.

Though the course lasts only two years, the high standard of admis-sion makes it possible to accomplish a great deal in that time, and to do the work thoroughly. The studies classed under " general subjects" em-brace a full course of French literature, and a continuation and comple-tion of the history course of the Lycées. Instruction in the English language includes the technical terms of seamanship, gunnery, and steam-engineering. The manner in which this instruction is given is peculiar and admirable. The lithographed *cours* of the professors, far from being a mere dictionary of terms, comprise connected treatises on these three subjects, in English; and excellent treatises they are, apart from their main object of teaching a foreign language, being clear, compact, and systematic. They, therefore, give practice not only in a nautical but in a general vocabulary, and in the construction of sentences, and the application of grammatical principles. The character of the exercise will be best shown by examples, a few of which are given in the appen-dix, taken from different parts of the work.*

The second division, known as the scientific subjects, comprises three groups of studies, astronomy and navigation, analysis and mechanics, and physics and chemistry. More time is given to these subjects than to the others, and the time is in general better arranged, the whole

---
* Note I.

forenoon being taken up with them. The course in the first group of studies, astronomy and navigation, begins with a short review of descriptive geometry and of spherical trigonometry. Astronomy is taught chiefly as subsidiary to navigation, but it receives a pretty full treatment, always from a mathematical point of view. The course in navigation is both theoretical and practical, and includes all that could well be taught in a stationary ship. There is a small observatory on shore, which is in the charge of this department. The hour before breakfast, on two days in the week, is devoted to observations and the solution of practical problems. The time is chiefly devoted, however, to a thorough foundation in the theory of navigation and to working out examples; the year on board the practice ship, after graduation, giving ample time to develop and perfect the exclusively practical part of the subject.

In the second group, analysis and mechanics, the course begins with a short review of certain subjects in algebra, such as fractional and negative exponents, the binomial formula of Newton, series, &c. Next comes a thorough course in analytical geometry, including higher plane curves. This is followed by the differential and integral calculus, including the method of least squares. After this, the class takes up mechanics, going through statics. This completes the course for the first year. The second year begins with the subject of differential equations, after which comes the rest of mechanics, embracing kinematics and dynamics. The second year's work includes an elaborate course in theoretical mechanics, and in mechanics applied to machinery.

Physics and chemistry are so divided that a part of both subjects is taught in each year. At present the Naval School does not possess the necessary means for instruction in either subject, such as apparatus, laboratories, &c., and the course in both is somewhat meager and inadequate. It is hoped that before long a laboratory will be provided. Of course it will have to be on shore, and its distance from the school will interfere seriously with its usefulness. At present some use is made of the laboratory connected with the pharmaceutical department at Brest; but this answers very ill for instruction in chemistry, and is of no use at all in physics. The lectures in the former branch are given up to general chemistry, with little view to its application to the naval profession; while many of the theories taught are obsolete, the dualistic formulas are retained, and the lectures in general do not represent the present condition of the science.

In the professional branches, the only regular *cours* delivered are in steam-engineering, ordnance and gunnery, and naval architecture. In the other subjects, and even in the first two of these, reliance is largely placed on practical exercises, supplemented by oral explanations. The lithographed *cours* form a series of elaborate works on the various subjects, and, with the manuals, afford all the necessary materials for imparting a thorough knowledge of the theory of all the branches. In seamanship, the *cours* contains a description of the parts of a ship, treated in a regular and systematic way, with the most general and elementary matters at the beginning; thus avoiding the faults of construction and arrangement which characterize most of the text-books on the subject. The *cours* for the second year is a complete sailmaker's and rigger's guide, and contains a general explanation and description of maneuvers. The problems of mathematics and mechanics involved in the subject, such as arise in lifting and getting on board heavy weights, in the action of sails and rudder on the motions of a ship, &c., are worked out in detail. As between theory and practice, however, by far the greater part of the time in this branch is given to the latter.

The course in ordnance includes the subjects of metals, the fabrication, testing, and inspection of guns; a full description of the three principal types or models in use in the French Navy, those of 1858–'60, of 1864–'66, and of 1870; carriages; powder; projectiles, and their manufacture; the complete discussion of the theory of motion of projectiles in air; the calculation of range tables, and the effect of projectiles on armor. Under the head of ordnance and gunnery (*artillerie*) come also instruction in stationary and movable torpedoes; the *cours d'infanterie*, which is a treatise on small-arms, principally the Gras rifle, together with the theory of firing (*étude théorique du tir*), mathematically treated; and a short explanation of topographical charts. The course is filled out by the manuals of the seaman-gunner, and of small-arms, and torpedoes.

The remaining professional subjects, steam-engineering and naval architecture, are treated with great fullness and thoroughness. With the exception of astronomical navigation, calculus, and mechanics, they are the most elaborate and skillfully-managed courses at the school. The instructors are a principal mechanician for the first branch, and two lieutenants of high scientific attainments for the second. Naval architecture includes ship-building, as well as the higher problems of naval architecture properly so called. The published lectures in both courses are well-digested and exhaustive works from a scientific and mathematical, as well as a professional point of view. (See Appendix, Note J.)

The practical exercises in the course are as follows:

1. Seamanship.
2. Great guns.
3. Howitzers, in boats and on shore.
4. Infantry.
5. Landing parties (*Exercices de débarquement*).
6. Boats, with sails and with oars.
7. Steam-engine.
8. Fencing.
9. Swimming (school in town).
10. Dancing (first class only).

The last three are considered as recreations. The seamanship exercises include long drills on Thursday and Sunday, in good weather. Instruction is given in signals. At certain times the corvette is under steam, and the students are then given practical instruction in the working of the engines, two being stationed at a time in the engine-room, and two at the fires. Seamanship instruction of a practical character, which does not involve getting a ship under way, is given on those afternoons when the regular exercise is put off by bad weather.

The system of instruction on board the Borda is one which has many advantages in the hands of able instructors, while with an inefficient staff it would inevitably go to pieces. In it, the instructor and the instructor's teaching are everything; the text-book nothing. As far as he has occasion to use them, the instructor makes his own text-books. The system is one that affords at the same time an excellent check and a powerful stimulus to the teacher. It is practically his class-room work that is published in the lithographed sheets, and the whole details of his teaching are therefore laid open to inspection. If he is incompetent he cannot conceal it, and if he devises new and original methods, if his studies are carried into the most recent and most advanced stages of scientific and professional progress, he gets full credit for it. The outside examiners are placed in a position to inspect not only results, but methods, and, if either are defective, to ascertain the most effectual remedy.

5.—DISCIPLINE—MODE OF LIFE—ROUTINE.

The discipline of the Borda is severe, even for a ship in commission. The supervision is so close and constant that there is very little opportunity of committing grave offenses without immediate discovery; and the list of punishments is calculated to preclude their repetition. The list is as follows :—

1. Reprimand (1) pronounced by an officer or professor; (2) by the second in command or executive officer.

2. Punishment squad, one hour a day, for three days at most.

3. Coventry (la police), not more than ten days.

4. Prison, not more than ten days.

5. Dark cell (cachot), not more than five days.

6. Suspension.

7. Expulsion.

Each offense, or rather each punishment, as representing an offense, has an influence on the conduct-mark. This mark is on a scale of 20. Demerits (points de punition) are given as a record of conduct, and the maximum number allowed in a term is 200. An absence of demerits gives a conduct-mark of 20, while the maximum gives a mark of zero. A student that receives the latter as his term-mark in conduct, is reported to the minister for dismissal. The number of demerits attached to each offense is fixed by a scale, according to the character of the punishment. Thus, every reprimand gives from one to two demerits; the squad gives two demerits, and one in addition for each day of punishment; coventry gives five demerits, and one for each day, and so upwards. Delinquents whose offenses cannot be suitably punished by the captain are sent to the guard-ship, where the Préfet disposes of them, either by a severer punishment, or by a report to the Minister of Marine, recommending their dismissal.

The character of the punishments is somewhat peculiar. Delinquents placed in the punishment squad are posted in line with a short interval between each man, and required to keep the position of the soldier without arms for one hour. If the number is large, they may be drilled for an hour. Coventry consists in isolating an offender from his companions and from everyone else during the hours of study, recreation, and meals. He attends lectures and practical exercises. He has his meals from the mess-table of the crew; and his allowance of wine is weakened with water. Delinquents undergoing imprisonment are allowed to attend the lectures and interrogations in scientific subjects only, but they are deprived of these also, if it is found expedient, as sometimes happens, to confine them on board one of the corvettes instead of the Borda, for the sake of greater isolation. When the fifth punishment is ordered, that of confinement in the *cachot*, the delinquent is taken out of his dark cell for one hour only in the twenty-four. This hour is employed in solitary exercise aloft (*exercice de gymnastique dans le gréement*). In both the prison and the *cachot*, delinquents' are allowed a blanket and their canvas blouse and trousers, but they sleep on the floor. Their meals are served from the crew's mess; but in the *cachot* they are reduced on the first, third, and fifth days of confinement to soup, bread, and water. Naturally, with such punishments, there is little difficulty in maintaining discipline on board the Borda.

Deprivation of liberty to go ashore does not exist as a separate punishment for offenses; but the privilege is taken away in consequence of very low marks, or of any of the five school punishments, incurred during the previous fortnight. A whole class may be occasionally deprived of liberty for a general infraction of the regulations.

Lieutenants and professors can only pronounce reprimands; all the other punishments, except the dark cell and removal from the school, may be inflicted by the commander, but the duration of the punishment is always fixed by the captain. Sectional inspections are conducted daily by the chiefs of sections, and a general inspection is held on Thursday by the commander and on Sunday by the captain; but informal inspections of all parts of the ship occupied by students, including also their chests, lockers, &c., are made at any time by the *adjudants*. Offenses are reported by any officer, from the commander down to the *adjudants* and instructing petty officers, under whose notice they come.

The students are not allowed to have any articles in their possession other than those authorized in the prescribed outfit, and they can keep nothing under lock and key. Any books, watches, rings, or other unauthorized articles that they bring with them when they first join the school are taken away, and only restored at their departure. They are forbidden to obtain or receive anything, from outside, even from their friends, but they may procure tobacco and such other small matters as they need at a little shop (*cantine*) on board. Novels and newspapers are especially prohibited, and the penalty for having a novel in possession is confinement in prison.

Formerly, there was a relation of authority and responsibility between the older and younger students, but this exists no longer. No monitorial authority of any kind is exercised, and positions of command in drills are purely temporary and cease when the drill is over. Orders are received from the chiefs of sections (lieutenants), and complaints in each section are made to them at inspection. If the complaint does not meet with attention, the matter can be laid before the commander or captain at Thursday or Sunday inspection.

The two classes go ashore on alternate Sundays once a month, the second class on the first Sunday in the month and the first class on the second. On these days the class not on liberty has practice for six hours in working ship on board the corvette, and on the other Sundays in the month the two classes are exercised either together or separately, using both the small vessels. Dinner takes place on board the corvette to save time. The students on liberty go ashore in charge of a lieutenant, who remains on duty during the liberty hours at an office provided for him at the *Etablissement des Pupilles*. He has, during this time, a general oversight of the students on shore, and he brings them off to the ship. The hours of liberty are from 11 a. m. till sunset, or till 7 p. m., when sunset is later.

No students are allowed to go on shore, even on liberty days, unless the privilege is asked for them by their *correspondant*. This person is a resident of Brest, selected by the parent or guardian of the student to act as his agent at Brest and to look, in a general way, after his interests. No officer of the school or contractor furnishing supplies to the school can be a correspondent. Correspondents cannot visit students in their charge on board the ship; indeed, this privilege is not accorded to any one; but a half hour, after the infantry drill on Thursday, is devoted to interviews, which take place on the drill-ground or in the building adjoining. In exceptional cases students may go ashore to visit their parents when the latter are temporarily at Brest, but this privilege is only granted on Sunday, and never, except in grave and peculiar circumstances, to pupils who reside in the neighborhood.

Each student receives on his arrival a number, which he retains throughout the course and by which he is known. It is placed on all his clothes, his desk, his books, and his hammock.

According to the table of studies already given, it has been seen that $62\frac{1}{2}$ hours a week are given to special exercises and 10 to general study.

The remaining 95½ hours which go to make up the total of the week are roughly divided as follows:

	Hours.
Sleep	57
Recreation	16
Meals	9½
Physical exercise	3
Inspections	2
Miscellaneous (including dressing, prayers, and religious service)	8
Total	95½

The students rise at five o'clock every day in the year. Three-quarters of an hour are occupied with dressing, prayers, and stowing hammocks; after which, the early morning study takes place for an hour and a quarter on five days in the week, varied on Thursday and Sunday by exercises in practical navigation. The study is followed by half an hour of exercise, either gymnastics or fencing. All this is done before breakfast. The routine of lectures, study, and exercises in the forenoon and afternoon has already been given. The aggregate number of hours of recreation during the week, 16, seems sufficiently large, but the programme is so arranged as to cut up the time into little scraps or recesses, making it hardly available for solid amusement. Thus there is a recess of fifteen minutes after breakfast, from 7.45 to 8; of 30 minutes, from 9.30 to 10, between the cours and study; of 30 minutes after dinner, from 12.30 to 1; of 15 minutes in the period of afternoon exercise, from 2.45 to 3; from 15 to 30 minutes after afternoon luncheon, between 4.30 and 5; and of 15 minutes between the evening cours and study, from 6 to 6.15. After supper there are no studies; recreation lasts from 8 to 8.45, followed by prayers and turning in at 9.

On Thursday and Sunday the programme is modified in order to give six hours in the afternoon on board the corvette; while the whole of Thursday morning is taken up with infantry drill on shore, and of Sunday morning with inspection and mass. As both these days are harder working days than the others, the hour for turning in is thirty minutes earlier.

There are four meals a day on board the Borda, as follows: Breakfast, 7.30 a. m.; dinner, 12 m.; afternoon luncheon, 4.15 p. m.; supper 7.45 p. m. The table is good, though exceedingly simple. The breakfast consists, after the French fashion, simply of coffee and bread and butter, while the afternoon luncheon (goûter) is of bread alone. Dinner is composed of soup, two dishes (plats)—one of meat, the other of vegetables—and dessert; and supper of meat, pudding or vegetables, and cake or sweetmeats. Half a pint of wine is allowed at dinner and at supper. On Friday the dishes of meat are replaced by fish. The regulation requires that the bill of fare shall be so arranged as to provide for the twelve meals per week, of which meat forms a part, four of boiled beef, two of roast beef, three of mutton, two of veal, and one of poultry.*

---

* The bill of fare for one day is given in the Appendix, note K.

Students are required to take a bath (*grand bain* or *bain complet*) once a month, and to take a foot-bath once a week. This somewhat infrequent ablution is supplemented in summer by sea-baths. The half hour devoted to the regular baths is that assigned in the programme to gymnastic exercise, just before breakfast. The uniform of *élèves* is the usual blue jacket and trousers; but it is not much worn on board ship, the customary dress being the white canvas blouse and overalls. This is worn even at studies and lectures. Blue is worn at mass and at inspections and on shore. When on liberty ashore students are required to wear a sword.

There is sick-call every morning and evening on board. In case of light illness students are placed in the sick-bay. Severer cases are sent to the hospital on shore, where a ward is specially reserved for the school. In the latter case the correspondent is immediately notified.

Religious service consists of prayers morning and evening, and mass on Sunday morning. The latter lasts half an hour, and attendance is required of all except the Protestants. The number of these is very small, there being five in 1878. They are allowed to have such service as they see fit by themselves in the sick-bay. An hour is set apart on two evenings in the week, during which the students who feel so disposed may visit the chaplain.

The general impression obtained by an examination of the Borda system is that it is one of extreme severity and repression. Not that the application of the system or the method of administration is harsh. On the contrary, the relations of the governing authorities and the students seem to be of the most amiable and cordial character. A spirit of subordination and a general desire to perform well the allotted tasks and duties is said to pervade the school; and though the punishments are severe they are infrequent. The essential features of the system are the close and constant supervision maintained over the pupils, and the prevention of violations of discipline, by the imposition of the heaviest penalties. Such offences as hazing or going ashore without leave are impossible. The constant watch, the isolation of the ship, and the absolute authority of the Préfet Maritime in and about the port effectually prevent any co-operation of outside persons in an attempt to go ashore without leave. The penalty for disobeying the orders of a sentry is the *cachot*, a penalty which no one who has once undergone it would care to run the risk of repeating. The features of the system that make the discipline so perfect, from a military point of view, are those which tell most severely on the students. In considering its effects upon the latter, however, considerable allowance must be made for peculiarities of national character and modes of thought. Certainly the colorless life of the Borda, its close confinement, its constant supervision, and the absence of all that gives charm or variety to existence would be intolerable to an American or English boy of the age of the French *élèves*.

6.—FEES AND ACCOUNTS.

As in other French Government schools, and as in most naval schools in Europe, the pupils in the Borda are required to pay the expenses attending their maintenance during the period of education. The regular fees are 700 francs a year for board (*pension*), and 1,000 francs for outfit (*trousseau*). The board is payable quarterly in advance; the outfit in two payments, 800 francs at admission and 200 francs at the beginning of the second school year. The amount for board may be paid either at the treasury in Paris or at the office of the receiver of finances in the departments, and the authorities at the school have no control of it, nor does it even pass through their hands. The other payment is made directly to the treasurer of the school, and is expended by him in the purchase of the articles prescribed by regulation. The weekly allowance of students and the cost of certain personal services of a minor character are also paid from this source. All the required articles are procured by the administrative officer of the school without the intervention of the student; and though he is allowed to keep the clothing brought from home that conforms to the regulations, no deduction is made on this account from the required deposit. So far from this, he is even obliged to give up all the money remaining in his possession after his arrival, which is deposited in the school treasury and credited to him on the books, being expended from time to time to meet any extra expenses that may be incurred for him. Any balance in his favor at the close of the course is turned over to him at the final settlement.

Exemption from the payment of the whole or part of the *pension* or the *trousseau*, or both, is granted to persons who are too poor to pay them, upon application. The application must be sent in by the 1st of August to the prefect of the department, by whom it is forwarded to the Ministry of Marine, with an attestation of the municipal council of the place where the applicant resides; a statement giving detailed information in regard to his means of support, the number of children or persons dependent on him, and an extract from the tax-list. The applications, with the other documents, are referred to the council of instruction of the Naval School; and, according to their decision, the whole or some part of the customary charges is remitted. A similar arrangement may be made with reference to the outfit at graduation, which costs 570 francs. The number of pupils receiving assistance (*boursiers*) in one shape or another is not limited, but it amounts to about one-fourth of the whole number at school.

The accounts of the school with the pupils or their families, and with the general disbursing officer of the Ministry of Finance, are in the charge of the council of administration, and more particularly of the commissariat officer. The charges against pupils are as follows:*

I. Ordinary expenses:

1. Outfit, including clothes, books, instruments, &c.

---

*These charges are all in addition to the *pension*, which goes directly to the treasury.

2. Personal service of certain kinds, including bootblack, barber, small repairs of clothing.

3. The weekly allowance of 1f. 25c (25 cents).

(All the above, except 3, are covered by the indemnity for *trousseau*, in the case of beneficiaries.)

II. Extraordinary expenses:

Injury to public property, and books or clothing lost or prematurely destroyed.

III. Optional expenses:

Lessons in fencing, dancing, or other accomplishments (*leçons d'agrément.*)
Pocket-money on the practice cruise.

The items chargeable on the accounts of the school against the naval appropriation are for the students' mess and washing, the indemnities for trousseaux, the extra pay of warrant and petty officers, and the small running expenses. All purchases are submitted to one of three boards of inspection. Of all these boards the senior lieutenant is the presiding officer, and the commissary is a member. The third member is the senior medical officer for the inspection of provisions, a professor for books and instruments, and a chief of section for articles of outfit.

7.—PRACTICE CRUISE.

The final practice cruise begins immediately after the close of the vacation following the two years' course in the Borda. Though the practice-ship is independent of the captain of the Borda, it may be considered as in some sense a part of the same general establishment. Both are under the direction of the Préfet Maritime, and the captains of both are members of the committee on improvement at the Naval School. The practice-ship is the Flore, a screw-steamer, first class, of 18 guns, and engines of 380 horse-power. The duration of the cruise is about ten months, from the 1st of October to the latter part of July. It is followed by a week of examination, after which the ship remains at Brest for six weeks preparing for another cruise.

The cruising-ground of the Flore is generally among the French West India Islands. The course of instruction includes the theory and practice of steam-engineering, gunnery, seamanship, navigation, hydrography, and landing drill. The students perform duty in turn as officers of the deck, under supervision, and work the ship. A thorough practical course is given in steam-engineering. The details of the programme are somewhat varied from year to year, but it always includes the subjects mentioned, and generally some others; among them, naval hygiene, taught by the surgeon, and drawing, by a master specially appointed. Last year, instruction was given in gymnastics by an ensign.

On passing the examination, the graduates of the Flore become *aspirants* of the first class, or midshipmen, and they are shortly after sent to sea in cruising-ships. They remain in this grade two years, making a total of five years from their admission to the service to their promotion to the grade of ensign.

CHAPTER XV.

THE ENGINEERS' SCHOOL (*Ecole d'application du Génie Maritime*).

The School of Engineers (constructors) at Cherbourg is simple in its organization, as would be expected from the small number both of students and professors. The classes are two in number, and are composed of from three to six members, averaging about four. The school is under the general supervision of the Préfet Maritime, and is inspected from time to time by the inspector-general of the corps of engineers. At its head is a constructing engineer as director. Changes in organization are considered by the committee on improvements (*Conseil de Perfectionnement*), composed of a vice-admiral as president, the inspector-general of engineers, the director and subdirector of the school, a director of naval construction or other superior officer of engineers, and a captain. The committee, with the exception of the subdirector, acts also as a board of examiners, and conducts the final examinations of each class.

The *personnel* of the school consists of the director, two constructing engineers of lower rank, and two civil professors, one of them for instruction in English, the other in freehand drawing. This is the whole teaching staff. The director has personal charge of one course of lectures. The English professor gives three lessons a week, of an hour and a half each, and the professor of drawing two lessons a week, of two hours each. The remaining courses are conducted by the engineer professors. The latter are appointed for five years, but may be retained at the school three years longer, upon the recommendation of the director. The engineer next in rank to the commanding officer is called the sub-director, and has the charge of carrying out the interior discipline of the school; but owing to the small number of pupils, the discipline is of the simplest character. A clerk and draughtsman complete the *personnel* of the establishment.

The cost of the school is insignificant, especially in comparison with the important object it fulfils—the professional education of a corps of accomplished naval constructors. A thorough preparatory training in physics and mathematics is given to the pupils in their course at the Polytechnic; all that remains to be done is to give the principles already learned the fullest and widest application of which the profession admits. The material wants of the school are largely supplied by the ordinary resources of the arsenal and dockyard at Cherbourg; and the teaching staff, as has been stated, contains a minimum of special teachers employed for this purpose alone. A small appropriation is made yearly, upon estimates offered by the *Conseil de Perfectionnement*, for the purchase of books and scientific journals for the library, and for lithographing and stationery. The expenditure of this sum is in the hands of the director, and is regularly accounted for by him. A full collection of models, plans, documents, and drawings is supplied by the department of naval construction; and beyond the small appropriation referred to, the school, as such, is not a source of expense to the government.

The course of instruction covers two years. The session begins in the month of November, immediately after the close of the final examinations; the exact date being fixed each year by the director. The course comprises theoretical instruction in lectures, for seven or eight months, and studies in the dockyard at Cherbourg (first year), and in the national engine-works at Indret (second year). The courses are finished on the 30th of June. The following subjects are included in the courses:

Naval construction.
Strength of materials.
Naval architecture.
Steam-engine.
Thermodynamics.
Technology.
Naval ordnance.
Regulation of the compass.
Accounts.
English.
Mechanical drawing.
Freehand drawing.
Ship and engine drawing.

As the profession of naval construction is a favorite branch among the higher graduates of the Polytechnic—coming usually after mines, and roads and bridges—the pupils are selected men, and distinguished in a high degree by earnestness, intelligence, and thorough scientific attainments, as far as they have gone. They enter the school of application at the age of twenty or twenty-one, an age which is most favorable for professional study. Besides the incalculable advantage of a sound preliminary training, they begin their professional course with a feeling of elation and encouragement derived from having already successfully tested their powers. Their number is large enough to keep up a generous emulation, while at the same time their instructors are able to give the closest attention to the wants of individuals; and hence they get all the benefits, with none of the disadvantages, of private tuition. While the system of instruction, by lectures and interrogations, is similar to that in other French institutions, the interrogations are much more frequent and personal, and each professor is his own *répétiteur*. The professional courses are marked by great originality of treatment, and text-books serve only a subsidiary purpose. At frequent intervals examinations (*interrogations générales*) are held, which have a certain weight in the final classification, though their object is as much that of giving a summary review as of testing knowledge. The hours of instruction are from half-past eight to half-past ten in the morning, and from twelve to five in the afternoon; and attendance is always required.

Practical instruction, or rather illustrative instruction, is given during the session, in visits to the arsenal and workshops, and to vessels making short trial trips. The real practice, however, is given during the summer. On the 1st of July the students are sent either to one of the naval ports or to Indret. Here they are attached to the different branches of work in the department of naval construction, and go through a regular course in the yard or the machine-shop. At Indret their attention is directed exclusively to the fabrication of engines, and all that is accessory to this branch of the profession. The authorities are directed to afford the students every facility, and to see that they pursue their work with assiduity. The director of the school makes tours of inspection during the summer to assure himself that the students are properly occupied; and the engineer professors are charged in turn with the direct supervision of their pupils, at the station to which they have been sent. This has the additional effect of giving the instructors an opportunity of refreshing and broadening their professional knowledge, and of keeping them familiar with the latest developments of professional science; and it enables them to obtain easily the draughts and other documents necessary for supplementing and revising their lectures.

Before he leaves the school in July each student receives from the director detailed instructions to serve him as a guide during his summer work. Copies of these instructions are also sent to the chief of service at the station where the student is to work. He is required to keep a journal containing full descriptions of the work performed, and of kindred matters coming under his observation, accompanied by draughts and sketches. These are examined by the construction officer under whose orders the student is placed, and finally handed in to the director of the school. On his return the whole of the student's work is examined and marked.

Second-year students return to Cherbourg on the 20th of September; and first-year students on the 10th of October, the intervening twenty days being passed by the latter in vacation. The time from the reopening of the school is passed in preparation for the final examinations, which begin on November 3. During this period the upper class has also a special course in ship and engine design.

Marks are given by the board of examiners at the annual examinations. The final mark of the student in each branch is determined by combining the mark of the board with that obtained in the school work during the course; the former having double weight. Each branch has its coefficient, which is combined in the usual way with the mark given in the subject; and the sum of the products, together with a mark for diligence and conduct, gives the final average of the student. The mark for conduct has a low relative weight. The following table explains the system of marking and gives the coefficients in each branch:

Subject.	School mark.	Relative weight of school mark.	Mark of examining board.	Combined mark.	Coefficient.	Final marks.
1 Ship-building........................	a	$\frac{1}{2}$ a	a'	$A = \frac{1}{2}a + a'$	8	8 A
2 Strength of materials	b	$\frac{1}{2}$ b	b'	$B = \frac{1}{2}b + b'$	8	8 B
3 Naval architecture.................	c	$\frac{1}{2}$ c	c'	$C = \frac{1}{2}c + c'$	9	9 C
4 Steam-engine	d	$\frac{1}{2}$ d	d'	$D = \frac{1}{2}d + d'$	9	9 D
5 Thermodynamics..................	e	$\frac{1}{2}$ e	e'	$E = \frac{1}{2}e + e'$	7	7 E
6 Technology	f	$\frac{1}{2}$ f	f'	$F = \frac{1}{2}f + f'$	7	7 F
7 Naval ordnance	g	$\frac{1}{2}$ g	g'	$G = \frac{1}{2}g + g'$	5	5 G
8 Regulation of compass...........	h	$\frac{1}{2}$ h	h'	$H = \frac{1}{2}h + h'$	5	5 H
9 Accounting	i	$\frac{1}{2}$ i	i'	$I = \frac{1}{2}i + i'$	4	4 I
10 English { first year ...	j	$\frac{1}{2}$ j	j'	$J = \frac{1}{2}j + j'$	3	3 J
11 { second year .	j	$\frac{1}{2}$ j	j'	$J = \frac{1}{2}j + j'$	3	3 J
12 Mechanical drawing { first year ...	k	$\frac{1}{2}$ k	k'	$K = \frac{1}{2}k + k'$	3	3 K
13 { second year .	k	$\frac{1}{2}$ k	k'	$K = \frac{1}{2}k + k'$	3	3 K
14 Ship and engine drawing.........	l	$\frac{1}{2}$ l	l'	$L = \frac{1}{2}l + l'$	8	8 L
15 Summer work { first year....	m	$\frac{1}{2}$ m	m'	$M = \frac{1}{2}m + m'$	4.5	4.5 M
16 { second year .	m	$\frac{1}{2}$ m	m'	$M = \frac{1}{2}m + m'$	4.5	4.5 M
17 Freehand drawing { first year....	n	$\frac{1}{2}$ n	n'	$N = \frac{1}{2}n + n'$	2	2 N
18 { second year .	n	$\frac{1}{2}$ n	n'	$N = \frac{1}{2}n + n'$	2	2 N
19 Diligence and conduct......... { first year .	o	$\frac{1}{2}$ o	o'	$O = \frac{1}{2}o + o'$	1.5	1.5 O
20 { second year .	o	$\frac{1}{2}$ o	o'	$O = \frac{1}{2}o + o'$	1.5	1.5 O
Final mark...........						Sum.

The order of seniority of the graduating class is fixed by the final classification, according to the system described; and in this order the engineer students are promoted to the grade of assistant engineer of the third class, as vacancies occur. The student who graduates from the school at the head of his class is sent, as a reward, to England, on a tour of scientific study and inspection. On his return he presents a report of his observations to the director of the school.

It will be observed that no provision is made in the course of constructing engineers for a cruise on board a ship of war. By the old regulations of 1865 constructing engineers were obliged to perform a certain amount of sea-service in the lowest grade before promotion; and they were also sent to sea in flag-ships in the higher grades. This regulation was abolished in 1876, as it was found that it removed engineers for too long a time from the duties of their profession on shore, without any proportionate advantage. The Minister of Marine still retains the power to send engineers to sea at his discretion; but, as in England, it is only done to a very limited extent and for short periods.

At Cherbourg, as at Greenwich, provision is made for the reception of construction students from civil life, either Frenchmen or foreigners, under the name of free pupils (*élèves libres*). The free pupils are required to obtain permission to attend from the Ministry of Marine and pass an examination for admission; and they receive instruction in the following courses:

Ship-building.
Strength of materials.

Naval architecture.
Marine engines.
Thermodynamics.
Technology.
Naval ordnance.
Regulation of the compass.

Every facility is given to the free pupils, except that the plans and documents in the school archives are not open to their inspection, without special authorization. They receive a diploma at graduation, stating the character and duration of their course of study, and the degree of capacity they have shown.

Officers on duty at Cherbourg are also allowed to attend the lectures at the school, upon receiving permission from the Préfet.

CHAPTER XVI.

THE TORPEDO SCHOOL (*Ecole des défenses sous-marines*).

The establishment at Boyardville has two objects, the training of officers and men for the torpedo-service, and the performance of experiments for the development and perfection of the materials of this branch of maritime warfare. The two functions are largely performed by the same officers. At the head of the institution, but still under the orders of the Préfet Maritime of the district, is a captain, who has general direction, and who also gives such courses of lectures as he sees fit. Under him is a commander, as senior executive aid, another commander in charge of the courses of superior officers, and a number of lieutenants, whose duties of instruction are divided between the junior officers and the instructing warrant and petty officers. The general principle prevails of making instructors of various degrees at the same time students of higher courses than they teach. Thus the lieutenants give lectures to the instructing warrant and petty officers, while the latter teach the men, under the supervision of the lieutenants, or of ensigns pursuing the officers' course. In general, instructors of all grades are selected from those who have been students, very often immediately after they have received their certificates; and in recognition of their special qualifications, they all receive extra pay while performing this duty.

Apart from the commissioned officers, the instructing force consists of two first masters (warrant officers), one belonging to the corps of gunners, the other to that of helmsmen. The latter has charge of the instruction of men in physics and in telegraphy, having an expert as assistant in the latter branch. Besides these, there are several warrant and petty officers, of the corps of gunners, helmsmen, and machinists, according to the number of pupils; but at least half of these must always be of the gunnery branch. The workshop is in charge of a principal mechanician or first master machinist.

The board of instruction (*conseil d'instruction*) is composed of the captain as president, the two commanders, the lieutenant in charge of the officers' course, and one of the lieutenants in charge of the course for men. It prepares programmes of study, and revises the official manual of torpedoes. The course of study is both theoretical and practical. Marks are given by the captain for the work performed by the officer students. This work consists of notes of lectures, practical experiments, and the preparation of essays or discussions on subjects relating to torpedoes. An officer unconnected with the school, either an admiral or captain, is detailed to conduct the examination at the close

S. Ex. 51——10

of the course, on the passing of which depends the certificate necessary for a torpedo-officer.

The course for superior officers (captains and commanders) lasts five months, beginning May 1 and November 1; that for lieutenants and ensigns is six months, beginning April 1 and October 1. Torpedo officers (*officiers torpilleurs*) stationed regularly at the seaports are sent once every three years for the five months' course. The course for warrant officers, petty officers, and men lasts six months, beginning January 1, April 1, July 1, and October 1, so that there are always at the school a division in the first half, and another in the second half of the course. The pupils in this course are selected by a board of officers appointed for the purpose at each port. The candidates must be men of good record as to conduct, and of a certain intelligence and aptitude; and they must belong to one of the three corps of gunners, helmsmen, and machinists, or to the seamanship branch (*manœuvre*). They are examined upon their arrival at the school, and sent back if found disqualified. At the end of the course they pass an examination, and receive certificates of their fitness for torpedo duty. The best men are promoted at the end of each course, the choice being determined by a board of officers (*conseil d'avancement*) at the school.

The practical researches and experiments with the torpedo are conducted by a board known as the *Commission permanente d'expériences*. It consists of the members of the board of instruction, with the addition of an assistant engineer (constructor), and a captain of marine artillery. The lieutenants in charge of the course of instructing warrant and petty officers are admitted with a consulting voice only. All officers pursuing courses at the school have the advantage of being able to attend discussions and witness experiments of the permanent commission.

A late decree (March 14, 1878) fixes the course for machinists in Whitehead torpedoes at four months, beginning January 1, May 1, and September 1, of each year. Still another decree (April 4, 1878) has added to the list of pupils in each course an assistant constructing engineer and a variable number of foremen and other officers of dockyard works (*maistrance des arsenaux*). The reason for this lies in the fact that the direction of naval constructions at the dock-yards is charged with the safe-keeping and delivery of torpedo materials; and it is therefore highly important that there should be officers connected with the corps of constructors who understand their properties.

CHAPTER XVII.

THE SCHOOL OF MACHINISTS (*Ecole théorique et pratique des mécaniciens*).

The school of machinists is situated at Toulon. It was established by the decree of February 13, 1879, and took the place of the two schools that formerly existed for a similar purpose at Brest and at Toulon.

The superior officers at the head of the school are a captain and commander. The instructors are composed of (1) professors of hydrography, for the scientific courses; (2) principal mechanicians and first master machinist, for the technical courses; and (3) a first master machinist, to direct the workshop instruction. The first are selected by the Minister of Marine, and the last by the commandant of the school. A novel feature is introduced in the appointment of the other class, in the requirement of an examination for the position of instructor. Examinations of applicants are held yearly at Toulon, by the permanent commission for the examination of machinists. Lists are kept of those who pass, and from these lists the instructors are selected. The subjects of the examination include arithmetic, algebra, geometry, elementary descriptive geometry, physics, mechanics, and the steam-engine. The instructors, or professors as they are called, are assisted by master or second master machinists in the capacity of *répétiteurs*.

The pupils at the school consist of firemen artificers (*ouvriers chauffeurs*), candidates for promotion to quartermaster machinists; of quartermasters and machinist pupils (*élèves mécaniciens*), candidates for second masters; and of second masters, candidates for first masters. The course for the firemen artificers is six months; for all the other classes one year. Competitive examinations for promotion are held at the end of each course. There are two "commencements" or dates of entry in each year, on May 1 and November 1. By this arrangement two classes in each category of candidates (except the six months' pupils) are always together at the school; one class in the first half, the other in the last half of the course.

Admission to the school is only obtained after passing a double examination: first, at the shore station or on board the ship where the candidate is employed for the time being, and, secondly, at the school itself. Twice a year the five Préfets Maritimes, and the commanders-in-chief of the various squadrons, send to the Ministry lists of the names of machinists of all grades whom they propose for admission to the school. The lists are drawn up after an examination of candidates in the arrondissement or squadron, conducted by a board composed of a line officer, a constructing engineer, and a principal mechanician. A good-conduct record, and a certain period of sea-service, the latter vary-ing according to the grade of the candidate, are essential qualifications for admission to the official lists; and the board graduates the candidates according to their professional capacity. The classified lists are sent to Paris twenty days before the half-yearly dates of admission, and from them the Minister of Marine makes a selection according to the number of pupils for whom provision is made at the school. The se-lected candidates are sent in detachments to Toulon, where they under-go the final examination for admission, conducted by the authorities of the school. Candidates failing at this examination are sent back to their divisions.

The method of instruction is similar to that at other French schools. Courses of lectures are given by professors, and full synopses of each lecture, covering all the material points, in fact sometimes the whole lecture itself, are delivered to the students. These are the well-known *feuilles autographiées* that have already been mentioned in connection with the Polytechnic School and the Borda. Further instruction is given and recitations are held on the subjects of the lectures by the *répétiteurs*, and marks are given at the recitations. Theoretical instruc-tion consists of the following subjects * :

Candidates for first master :
1. Arithmetic, elementary algebra, geometry, and plane trigonometry.
2. Mechanics and physics.
3. Theory and description of engines.
4. Management of engines.
5. Repair and preservation of engines.
6. Erection of engines.
7. Regulation of engine-work.

Candidates for the grade of second master, and machinist pupils, have a course in the same subjects, omitting algebra and trigonometry. Candidates for quartermasters omit also subjects 2, 6, and 7. This grade, as well as that of second master, is divided into two branches, the theo-retical and the practical; and the course for candidates for the latter branch is more limited than the other. For example, candidates for practical quartermasters have no lectures at all, and those for practical second masters have a special limited course in mathematics, and omit the lectures on the management and the erection of engines altogether. For all students, there are classes in mechanical drawing and in shop-work; and one day in the week is devoted to visits made by the stu-dents in company with their instructors to ships of the reserve, to ships in construction or making trial trips, and to workshops.

The examinations for promotion are competitive and are conducted by a board composed of a captain, a commander, a constructing engineer, a hydrographic examiner or professor, and a principal mechanician. The members of the board are appointed for two years, but they continue to

* Detailed programmes of the course are given in the Appendix, Note L.

perform during the period such other shore duty as may be assigned to them. While actually conducting the examinations they have the privileges and emoluments attached to special service.

The examination consists of three parts—

(1) A piece of manual work, in boiler-making, forging, or fitting. This is not required in the examination for the highest grade.

(2) A scale-drawing, of objects more or less difficult according to the class examined, as follows:

Candidates for quartermasters, a simple part of the machinery, as a crank, cock, or beam.

Candidates for second masters, and machinist pupils, an apparatus or complex part of the machinery, as a piston, pump, or donkey-engine.

Candidates for first masters, a complete engine or boiler.

(3) An oral examination, on the subjects of the theoretical course. The exact subjects of the examination may be seen in the detailed programmes of the courses, given in the appendix. Each candidate draws a question from each chapter of the programme for his grade; and a supplementary question, also selected from the programme, is given by the president. This somewhat imperfect form of examination is supplemented in the case of candidates for the highest grade by a written paper on a single question given to all the candidates. The candidates of all grades also answer questions on the management of engines, taken from the supplementary programmes. No limit is fixed to the number of these supplementary questions; and the mark of the candidate in the branch of management of engines is obtained by combining the mark for his first answers with those of the supplementary questions. For machinist pupils, all the questions are selected by the commission, and the two questions are given on each of the first two chapters. Candidates for the practical masters and quartermasters pass a limited examination, based on the limited course pursued at the school.

The system of marking is somewhat complicated, and it will hardly be necessary to go into its details. The drawing, manual work, and each theoretical subject count equally in determining the final mark. Additional marks (*points supplémentaires*) are given in recognition of remarkable professional aptitude, and of previous good service. Upon the arrangement of the candidates in the final class-list depends their seniority, and therefore the order of their promotion ; but those who fall below a certain standard—and quite a high standard—are not entered at all upon the lists for promotion.

The extent and character of the course may be seen from the programmes of study. These programmes, as they stand, contain the questions given in the examinations, each paragraph representing a question, any one of which may be drawn by a candidate. The programmes may therefore be depended on as an exact statement of the course of study.

CHAPTER XVIII.

THE SCHOOLS FOR COMMISSARY PUPILS (*Cours d'administration des élèves commissaires*).

The schools for commissary pupils are at the various naval ports, whither the pupils are sent upon their admission to the service. The instructors are a professor and assistant professor at each station, both of whom are officers of the pay or commissariat corps of the navy. The course lasts two years, and the session begins on the 3d of November in each year and ends on the 30th of September following. During this period instruction is given on at least three days in the week. Besides the courses in naval administration, pupils are required to pursue the course of study in the English language, which is opened at all the naval ports for the benefit of officers generally; unless they can give adequate proof of a thorough acquaintance with either English or Spanish. The courses in naval administration are as follows:

I.—GENERAL VIEW.

1. The Navy; its purpose; the mercantile marine and its relations with the Navy; French establishments abroad.
2. The ministry of marine and of the colonies.
3. General organization of naval arsenals and other establishments.
4. *Organization of the colonies.
5. Composition and organization of the various corps in the navy.
6. Legal status of an officer of the Navy.
7. Recruitment of the naval forces.

II.—MARITIME JUSTICE. †

III.—COMMISSARIAT SERVICE.

1. The commissary-general.
2. Maritime inscription.
   a. Organization, classification, drafts.
   b. *Maritime police.
   c. *Right of domain in territorial waters.
   d. *Coast fisheries.
   e. *Wrecks.
   f. Naval pensioners (*Etablissement des Invalides de la Marine*).
   g. *Pensions and compassionate allowances.
3. Duties of commissary officers in connection with surveys or boards of audit, in relation to
   a. Staff officers, foremen of dockyard works, and other officials.
   b. *The administration and accounts of the different corps.
4. Armaments.
5. Administrative service at sea.

---

* The starred subjects are those taught during the second year of the course.
†This subject enters only into the examination for higher grades.

6. Stores.
7. Provisions.
8. Hospitals and prisons.
9. Public works.
    a. Laborers.
    b. Accountability for material (1) on shipboard, (2)* in the arsenals.
    c. Purchases.
10. Organization of central offices of sous-arrondissements.
11. Financial accountability and details of expenditure.†
12. * Accountability for provisions and stores.

Commissary pupils attend the courses for both years; those of the first year only are attended by the assistant commissaries appointed from ensigns and from graduates of the Polytechnic School.

The courses of the second year are practical as well as theoretical. During this year it is the special charge of the assistant professor to instruct his pupils in the application of principles and methods to the actual duties of the corps. Written exercises are performed by the students under his direction, bearing upon every detail of the commissariat service, especially at sea. The professor is required to conduct the pupils to the various workshops and storehouses to instruct them in the details of business connected with the reception and inspection of stores, supplies, and equipments, and every facility is given him in the performance of his work. Outside of lecture hours the pupils perform subordinate duties in the various branches of the commissariat department of the station; in which they are subject to quarterly transfers to familiarize them with all the parts of their profession.

Interrogations take place frequently, and examinations are held at the end of each year, before a board composed of the commissary-general of the marine, an officer of inspection, and three commissariat officers. The examinations are both written and oral; marks are given, and the final classification determines the order of seniority. Fifty per cent. of the maximum aggregate is required in order to pass, and candidates who fail are allowed to go over a second time the course for the year.

It should be added that complete special libraries, composed of works pertaining directly or indirectly to matters connected with the commissariat service, are to be found in each port, and that they are placed at the disposal of the students.

---

* Second-year course.
† This subject enters only into the examination for higher grades.

CHAPTER XIX.

GENERAL CHARACTER OF THE FRENCH SYSTEM.

It was remarked by Captain Hore, the naval attaché of the English embassy in Paris, in his testimony before the Committee on the Higher Education of Officers, in 1870, that the English had no system of naval education, and that the French had too much system. That there is much truth in the first part of the observation will be acknowledged by every one who has examined the matter at all, and Captain Hore, being himself an English officer, was in a position to know. But his judgment upon the French system must be taken with some allowance, and certainly his testimony before the committee does not indicate such an acquaintance with the subject as would entitle his opinion to great weight.

The fault of "too much system," if it means anything, means a sacrifice of results to methods, an effort which looks rather to the perfection of the machinery than to the work done by the machine. It is a fault very commonly charged to French methods of administration, and one to which, perhaps, they are largely open. It is a necessary consequence of extreme centralization, and it affects to some extent the system of public education, including both the secondary and the professional schools. Its injurious effects are felt, however, rather in minor details and in exceptional instances than in the general result. In most respects the public-school system, although highly organized, is sufficiently elastic to meet the wants of individual cases. As to the training-schools for this or that professional service, and especially for a military service, it is doubtful whether a flexible system is as productive of good results as a more rigid one; and in the French naval schools there is certainly in matters of theoretical and practical instruction much greater flexibility than is to be found in those of almost any other nation. The system of oral teaching, without the restrictions of a text-book and supplemented by individual explanation and interrogation, cannot be other than an elastic one. With regard to discipline, however, it must be confessed that if French boys are at all like other boys, a great many rules might be relaxed with direct and positive benefit.

The broad features of the French system of naval education may be readily recognized from the detailed description that has been given. They consist in a unity of purpose underlying the whole plan; a rational organization, with a distinct perception of the ends in view, and an adaptation of means to reach the ends proposed; the exaction of a high standard of preparatory training; and great originality, freedom, and thoroughness of instruction. Looking at the details, we find, in the case of line officers, a system of local examinations for admission, competitive in character, with requirements based on the programmes of the best schools in the country. These are followed by a three years' course of theoretical and practical training. The first two years are passed in a stationary ship, with all the disadvantages that such a school-house entails—disadvantages in this case even greater than in that of the Britannia. The only compensating advantage is that of making possible an excessively rigorous discipline, an advantage of more than doubtful character. The course of theoretical instruction is the fullest and most advanced *required* course pursued by the cadets of any navy in the world except that of Germany, and the practical and professional branches receive an ample share of time and attention, although the course in these subjects is rightly considered as only preparatory to the work of the practice-ship. It is, nevertheless, extensive and thorough, including frequent exercise in the details of seamanship, gunnery, navigation, and the manipulation of engines. Following the two years' course comes the third and final year in a sea-going practice-ship, with review and completion of the course, in the theory and practice of subjects purely professional.

The other corps of the service are as well taught as the line in the particular duties of their several professions; for it may be taken as a cardinal maxim of professional education in France that a man cannot be expected to know how to do a thing by a process of inspiration or intuition, or even by "picking up," but that he must be *taught* to do it. Hence, they have a thorough course of instruction for the men who are to build their ships, to fight their ships, to govern the employment of the motive power of their ships, to conduct the details of internal administration on board their ships. The constructors have a four years' course: two years at the Polytechnic, the first school of mathematics in the world, and two years at their special school of application, with practice in the great ship-building and engine-building establishments. The engineers or mechanicians have a series of courses and examinations preceding each promotion, whose extent and character leave little to be desired for this branch of the profession. Finally, the administrative officers or commissaries are taught effectually the principles, the laws, and regulations which are to govern them in the future exercise of their duties.

It will be seen that the education given to officers in general is of a high and extensive character. This supplies the want of special subsequent training to some extent, but not wholly. There is a decided need at present of facilities for higher education in the branches which particular inclinations may lead individual line officers to take up. It is not unlikely that such a higher college, similar in purpose to the half-pay courses at Greenwich, may in time be established; though there is by no means the same necessity for it that exists in England.

PART III.

GERMANY.

CHAPTER XX.

PERSONNEL OF THE GERMAN NAVY.

The executive or line officers (*See-Offiziere*) of the German Navy are divided into three classes: The staff of the Admiralty, whose duties are connected with organization and administration; the naval staff, composed of officers occupied with some special branch of the profession; and the sea-going officers, whose duties are on shipboard and at naval stations.

The grades of officers, with the numbers in each grade, are as follows:

LINE (*See-Corps*).

Admiral	1
Vice-admiral	1
Rear-admirals	3
Captains	20
Commanders or captains of corvettes	45
Lieutenant-commanders	74
Lieutenants	128
Sub-lieutenants	128
Midshipmen (*See-Cadetten*)	100
Cadets (about)	50

MECHANICAL ENGINEERS (*Maschinen-Ingenieur-Corps*).

Chief mechanical engineers	2
Mechanical engineers	6
Assistant mechanical engineers	12

The officers of this corps are charged with the management of engines on board the large ships.

PAY CORPS.

Paymasters	26
Assistant paymasters	26

MEDICAL CORPS.

Surgeon-general	1
Chief surgeons	59
Surgeons	19
Assistant surgeons, first class	16
Assistant surgeons, second class	16

ENGINEER CORPS.

I. Division of naval construction:

Directors	3
Chief engineers	4
Engineers	12
Assistant engineers	11

II. Division of marine engines:

Directors	3
Chief engineers	4
Engineers	11
Assistant engineers	11

III. Hydraulic works:

Directors	2
Chief engineers	3

There is also the marine infantry (*See-Bataillon*), and there was until recently the marine artillery; but the last has lately been abolished, and its place is to be taken by the four companies of seamen-gunners.

The total force of petty officers and seamen make up the two seamen's divisions, the first division being stationed at Kiel and the second at Wilhelmshafen. There are also two dockyard divisions, in like manner stationed at Kiel and Wilhelmshafen.

The preliminary steps in the career of a line officer in the German Navy begin with the application of his parent or guardian for an appointment. These applications are made to the Admiralty in August and September of each year. In the following spring the applicants present themselves at Kiel for the entrance examination. The examination is not competitive, but the standard for admission is high, involving not only the attainments called for by the questions given, but the previous completion of a certain course in the public schools—schools which are unsurpassed by any of their kind in the world. The successful candidates are appointed cadets in the Navy, and embarked immediately on board the cadets' practice-ship. The cruise lasts six months, and the time on board is fully occupied by training in practical duties; no theoretical instruction being given, except such as is necessary by way of explanation.

At the end of the practice-cruise, the cadets return to Kiel and join the cadets' division at the Naval School. Here they have a course of theoretical instruction in professional and scientific subjects. The duration of the course is six months, from October 1 to March 31. It is followed by the second of the important examinations which all officers must pass. These are four in number. The first is the entrance examination; the second, which follows the cadets' course at the Naval School, is called the midshipmen's examination, as it is followed by the promotion of proficient cadets to that grade; the third and fourth, known respectively as the first officers' examination, and the officers' professional examination, occur later in the career of officers.

After passing the examination, the newly-promoted midshipmen are sent for a month to the gunnery-ship at Wilhelmshafen. Early in May they are divided among the ships of the iron-clad squadron, where they remain on service during the greater part of the summer. In September or October they are ordered back to Kiel, and embarked on board the midshipmen's school-ship, of which there are two always in commission, one of them leaving Kiel every year. The cruise lasts two years.

The ship goes around the world, but most of the time is generally spent on the coasts of China and Japan. Instruction is given both in lessons or recitations and exercises, and comprehends a very full programme of theoretical and practical study. The regular officers of the ship are the instructors.

At the end of the cruise the ship returns to Kiel and the midshipmen undergo their third examination, called the first officers' examination, preparatory to promotion to the grade of sub-lieutenant. Upon passing this examination they are subjected to another and peculiar test of fitness. Their names are proposed, one by one, for approval, to the officers attached at the time to the naval station at Kiel. These officers, whose number is always considerable, vote as at any other election; and an unfavorable result puts an end to the candidate's career.

After passing this test officers are promoted to sub-lieutenants as vacancies occur. They return to the Naval School (without waiting for their promotion) and go through the officers' course. This begins in November and lasts a year. It includes professional and scientific subjects, among them steam-engineering and fortification, besides all the others which were taught in the cadets' course. In these last, the course is far more advanced, having progressed by regular and systematic steps. The officers' course at the Naval School is closed by the officers' professional examination, with which the compulsory education of officers comes to an end.

A still higher course of an extensive and elaborate character, and lasting three years, is open to voluntary students at the Naval Academy. This institution, though united with the Naval School and under the same direction, has its specific object distinct from the other, and its students, being officers of higher rank and attending the academy voluntarily, are subject to separate regulations.

The steps in a naval officer's education may therefore be summarized as follows:

August, September: Application of parent or guardian.
    First year:
April.—Examination for admission.
April to September.—Cadets' practice-cruise.
October to March.—Naval School: Cadets' course.
    Second year:
April.—Midshipmen's examination: Promotion to midshipman.
April to May.—Gunnery-ship.
May to September.—Iron-clad squadron.
    Third year, fourth year:
September to September (two years).—Midshipmen's practice-cruise.
October.—First officers' examination.
      Election at Kiel.
      Promotion to sub-lieutenant (by vacancies).
    Fourth year, fifth year:
November to November.—Naval School: Officers' course.

Fifth year:
November.—Officers' professional examination.
Total period of education, five and a half years.
Subsequent education:
Voluntary three-years course: Naval Academy.

PROMOTION OF SEAMEN TO THE CORPS OF OFFICERS.

Seamen of the German Navy who desire to qualify themselves for advancement to the corps of officers may, if they have shown special fitness, be recommended for promotion by the commandant of the seamen's division to which they belong. They are then required to send in the papers presented by applicants for admission as cadets, and, in addition, a certificate from a commanding officer (or merchant captain) with whom they have served, as to their conduct, acquirements, and capacity, and a statement that they have served twelve months at sea in a man-of-war or merchant vessel. The examination for admission must be held before the candidate is twenty years of age, and it is governed by the same regulations as in the case of cadets. After the examination the career of seamen advanced to the corps of officers is the same as that of the officers themselves.

CHAPTER XXI.

ADMISSION.

Application for permission to enter the Navy as cadet is made at the Imperial Admiralty in Berlin, in the months of August and September of the year preceding the entrance examination. The application must be accompanied by a number of papers, such as certificates of birth, religious creed, confirmation, &c.; a full narrative of the life of the applicant, written by himself and duly attested, stating the schools he has attended, his course of study, changes of residence, illnesses, German and foreign works read by him, and other minute facts; school diplomas; and the health certificate of a military or naval surgeon. The parent is also required to give a bond to make the deposit necessary to procure the first outfit of his son, to pay for his subsequent outfits, and also to make him an allowance up to the time of his promotion to the grade of lieutenant. The allowance is fixed at 30 marks (about $7.50) per month. It is paid for the first six months in advance, and afterwards in quarterly payments. It is not paid directly to the cadet or midshipman, but to the treasurer of the Naval School; and it is disbursed by the paymaster who h   for the time being, the accounts of the cadet. In certain exceptional cases, as in that of midshipmen who contract debts on the practice-cruise, it may be paid, like the regular salary, to an officer designated by the captain, and the midshipman has no control over it for the time being, except according to the discretion of the officer who has it in trust.

Candidates must be under seventeen years of age at the date of admission. Exception is made, however, in the case of *graduates* (*Abiturienten*) of the Gymnasia, Realschulen of the first-class, and similar institutions (that is, institutions of an equally high standard), who may be admitted up to the completion of their nineteenth year. No minimum age of admission is prescribed, but a limit is practically fixed by the regulation which requires candidates to present a certificate from a gymnasium of fitness for the upper second class, or to show in the examination their qualification for it. To do this, the candidate must have passed through the lower classes of the schools, which he could hardly accomplish before his fourteenth or fifteenth year.

The examination for admission, like the later examinations to which officers are subjected, is conducted by a special board of examiners (*See-Offizier und Cadetten Prüfungs-Commission*). The board is appointed by the Minister of Marine (*Chef der Admiralität*), and its proceedings are governed by minute regulations. Besides the entrance examination, it has charge of the midshipmen's examination, and the first and the final examination for officers. It has no duties of instruction. The latter are performed entirely by the regular officers and professors of the schools, and it is the object of the system to make the examiners a distinct body. The board of examiners holds therefore a very important place in the German system of naval education. As it is governed by similar regulations in the conduct of all the examinations, it may be well to give here a general outline of the system.

The board is composed of a president, examiners, secretaries, and officers to do proctor's duty in written examinations. The president conducts the meetings of the board, gives special directions in regard to examinations, and is present at the oral examinations. The examiners conduct the oral examinations, and set the written papers. Three times the number of papers required are made out in each subject by the examiners, from which the president of the board chooses the allotted number. The examiners mark the work-papers. Candidates obtaining, or seeking to obtain, improper assistance, are rejected without further formality.

In addition to the prescribed subjects, candidates, at all the examinations, are allowed to offer one modern language, as an extra subject. In this case they receive an examination in the language. If they receive a mark of above 55 per cent. in the extra, it goes to increase their final mark, unless this final mark is below the passing standard; in which case the extra does not count at all. In no case does it count so as to diminish a final mark.

The oral examination is intended to complete and supplement the written, in order that the examiners may determine, with greater accuracy, the merits of a candidate. At least ten minutes must be allowed to each candidate.

The system of marking at the examinations is that employed in all the educational establishments of the Navy, and has some peculiarities that are worthy of notice. The scale of marks ranges from 1 to 9, and a general average of 5 is required in order to pass. A mark of 5 is also required in certain designated subjects. The time occupied in writing a paper is considered in marking. The marks for the oral and written examinations are combined with equal weight. To fix the relative weight of subjects in the final calculation, they are divided into three classes, according to their importance. Each class has its coefficient; that of subjects of the first class being always 3, of the second 2, and of the third 1.

In computing the final mark of a candidate, the marks for separate subjects are not taken directly as the basis of the computation, but the difference by which each mark exceeds or is less than 5, the passing mark. All marks greater than 5 give plus quantities, and all marks less than 5 give minus quantities. These differences, plus or minus as the case may be, are then multiplied by their respective coefficients, and the algebraic sum of the plus and minus products constitutes the final mark. The result may be a plus quantity, a minus quantity, or zero.

Candidates are grouped together, according to the result, in five classes. Those whose final mark is a minus quantity are designated as *not passed*. Those whose mark is zero, and who have obtained the required mark in the separate subjects in which a standard is exacted, are *satisfactory*. Above this limit candidates are classed as *good*, or *very good*, according to the degree of excellence. Candidates whose final mark is zero or above, but who have failed in one or more of the separate subjects in which there is a passing standard, are classed as having *passed with conditions* (*bedingt bestanden*). The last may be recommended for re-examination.

At the conclusion of each examination, the board makes up its report with a general merit-roll and recommendations in the cases of candidates who have failed. It also issues certificates to those who pass. The recommendations of the board decide, finally, the disposal of all doubtful cases.

The examination for admission is held annually at Kiel, in the month of April. Four weeks before the date fixed for the entrance examination, the papers sent in by applicants for admission are sent to the president of the examining board. Candidates are first subjected to a medical examination by a naval surgeon, in presence of the president and recorder of the board. The results of this examination are forwarded to the Admiralty, but they have no bearing upon the mental examination.

The requirements for admission in the mental examination are as follows:

A. FIRST CLASS: COEFFICIENT, 3.

I.—LATIN.

The examination is based on the course of study in the lower second class of a Gymnasium or Realschule of the first class, and includes the authors usually read at schools in this and the preceding classes, together with written translation from Latin into German and the grammatical analysis of passages.

II.—GERMAN * (1½ hours).

Preparation of an essay on a simple subject, without errors in spelling or grammar, and showing a certain facility in expression.

III.—MATHEMATICS * (2½ hours).

1. The whole of arithmetic, and algebra through logarithms and exponential equations.
2. Elementary plane geometry.
3. Trigonometry, plane and spherical, including the discussion of trigonometrical functions, the deduction and application of formulas, and the determination of the areas of right-line figures and segments of circles.
4. Elements of stereometry.

B.—SECOND CLASS: COEFFICIENT, 2.

IV.—PHYSICS.

Elements of physics and mechanics.

* In all the programmes of examination, given under the head of German schools, starred subjects are those in which a mark of 5 is required in order to pass.

V.—GEOGRAPHY, PHYSICAL AND POLITICAL.

VI.—HISTORY.

History in general, and the history of Germany in particular, with special reference to the growth of its territory and the development of its constitution, and to the principal events of the most important wars since the middle of the 18th century.

C.—THIRD CLASS: COEFFICIENT, 1.

VII.—FRENCH AND ENGLISH.

Tolerable fluency is required in reading and translating easy passages into German, and *vice versa*, with some readiness in grammatical analysis.

VIII.—DRAWING.

The candidate must produce a freehand drawing of his own, duly attested, or in default of this, is required to draw from models or objects set by the examiner.

If the candidate has the certificate of fitness for the upper second class of a gymnasium, or realschule of the first class, or if he has passed satisfactorily the first-class course in the Cadetten-Haus at Berlin, he is exempt from the examinations in Latin, German, and history. In obtaining the final mark in such cases, the subjects in which examination is waived receive a mark of 5, satisfactory; but candidates are always at liberty to obtain a higher mark by passing the examinations, if they desire.   •

If the candidate has been graduated with a diploma (*Abiturienten-Zeugniss*) from a gymnasium, a realschule, or other similar institution of learning, he is entirely exempt from the entrance examination, provided the diploma classes him as *good** in mathematics. In the absence of this qualification he must pass an examination in mathematics before the board. The board decides according to the result of the examination, and sends to the Admiralty an abstract of the proceedings in relation thereto, in which it is stated finally whether a re-examination may be granted. The Minister makes his decision in regard to the appointments on the basis of this report.

* This term has a definite official significance in the German school system.

CHAPTER XXII.

CADETS' PRACTICE-CRUISE.

Cadets who have received appointments are arranged provisionally according to the result of the examination, those who come in with graduates' diplomas being ranked first according to age. All the cadets are embarked in April, on board the cadets' training-ship, and receive there their first training as seamen and officers. The training-ship makes a cruise during the summer, and returns to port about the end of September. Those cadets who, during the cruise, appear unfit for the naval service, from want of capacity or otherwise, are to be reported as soon as possible to the Admiralty by the commanding officer, after consultation with the other officers. On the basis of this report, the minister finally orders their discharge from the service. The time passed on board the training-ship, however, does not in these cases count as service-time towards the performance of obligatory military service, required by the German laws of all citizens.

The object of the six months' practice-cruise, with which the German cadets begin their professional education, is to discover their physical aptitude for the service, and to give them the needful elementary training in the practical duties of their profession. Hence there is no theoretical instruction, as such, on board the school-ship.

The course may be divided into five branches, which are pursued as follows :

1. *Seamanship.*—This includes the first instruction in the standing and running rigging and sails, and practice in setting up and taking down rigging, sending up and down upper masts and yards, setting, reefing, furling, and taking in sail, and working ship under sail.

2. *Gunnery*, including the guns and carriages on board the ship, their equipments, and ammunition. In exercises, the cadets are shifted to different guns, and to different numbers, so as to give every variety of practice. At quarters they have their regular stations. Firing-practice is so arranged that each cadet has two firings with the 8-c. m. gun, and two with the 12-c. m. or 15-c. m.; one of which is with the fixed target, and the other with the floating target at sea.

3. *Navigation.*—The cadets have some instruction of a practical and elementary character in the use of reflecting instruments and charts, and they are taught the use of the compass, log, and lead.

4. *Official duties* (*Dienst-Kenntniss*).—Manual and use of small-arms; landing parties ; infantry drill on shore.

5. *Rigging, steering, and sailing boats.*

In addition to this programme, the cadets, when opportunty offers, in going into or out of harbors, or passing remarkable objects, receive informal instruction orally from the officers, partly with the object of training their powers of observation. The small number of the cadets, generally not more than 40, makes it possible for the officers of the training-ship to give constant attention to their wants and frequent personal instruction. One officer, called the "cadets' officer," is specially detailed to superintend the military and nautical training of the cadets, to keep the liberty-lists and punishment-lists, and to examine the journals kept by the cadets. The cadets are arranged in four watches. Each of them serves in turn as cadet of the watch, and during his tour receives instruction in the duties of the position from the officer of the watch. The other cadets perform the regular duties of seamen of the watch. In exercising with sails, they perform duty aloft, to a moderate extent at first, and with the assistance of the petty officers and seamen of the ship. The latter are gradually separated from them, as they become more proficient. At quarters, they are formed in gun's crews, as soon as they have learnt the drill sufficiently.

The daily routine should be carefully examined, as it represents one of the most systematic and thoroughly-digested courses of practical instruction of its kind in existence, and one thoroughly adapted to the age, training, and requirements of the cadets.*

The practice-ship used for the cadets is the Niobe, a sailing-frigate. She is armed with 14 guns, 4 of 15 c. m., 6 of 12 c. m., and 4 of 8 c. m. All are Krupp breech-loading rifles, except two of the last, which are howitzers. The embarkation on board the school-ship takes place about the middle of April. The first four weeks are spent in the harbor of Kiel, learning the elements of the profession, and the routine of duty on shipboard. The morning is generally occupied with infantry drill on shore, the afternoon with instruction upon the sails and rigging, with boat-sailing, or with target-practice with small-arms on shore. At the end of this time an examination is held on board the ship by the commander-in-chief of the station at Kiel.

For the following week the cadets have practice in working the ship in and before the harbor; and as soon as they have acquired some familiarity with this exercise the real cruise begins. This lasts till late in September, at which time the training-ship returns to Kiel. The cruise extends to various points in the Baltic and the North Sea, including perhaps the Channel, where the ship may put in to the important English ports, such as Portsmouth and Devonport. In July, 1878, the Niobe was at Dartmouth at the closing of the semi-annual session of the Britannia.

During the cruise the routine at sea or in port is followed, as the case may be. The former includes exercises with sails on four mornings in the week, daily instruction in seamanship, navigation, and official duties, and practice in reefing, furling, loosing sail, &c.; instruction three times

---

* Appendix, Note M.

a week in the use of the lead, compass, &c.; sailmaking, knotting, and splicing, twice a week; exercise with small-arms and great guns. In port, the routine includes exercises with sails five days in the week; with great guns, three days; boat sailing and steering daily, and small-arms occasionally.

The captain of the training-ship makes weekly reports to the Admiralty of the progress and conduct of cadets. The usual punishments are reprimand, extra watch duty, confinement, and deprivation of leave; but in cases of persistent misconduct the cadet may be removed immediately from the service on the application of the captain and officers of the training-ship.

The cadets sleep in a room built for them on the gun-deck. Here they have their lockers, and the room serves also for study and mess room. Seamen are detailed to act as servants, take care of clothing, &c., the allowance being one man to four cadets. Seamen are also detailed as assistant stewards and waiters. The cadets' mess is in charge of the cadets' officer and the paymaster, who are responsible to the captain.

A certificate of service from the captain and officers of the cadets' practice-ship is given to the cadets at the beginning of September. The certificates, together with a provisional rank-list based on the degree of application shown by each cadet, are forwarded by September 15 to the Admiralty, which thereupon orders qualified cadets to attend the cadets' course of the Naval School at the beginning of October. Before this assignment, the captain of the training-ship requires them to take the oath of service.

CHAPTER XXIII.

THE NAVAL ACADEMY AND SCHOOL: GENERAL ORGANIZATION.

The Naval Academy and Naval School (*Marine Akademie und Schule*) form really two establishments united under one government. The Academy is devoted to the higher education of officers who have shown marked ability, and who come as voluntary students for three years. The studies and special regulations of the Academy will be taken up later. The School, on the other hand, is attended by midshipmen or acting sub-lieutenants, and by cadets, and its courses are compulsory for all officers. The School is divided into two parts, the officers' division (*Offizier-Cœtus*), and the cadet's division (*Cadetten-Cœtus*). To the officers' division belong acting sub-lieutenants (*Unter-Lieutenants-zur-See ohne Patent*) and midshipmen, who have finished their second practice-cruise, and have passed the first naval officers' examination, and who are preparing for the officers' professional examination. This course lasts eleven months. The cadets' division contains cadets who have finished their first cruise in the cadets' practice-ship, and who are preparing for the midshipmen's examination. The course extends over the six winter months.

The institution comprising the Academy and School is under the supervision of the commander-in-chief at Kiel, in matters of command and discipline; but in all that pertains to instruction and maintenance, it is directly under the Admiralty. The government of the institution stands in the same relation to the commander-in-chief as that of a naval garrison. The military head of the establishment is a naval officer of high rank, known as the director of the Naval Academy and School.

The direction of all matters relating to instruction rests with the committee on studies (*Studien-Commission*). This committee is composed of five members, appointed by the emperor. It consists of a rear-admiral as president, three staff officers of high rank, one of them the director of the Academy and School, and one professor in the University of Kiel. The duties of the committee are not merely advisory. On the contrary, it has actual charge and direction of the whole course of instruction given at the institution; and in matters which fall within its province, the director is required to carry out its decisions. It arranges the programme of studies, and takes cognizance of the work performed by instructors and students, being directed to keep in view the diligence of the former, as well as the progress of the latter. It nominates the candidates for appointment as instructors. Its members frequently attend the lectures and recitations, and the results of their observations are considered in meetings of the committee. Where reform is called for, changes in fundamental regulations are proposed to the Admiralty. The instructors have, however, the right to propose changes in the course, which are then considered by the committee; and in drawing up the course, and arranging the distribution of time, the committee acts upon a very rough scheme of instruction proposed by the instructors through the director. This may be modified if the committee sees fit.

The committee has extraordinary powers in regard to examinations, acting largely as a check upon the board of examiners. The papers set at the examinations of cadets and officers are submitted to its inspection, and may be altered or replaced according to its discretion. It delegates members to attend all oral examinations, including the quarterly examinations in the Academy and School. It examines all the marks given at examinations, and transmits the result of its inspection to the president of the examining board. It sets the papers for applicants for admission to the voluntary courses of the Naval Academy, inspects the marks given by the examiners, and delivers an opinion thereupon to the Admiralty, with recommendations as to the application. It decides at the close of each year what students of the Academy shall be allowed to continue their studies; and sends full reports to the Admiralty of the work done both in the Academy and School, giving a critical statement of the progress and performance of the students, and of the ability shown by the instructors, the methods of instruction pursued, and the particular results attained. To these reports, the director adds a statement of the progress of each student, specifying the subjects in which he has been particularly successful, and stating whether he has shown marked aptitude for any professional specialty.

The committee has supervision of the library and scientific collections, and it directs the expenditure of appropriations for these purposes, on recommendations from the director and the instructors.

Either the committee as a whole, or any member of it, is authorized to make suggestions or censures directly, either to the students of the School and the Academy, or to the instructors. Decisions and recommendations regarding the direction are sent to the Admiralty, but those concerning any part of the personnel are made to the director.

The director of the Naval Academy and School is the head of the personnel, military and civil, of the establishment, and has the powers, as to discipline, punishment, and granting leave of absence, of the colonel of a regiment. He has general guidance and supervision of the instruction, as well as of the interior service, and is responsible for the thoroughness of the work done by teachers and students. He arranges the details of the programme of studies, and the regulations for instructors, in accordance with the decision of the committee on studies. He has the power of suspending or detaching students guilty of gross breach of duty; and the return of offenders depends on their conduct

during the period of suspension. With him is associated the direction
officer, who has the disciplinary powers of the commander of a battalion, and upon whom devolve the supervision of the interior service,
and the special charge of discipline. In deciding on military questions
and matters of administration, the director acts through the direction
officer.

Three naval officers are attached to the School, called inspection officers, to assist the direction officer in the details of executive duty. To
them is intrusted the direct supervision and inspection of the students
quartered in barracks, and the particular enforcement of regulations.
The system of discipline is vigorous, but not severe. The students are
treated like men rather than boys, and are not repressed by unneces-
sary restraints. The liberal character of the German regulations forms
a strong contrast to the repressive systems prevailing in France and
Italy. The inspection officers are enjoined to establish between themselves and the students the relation of comrades, without at the same
time sacrificing their authority; and they are to study the peculiarities
of mind and character of those in their charge. At the close of each
month a meeting of the inspection officers is held, at which they make
a formal statement of their opinion of the character and aptitude of
each student.

Each inspection officer has assigned to his particular charge one or
more of the cadets' quarters, to which he gives special attention. All
petitions from students in his inspection go first to him. He sees that
the study-hours are properly occupied, and makes frequent inspections
to insure order and cleanliness in his buildings. He has charge, either
in his own inspection or as officer of the day, of the under-officers,
watchmen, writers, waiters, and house and section-room service. No
authority is given him to punish offenders. A breach of discipline requiring punishment is reported to the director. The inspection officers
are on duty by turns, for twenty-four hours at a time, as officer of the
day, with the usual duties of this position.

In addition to the three inspection officers, there is a fourth officer,
who is detailed as a sort of adjutant or military secretary (*Bureauchef*),
and who has charge of the office duties and the care of the library and
scientific collections. The pecuniary administration is in charge of a
committee (*Kassen-Commission*) consisting of the direction officer, the
Bureauchef, and the paymaster.

Instructors are of three classes, as follows :

(1) Civil instructors, who are imperial officials, detailed for duty at
the School or Academy.

(2) Officers of the military or civil branches of the Navy, ordered
exclusively for instruction.

(3) Other officers, civil functionaries, and professional men of distinguished attainments, who are employed to give instruction in addition
to such other occupations as they may already have. Instructors of the

first two classes have a fixed number of lectures to give, and hours of instruction to fill, and for extra work they receive extra pay. Instructors of the third-class (*Honorar-Lehrer*) receive special fees, according to the character and amount of the instruction they give.

In the courses at the Naval School, learning by rote and mechanical methods of teaching are rigorously excluded, and the practical wants and necessities of the profession are kept constantly in view. The object of the instructors is always to give students the most thorough understanding of principles, and this end is accomplished by frequent reviews and recitations, setting special problems, and close criticism of students' work. A lesson lasts one hour and a half, and the number of pupils in a section is limited to 25.

Practical instruction is given in visits of inspection on shipboard, and in the dockyards. There is also a course in practical surveying, near the close of the session, for those students who received no instruction in this branch while on the midshipmen's practice-cruise. The students of both courses have instruction in fencing and gymnastics, and the cadets have also lessons in dancing.

Quarterly examinations, both written and oral, are held, and marks given at these examinations are entered in merit rolls. The marks for the four examinations are combined at the end of the course, and give the standing of the student. The merit rolls together with the three best and three worst exercises in each subject are sent to the committee on studies.

The discipline of the School, under the inspection officers, is carried out by students, selected for special duties and responsibilities. These are (1) the superintendent of service, for the whole body of students, (2) the class-superintendent, and (3) the superintendents of quarters. All of them are named by the director of the school, usually according to seniority. The duty of these " cadet officers," as they may be called, is to enforce regulations, order, and discipline among the students under their charge. To this end, they are given adequate authority, and they are responsible for irregularities that they fail to report. Each of the students' quarters contains a study or living-room and two dormitories. In the former the students are occupied during the study hours, and discipline similar to that of a class-room is preserved by the superintendent of quarters.

The mess apartments are divided into three parts, devoted respectively to meals, to amusement, and to smoking. One of the inspection officers has charge of the mess in conjunction with a caterer chosen by the students. This officer has general supervision of accounts and expenditures, and receives all complaints and requests. He examines the students' wine account, and their accounts for extras with the steward, which are settled every month; and the steward is forbidden to open any credit beyond this. The officer of the day has oversight of the quantity of wine drunk at the table, and keeps it within proper limits.

Beer only is allowed at breakfast, wine and beer at dinner. The use of spirits is forbidden.

The mess hall and the other rooms connected with it may be used for music or games during the hours of recreation; but no games of chance are allowed.

All the cadets in attendance at the School are quartered in the school-buildings, and as many midshipmen as the space allows. The rest of the midshipmen and the officers provide their own quarters, but they must dine at the officers' mess. Those quartered in the buildings have all their meals at their own mess.

The library and reading-room are at the disposal of the students. They are required to attend divine service on Sundays and high church-festivals, the Protestants in the garrison church and the Catholics in their own church. They are marched to and from church in charge of an inspection officer.

Liberty to go outside the school limits is given during afternoon recreation on working days; and on holidays after service or inspection till 10 p. m. Permission at other times and for longer periods is given only to students whose conduct and diligence warrant it. Requests are forwarded at muster through the inspection officer to the director, and the liberty may extend to permission to attend the theater till 11 p. m., and social entertainments, where an invitation has been received, to a later hour. In all such cases the janitor notes and reports the hour of return.

The care of the students' clothing and quarters is in charge of the servants, who are enlisted men detailed for this duty.

For cadets the light punishments consist of extra drills, special dress for muster, suspension from duty, and the requirement of being in their rooms at a certain hour. The heavy punishments are confinement to the buildings, with required attendance at lectures, and performance of regular duties; confinement in quarters, during which the student is not allowed to leave his building, except for meals, musters, lectures, &c.; and confinement in garrison, during which the student loses, for the time, his part in the instruction. The punishments for midshipmen consist of formal censure and confinement of the three classes mentioned for four weeks. For officers the only punishments are censures, and confinement to the offender's apartment for two weeks.

ROUTINE.

6 a. m.—Reveille.
6.30 a. m.—Study.
7.35 a. m.—Breakfast.
8 a. m. to 12.30 p. m.—Three forenoon recitations.
12.45 p. m.—Muster, dinner, and recreation.
2.30 to 4 p. m.—Afternoon recitation.
4 p. m.—Recreation; instruction in gymnastics and dancing.
6 p. m.—Study; voluntary on Saturday.
8 p. m.—Supper and recreation.
10.10 p. m.—Lights out.

CHAPTER XXIV.

NAVAL SCHOOL.—CADETS' DIVISION.

MIDSHIPMEN'S EXAMINATION.

The direct object of the course for the cadets' division of the Naval School is the preparation of the cadets for the midshipmen's examination; and it furnishes the first theoretical instruction received by junior officers. It embraces the following subjects:

I.—NAVIGATION.

Outlines of astronomy; stars of the first magnitude and their distances; ocean and wind currents; arrangement and use of the log-line and glass; calculation of course and distance, and all allowances to be made for currents; construction and use of charts; ability to plot a ship's position from course and distance, from latitude and longitude, and from bearings and measurement of angles; conversion of time; calculation of latitude from meridian altitude of sun, moon, stars, and planets; ready use of the sextant and octant, and finding the index-error; construction and use of the artificial horizon; barometer, thermometer, chronometer, and the different compasses; correction of observed altitudes for semi-diameter, parallax, dip, and refraction; arrangement and use of Bremicker's nautical almanac; setting up of the binnacle and azimuth compasses, and obtaining the deviation.

II.—SEAMANSHIP.

Nomenclature of the various parts of a ship not included under ship-building; masts and rigging, their arrangement, application, and uses; loosing and furling, making and taking in sail; action of wind on sails and ship; action of the rudder; simple manœuvers under ordinary circumstances, with words of command; the national flags of maritime States, the system of signals in the Imperial Navy, and the International Code, including semaphore signals; the daily boat-service; salutes and words of command.

III.—GUNNERY.

Classification and nomenclature of guns and small arms; materials and principles of construction of guns and carriages; composition and elements of gunpowder, and general notice of the manufacture of those kinds in use in the German Navy; ignition, inflammation, and combustion; examination and testing of powder; indications and treatment of spoilt or damaged powder; preservation of powder and ammunition in magazines, on shore and on shipboard; classification and nomenclature of ammunition, fuses, and primers; elementary principles of firing.

IV.—LAND TACTICS.

Different kinds of troops; formations and evolutions of a company of infantry, with special reference to skirmishing; effect of the ground upon methods of fighting; principles for the guidance of a ship's company landed for fighting (except the use of boat-guns and field-pieces).

V.—MATHEMATICS.

Thorough knowledge of lower mathematics, and facility in calculation, leaving out the higher equations and series; plane and spherical trigonometry and stereometry

VI.—NATURAL PHILOSOPHY.

Fundamental principles of chemistry, with particular reference to the processes entering into gunnery and nautical science.

VII.—OFFICIAL DUTIES (*Dienst-Kenntniss*).

Preparation of official reports pertaining to midshipmen's duties; regulations governing the duties of midshipmen on shipboard; naval discipline on shore and at sea; organization of the Army and Navy.

VIII.—SURVEYING, OR TOPOGRAPHICAL DRAWING.

IX.—ENGLISH AND FRENCH.

Fluent reading and tolerably ready translation of English and French into German and *vice versa*; readiness in oral and written expression.

At the close of the course for cadets a faculty-meeting of the director and instructors of the Naval School decides what students may be permitted to attend the midshipmen's examination. Their report, together with the director's report on conduct, is transmitted to the Admiralty for approval. Those cadets who receive permission from the Admiralty are thereupon examined for promotion to the grade of midshipman.

The midshipmen's examination covers the whole course of instruction pursued by the cadets and is conducted by the permanent board of examiners. The subjects are arranged according to the following schedule:

FIRST CLASS: COEFFICIENT, 3.

I.—NAVIGATION:* 3 hours.

Six papers or problems are set in the examination, of which two are descriptive and four are practical problems.

II.—SEAMANSHIP:* 3 hours.

Three papers are given: one on general knowledge of the ship, one on maneuvers, and the third on service on board ship.

III.—GUNNERY: 2¼ hours.

IV.—MATHEMATICS: 2 hours.

Three papers are given: in algebra, stereometry, and spherical trigonometry.

SECOND CLASS: COEFFICIENT, 2.

V.—LAND TACTICS.

Two papers, of half an hour each, on infantry tactics and fighting on shore.

VI.—OFFICIAL DUTIES.

Three papers, of half an hour each, on (1) naval and military organization, (2) general regulations (including official correspondence), and (3) laws of discipline.

THIRD CLASS: COEFFICIENT, 1.

VII.—NATURAL PHILOSOPHY (CHEMISTRY): 1 hour.

VIII.—SURVEYING.

Drawings made by the students during the term are marked as a part of the examination.

---

* Mark of 5 (55 per cent.) required.

IX.—ENGLISH AND FRENCH.

The oral examination in navigation lays special stress on the use of the sextant, octant, charts, compasses, nautical almanac, &c.; in languages it is chiefly devoted to reading aloud and conversation. Cadets who fail at the midshipmen's examination may be allowed, upon recommendation of the board, to go over a second time the course for the year, beginning with the cruise in the practice-ship and ending with the course at the school. In order to do this they must join the class of newly-entered cadets, and they lose a year's seniority. In cases of idleness, want of ability, or bad conduct, the board will not recommend the retention of a cadet who fails, and in the absence of such a recommendation the cadet is invariably discharged. A third trial is never granted by the board, and therefore a second failure causes immediate dismissal. In such a case only the time of service which has elapsed since taking the oath counts toward the performance of obligatory military service.

Cadets who have passed the midshipmen's examination receive certificates of proficiency from the Naval School. Those who show extraordinary proficiency are proposed for the special approbation of the Emperor. At the same time with the proceedings of the board, a tabular statement of the results of the examination and a list of the cadets graded according to conduct are sent to the Admiralty. The order of standing as midshipmen is made out on the basis of these documents, combined with the certificates and provisional arrangements of the training-ship, and cadets can be proposed for promotion as acting or supernumerary midshipmen in the order of their seniority, as fixed by this process.

CHAPTER XXV.

MIDSHIPMEN'S PRACTICE-CRUISE.

Cadets who have received the certificate of proficency at the midshipmen's examination are sent immediately to Wilhelmshafen for training on board the gunnery ship. The course, which is mainly practical, lasts one month. At its close they are ordered to the squadron of evolutions, in the ships of which they make the summer cruise. The squadron is composed usually of three or more large iron-clads, and cruises for several months in the Baltic and North Sea. Returning from this cruise about September, the midshipmen are embarked at Kiel in the midshipmen's school-ship for a two years' cruise. Here they receive not only a thorough practical training for service as sub-lieutenants, but also an extended course of theoretical instruction to prepare them for the first officers' examination (*Erste See-Offiziers-Prüfung*).

As the midshipmen's practice-cruise lasts two years, and as there is a class ready to go out every year, two ships are set apart specially for this service. They are screw-steamers of modern type, fully rigged, so that they are available for sailing or steaming. They are supplied with a full battery of breech-loading rifles, and with all the appliances and equipments necessary for a training-ship. Among these is a complete library of professional works and works of history, poetry, and fiction in French, German, and English. The professional list embraces the latest works in all branches, including several American books. There is a full supply of instruments connected with steam-engineering, navigation, hydrography, and land surveying, for the midshipmen are landed from time to time for practice on shore. The ship is also supplied with a torpedo outfit. The cruise generally extends to China. The practice-ship for the cruise of last year was the Leipzig, a screw-corvette of 12 guns and engines of 700 horse-power.

The routine of study (*Stunden-Plan*) is made out by the captain of the school-ship and approved before the beginning of the cruise by the Admiralty. Slight changes may be made for weather or other emergencies, but more extensive changes on the cruise must be sanctioned by the Ministry. Semi-annual examinations are held on the work of each semester, and annual examinations at the end of the year. The captain is present at oral examinations. The same marking system prevails as at the Naval School. The captain transmits marks and reports to the Admiralty, and gives the necessary warnings to cadets who fail to reach the standard. The captain attends the class instruction from time to time, and forms an opinion of the value of the instruction given. He appoints the *cadets' officer*, who has charge of the mess funds, clothing and outfit, and allowances, and of all private interests of the midshipmen. In foreign ports midshipmen are organized in parties to visit places of interest. They are encouraged to go to balls and social entertainments, and are given every possible opportunity of conversing in foreign languages. The punishments are similar to those of the cadets.

For duty on board, the midshipmen are permanently divided in four watches, under the four division officers. These watches are rearranged every quarter. One midshipman acts as sergeant of the division, and the duty is taken in turn. Journals are kept by the midshipmen, and inspected daily by the division officer. In the journals they are required to make sketches of such objects and places as may be designated.

When the ship is under steam the midshipmen do watch duty in turn in the engine-room, where they receive instruction from the machinist in charge. At the end of this watch they have to make a report to the officer of the deck of the number of revolutions made, the position of the manometer and vacuum-gauge, the quantity of coal consumed, and the temperature and saturation of the boiler.

When the ship is under sail the midshipmen are stationed in turn as midshipmen of the tops. In port, each acts in turn as signal officer. They act by turns as gun-captains at great-gun exercise, and they are sent in charge of boats.

According to the daily routine midshipmen turn out at 6.30 a. m. From 8 to 9 they have instruction, or superintend the cleaning of the guns, after which they have instruction and exercises till dinner. In the afternoon, from 2 to 4, they again have instruction, and after the evening muster, at 4.30, they take part in the general exercises, either as topmen or at their regular stations. The evening is generally devoted to study, but on two evenings in the week they have fencing and gymnastic instruction.

The arrangements for living, mess, &c., are similar to those in the cadets' school-ship. A room is assigned to the midshipmen on the gun-deck, in which they eat, sleep, and study. The senior cadet in each division performs in turn, for a week at a time, the duties of officer of the day (du-jour-Dienst), as far as the mess and study periods are concerned, and is responsible for quiet and order in the steerage.

The mess is managed by a board, composed of the cadets' officer, paymaster, and two midshipmen, elected by their companions for a term of three months. The steward is engaged by contract at the beginning of the cruise, and two seamen are detailed as waiters.

Midshipmen sleep in hammocks. The personal service of the midshipmen is performed by seamen who volunteer for the duty, and who receive special compensation and exemption from all ship's duty for certain hours in the morning. One servant is allowed to two midshipmen.

Midshipmen are forbidden to contract debts; and if a midshipman is negligent and thriftless in pecuniary matters, the captain, as an extreme measure, may turn over his pay and his private allowance (Zulage) to the officer of his division, as a trust, to be administered according to the officer's discretion.

An amusement fund is formed by the paymaster out of sums reserved
S. Ex. 51——12

from the private allowance, and from other sources, which is expended in excursions made from time to time by parties of midshipmen.

The theoretical course of instruction includes navigation, steam-engineering, seamanship and naval tactics, gunnery and torpedoes, official duties and organization, ship-building, English and French. Navigation, gunnery, and engineering are taught by the navigating and gunnery officers, and by the engineer, respectively. Instruction in the direction and care of the ship's stores is given by the paymaster, and the other subjects are assigned to the various officers of the ship according to their qualifications. No officer is given more than two subjects of instruction. In order to give to instructing officers the needful time to prepare their instruction, they may be relieved of the deck, weather permitting, by the sub-lieutenant of the watch, from 8 a. m. to 12, and from 12 to 4 p. m.

Extra pay is given to officers engaged in duties of instruction on board the school-ship, at the rate of 2, 2.50, or 3 marks (mark = 25 cents) a day, according to the subject taught. The aggregate of these "supplements" is limited to $618 per annum, representing 912 hours of instruction, of which 96 hours are given to fencing. This leaves 816 hours, which may be taken as the total time devoted annually to theoretical instruction on board the practice-ship, an average of nearly fifteen hours a week through the whole year.

The hours of instruction are as follows:

FIRST SEMESTER.

Hours a week.

Navigation	4
Steam engineering	3
Gunnery	2
Ship-building	1
Seamanship	2
Official duties	2
English	1
French	1
Total	16

SECOND SEMESTER.

Navigation	3
Engineering	3
Gunnery	1
Torpedoes	1
Ship-building	1
Electives	1
Seamanship	2
Official duties	2
English	1
French	1
Total	16

THIRD SEMESTER.

Hours a week.

Navigation	3
Engineering	3
Gunnery	1
Torpedoes	1
Ship-building	1
Electives	1
Seamanship	2
Official duties	1
English	1
French	2
Total	16

FOURTH SEMESTER.

Navigation	3
Engineering	3
Gunnery	1
Torpedoes	1
Ship-building	1
Electives	1
Naval tactics	1
Organization	2
English	1
French	1
Total	15

At the close of the second and fourth semesters annual examinations · are held in navigation, engineering, and gunnery; at the close of the third semester are the final examinations in seamanship and official duties ; and at the close of the fourth, the final examinations in torpedoes, ship-building, naval tactics, and naval administration.

The programme of studies is arranged with reference to the regular ship duties of instructing officers, so as to work as little inconvenience as possible. At the same time, it is always kept in view that the ship is a school-ship, and that the first object of the cruise is to give regular and thorough instruction, practical and theoretical, to the midshipmen. The system is therefore free from the objections to instruction on shipboard in the English service, where the attempt is made to have a school on board the cruising ships of the fleet, and where, of course, instruction must become a secondary consideration, beside the all-important demands of actual service ; where, moreover, midshipmen are considered not only as students, but as officers having regular duties to perform.

The arrangement of studies is such as to give three hours of theoretical instruction a day, one in the morning and two in the afternoon. The· details of the theoretical course are as follows :

NAVIGATION.

FIRST SEMESTER : 4 hours a week.

The course begins with a short review of plane navigation, including compasses, deviation, log and lead, and charts. The subjects taken up in the course are nautical surveying, with the instruments on board the ship; preparation of charts; calculation of position, construction of nautical instruments, chronometer, sextant, circles, compasses, &c. ; setting up of the binnacle and azimuth compasses, and determination of deviation; practice in observing with the sextant and with Pistor's circle; arrangement of the nautical almanac; correction of harbor surveys; and correction of the chronometer error by hour angles of the sun.

SECOND SEMESTER : 3 hours.

Astronomical navigation ; review of the relations of hour angles and time; calculation of the meridian passage and altitudes of the fixed stars ; of the hour angle and chronometer rate, from observations by the midshipmen of the sun, moon, planets, and fixed stars.

THIRD SEMESTER : 3 hours.

Determination of chronometer error by equal altitudes of sun ; preparation and use of chronometer tables; latitude and longitude by Sumner's method; calculation of latitude by altitude of pole star, by ex-meridian altitude, by double altitudes; rising and setting of constellations; calculation of variation by sun's amplitude or azimuth.

FOURTH SEMESTER : 3 hours.

Calculation of time of high and low water; meteorology and geography of the sea ; winds and currents; laws of rotary storms ; general review.

SEAMANSHIP AND NAVAL TACTICS.

FIRST SEMESTER : 2 hours.

Putting in and taking out masts and bowsprit ; rigging shears; getting up and taking down tops, caps, topmasts, yards, and spars; setting up standing rigging ; reeving off running rigging ; bending sails; arrangement and use of the anchor and cables ; securing the anchor, gearing used for this purpose; shipping stores; getting ready for sea.

SECOND SEMESTER: 2 hours.

Boats and their management; striking topmasts and lower yards; shipping guns and other heavy weights; effect of wind on the sails; maneuvering under various circumstances of wind and weather; getting under way, anchoring, tacking, wearing, heaving to, bracing back, and scudding; furling and reefing sails in a stiff breeze; anchoring on a lee shore in a gale; hoisting and lowering the screw.

THIRD SEMESTER: 2 hours.

Clearing ship for action; getting a ship off when aground; setting up jury-rigging; repairs of masts and yards; docking ship; shipping and unshipping rudder; jury-rudder; rules of the road; exact knowledge of German signals, of the international code, and of the semaphore signals; review.

FOURTH SEMESTER: 1 hour.

Naval tactics, including fundamental knowledge of evolutions, the relative fighting qualities of ships in an engagement, the attack of fortified and unfortified places and coasts, and the protection of harbors and mouths of rivers.

GUNNERY.

FIRST SEMESTER: 2 hours.

Principles of construction and use of guns; carriages, equipments, and small arms, on board the school-ship; inspection of ordnance materials; exact knowledge of the detail drill, with the guns on board, of the range-tables, and of the regulations for the inspection and care of the guns, before, during, and after use.

SECOND SEMESTER: 1 hour.

Review of the whole subject of gunpowder, its preparation and inspection, ignition, combustion, and energy; regulations for the transportation, bringing on board, and care of powder and other ammunition; preparation of fuses; filling cartridges and shells; arrangement of magazine and shell-room; use of boat and field guns (taught in connection with landing parties).

THIRD SEMESTER: 1 hour.

Special exercises, as hoisting in guns on covered decks, shifting guns to fire ahead; pivot guns; service of guns in saluting; special regulations for naval guns not on board the school-ship.

FOURTH SEMESTER: 1 hour.

Foreign guns and small arms, with practical applications of instruction in visits to foreign men-of-war; general review.

TORPEDOES.

SECOND SEMESTER: 1 hour.

Explosives; loading, filling, and unloading torpedoes; fuses, their preparation and adjustment; contact-apparatus; properties of dynamite and gun-cotton.

THIRD SEMESTER: 1 hour.

Defensive torpedoes; planting and taking up; avoidance of enemy's torpedoes; construction and use of offensive torpedoes.

FOURTH SEMESTER: 1 hour.

Laying and inspection of electric torpedoes, inspection and care of wires; measuring instruments; review.

STEAM-ENGINEERING.

FIRST SEMESTER: 3 lessons.

Laws of steam; working of engines in general; different propellers; different types of boilers and their details; causes and prevention of foaming; management of the boiler, the last in connection with practice in the fire-room.

SECOND SEMESTER: 3 lessons.

Management of an engine; starting, stopping, backing; object and different methods of expansion; expansion apparatus; reversing gear; methods of distribution of steam, in connection with practical exercises in the engine-room; different ways of reversing.

THIRD SEMESTER: 3 lessons.

Principles and working of different condensers; use and disposition of detail parts, as cylinder valves, cylinder cocks, valve-chest cocks, condenser valves, oiling apparatus, waste cock, hand-reversing gear, manometer, counter, and indicator; calculation of horse-power and of tension; indicator diagram.

FOURTH SEMESTER: 3 lessons.

Management of the engine in action; preservation of the engine and boiler; putting in, hoisting, coupling, and uncoupling screw; distilling apparatus; different kinds of fuel, their inspection, and the indications of their quality; care of coal in bunkers; use and arrangement of all the pumps on board; nomenclature of the engine and boiler and their parts in English and German.

SHIP-BUILDING.

FIRST SEMESTER: 1 lesson.

Ship's frame, timbers, planking, diagonal bracing, &c., in wooden and iron ships.

SECOND SEMESTER: 1 lesson.

Frames employed in the construction of iron ships, armored ships, composite ships and iron ships sheathed in wood.

THIRD SEMESTER: 1 lesson.

Construction of masts, tops, spars, rudders, capstans, and boats.

FOURTH SEMESTER: 1 lesson.

Survey of ships; means of preservation of material employed in the construction of ships; small repairs performed on board ship; stowage, and its effect on the stiffness, stability, and motions of the ship under steam and under sail; general review.

OFFICIAL DUTIES.

FIRST SEMESTER: 2 lessons.

Organization of the Army and Navy on a peace and war footing; details of Naval organization as contained in orders, regulations, and instructions.

SECOND SEMESTER: 2 lessons.

Official and other relations of officers; garrison-duty; routine of office-work, and preparation of official reports and dispatches; duties of first lieutenant, navigator, officers of the watch, and inspection officers.

THIRD SEMESTER: 2 lessons.

Discipline, punishments, military courts; legal status of military persons; disputes of officers; courts of honor (*Ehrengerichte*).

FOURTH SEMESTER: 2 lessons.

Details of administration, especially in relation to the care of stores and materials, and the duties of boards of inspection and survey, and auditing boards.

ENGLISH AND FRENCH.

One lesson a week in each language during the course, including exercise in writing from dictation, writing letters, conversation, oral and written translation from German into the foreign languages and *vice versa*.

Elective subjects are international law, hygiene, &c.

The course of practical instruction is as follows:

Navigation.—Practical instruction in navigation consists of daily observations, calculations, and nautical surveying. Midshipmen keep a navigation book, in which are entered all their observations of the sun, moon, stars, &c. These books are regularly examined and corrected by the navigating officer. When opportunity offers, an excursion of several days is made, to survey part of a bay or harbor, the midshipmen being in charge of the navigating officer and one or two others detailed to assist him. At least one such excursion is made during each cruise, and during its continuance all other instruction is suspended.

Seamanship.—Practical instruction in seamanship is given partly by the officer of the deck to the midshipmen of the watch, and partly in separate exercises. During the latter the midshipmen are stationed as mizzen-topmen, and, in case of a large number, a few are stationed in the gangways. The exercise consists in sending up and down light masts and yards, loosing, furling, reefing, and shifting sail. This exercise may also be carried out in connection with maneuvers. It is held at such times as the captain may direct, and more or less frequently, according to the proficiency of the midshipmen. The test of proficiency is that they shall be able to perform the exercises with as great speed and precision as the crew. Practice in boat-sailing takes place whenever opportunity offers. Towards the latter part of the cruise, each midshipman takes the deck in turn, and works the ship, under the supervision of the officer of the watch, but giving the orders himself.

Gunnery.—Midshipmen are frequently exercised at the guns, with broadside and pivot carriages, and with howitzers. The exercise includes firing-practice. Their stations at the guns are changed from time to time, and they are exercised sometimes by themselves, and sometimes with the crew. The preparation of pyrotechnic materials is included in the instruction. Finally, the captain is expressly ordered to lose no opportunity of sending the midshipmen in charge of the gunnery officer on board foreign ships of war, to examine their ordnance.

Steam-engineering.—For practical instruction in steam-engineering, including the management both of fires and engines, midshipmen are divided into small parties. Instruction in the fire-room lasts two hours at a time, and is in charge of the leading fireman of the watch. The instruction is kept up until every member of the class shows complete understanding of and proficiency in the work. In the engine-room, midshipmen are in the charge of the engineer. Here they learn to work the engine and to manage the reversing-gear in maneuvering the ship. They become familiar with the uses of different valves, gauges, cocks, levers, and other detail parts of the engine. In addition to this, they have watches in the engine-room, as already described, a duty which they perform in turn. The greatest attention is paid to this branch of instruction, and every effort is made to familiarize the midshipmen with the practical use of the machinery of the ship. Parties are sent in charge of officers to examine the engines of foreign men-of-war; and at suitable times midshipmen are placed by themselves for several successive hours in charge of the engine or of the fire-room.

Infantry and other drills.—Instruction in fencing includes both broad and small sword. In infantry tactics and the use of small arms, instruction is given by forming landing parties, at which the midshipmen are detailed as non-commissioned officers. They have also the manual drill on shipboard, and frequent target-practice with rifles and revolvers. In the latter a high standard is required.

On the return of the practice-ship, the commanding officer sends to the Admiralty a full report of the proficiency of the midshipmen, and a provisional rank-list. The midshipmen receive certificates of service, drawn up by the captain with the assistance of the officers, containing a statement of their character, fitness for the service, and general scientific attainments. The reports and certificates state particularly whether the midshipman is out of debt and whether he is considered worthy of admission to the service as an officer. The midshipmen who receive a favorable certificate are ordered to Kiel to pass the first officers' examination.

The first officers' examination (*Erste See-Offizier-Prüfung*) is preliminary to promotion to the grade of sub-lieutenant, and covers, in general, the ground gone over in the course on board the midshipmen's practice-ship. The subjects are classified as follows:

FIRST CLASS: COEFFICIENT, 3.

1. *Navigation.**—Eight papers are given, of which five are practical and three descriptive. The examination lasts, in the aggregate, eight hours. It calls for thorough ability to do a day's work and increased facility in dealing with the subjects required in the midshipmen's examination, and it covers the programme of the course in the school-ship.

2. *Seamanship and naval tactics.**—Time allowed, six hours.

SECOND CLASS: COEFFICIENT, 2.

3. *Gunnery.*—Six papers, two of which are on torpedoes. Six hours are allowed for the whole.

4. *Steam-engineering.*—Three papers of one hour each; one on the erection and arrangement of engines; one on their working, and one on their manipulation.

5. *Official duties.*—Three hours.

THIRD CLASS: COEFFICIENT, 1.

6. *Ship-building.*—Three papers of one hour each; one on construction; one on the dynamics of ship-building, and one on materials of construction.

7. *French and English.*

The oral examinations are similar to the earlier ones, but more advanced in character. At the close of the examination the usual reports, recommendations, &c., are sent in by the board. Only one re-examination may be granted in case of failure, and this only for special reasons.

Midshipmen who pass the examination are thereupon subjected to a still further test of a remarkable character, which it is believed is peculiar to the German service. An election is held at Kiel, the naval sta-

* Standard of 55 per cent. required.

tion of the Baltic, to determine the fitness of each midshipman for the grade of sub-lieutenant. In this election all the officers attached to the station have a vote.

Upon the request of the captain of the school-ship, the Commander-in-Chief of the station is required to summon all the officers on duty, at an appointed time, to hold the election. The names of the midshipmen are then submitted, in the order of their seniority, for election to the grade of sub-lieutenant without commission. In case of a difference of opinion as to the merits of a candidate, the following rules are observed:

1. If the majority of votes is against the candidate his career in the Navy is finished, and the next candidate is proposed without further formalities.

2. If a minority of votes, or only a few individual votes, are cast against the candidate, the officers so voting are required to give the reasons for their opinion in writing; and this vote of the minority, with the indorsements of the Commander-in-Chief of the station and the Minister of Marine, is to be appended to the general report or petition (*Gesuchsliste*) addressed to the Emperor. In this report all those who have been favorably passed upon in the election are proposed for acting sub-lieutenants; and they become full sub-lieutenants as vacancies occur.

CHAPTER XXVII.

NAVAL SCHOOL.—OFFICERS' DIVISION.

OFFICERS' PROFESSIONAL EXAMINATION.

The acting sub-lieutenants (*Unter-Lieutenants zur See ohne Patent*) are ordered by the Admiralty to attend the officers' course at the Naval School. Midshipmen who have passed their first officers' examination, and are awaiting promotion, are ordered there at the same time. The course of instruction begins about the 1st of October and closes at the end of August in the following year. It completes the theoretical education of officers and prepares them for the professional examination (*See-Offizier-Berufs-Prüfung*), held annually at Kiel in September by the board of examiners. The first half of the session is devoted rather to purely scientific study, and the last half to professional branches. The dates of the beginning and end of the course are fixed by the Admiralty, and may be varied according to circumstances.

The organization of the Naval School for both the officers' and cadets' divisions has been already described. The course of the former includes the following subjects:

I.—NAVIGATION.

Exact knowledge of the construction and skill in the adjustment and use of instruments; the correction of errors, knowledge of the theory, and increased readiness in determining the time of culmination and the real and apparent rising and setting of fixed stars, &c., and in calculating longitude by chronometer with the necessary observations for error and rate; preparation of tables for the correction of the chronometer; readiness in observing and in calculating local time by the different methods; arrangement and calculation of complete observations of longitude by lunar distances from the sun, stars, and planets, with or without measured altitudes; calculation of latitude according to the methods previously given, and also by observation of simultaneous altitudes of two fixed stars; certainty in determining the variation and deviation of the compass and the causes of the latter, both on land or at sea; methods of coast surveying; execution of a survey by the use of the plane table, and preparation of charts and sketches; knowledge of sailing directions in their dependence upon prevailing winds and currents; acquaintance with the marked features and depth of water of German harbors, with the shoals, lighthouses, and beacons of the Baltic, the Sound, the Great Belt, Cattegat, and the North Sea, as well as those near the usual routes to the English Channel.

II.—LAND TACTICS.

Employment of different kinds of troops in fighting; the tactical formations and evolutions of an infantry battalion, especially the company column and skirmishing; general principles of scout and picket service, on the march and in camp; influence of the contour of land upon modes of fighting; tactics of the Naval Brigade (*gelandete Schiffsmannschaft*); and employment of boat and field guns in landing parties.

III.—GUNNERY.

Extended review of the subject of gunpowder and ammunition, in general; guns, carriages and equipments, small arms, and the care and use of ordnance, especially of rifled guns; general study of foreign ordnance; thorough knowledge of the working of ordnance, and the attendant circumstances, as, the form of the trajectory of solid shot, and of scattering projectiles, with or without reference to the resistance of the atmosphere, and rotation; preparation and use of range tables, and graphic representation of the path of projectiles; firing at long range; effect of shot, &c.; effect of firing on the gun itself; recoil and its prevention; different kinds of firing, according to the gun, projectile, charge, form of trajectory, direction, and object aimed at; employment of different methods in sea and coast fighting; condition and effectiveness of foreign naval and coast artillery; firing with small arms.

IV.—SHIP-BUILDING AND NAVAL ARCHITECTURE.

Technical knowledge of the properties of materials used in ship-building, their care and preservation; principal systems of ship-building, their advantages and disadvantages; structural parts of a ship, and details of construction of wooden and iron ships; calculation of displacement; ship-draughting; application of statical and dynamical laws to determine the lines of a ship, and the distribution of weights, stowage, &c.; effect of the form and distribution of weights, on the character of the ship as a gun-carrying machine, while stationary or in motion; rolling moments and pitching moments; preparation of sail-draught, with necessary calculations; action of sails; effect of stowage on the efficient working of the ship; relation between the ship and engine; docking ship, and inspection of hull, especially in iron ships.

V.—STEAM-ENGINEERING.

Physical laws relating to steam; temperature, elasticity, density, condensation, and expansion; principles of construction and working of different kinds of engines and boilers; different systems of propulsion; safety apparatus, and boiler equipment in general, with governing principles; determination of the saturation; expansion apparatus; apparatus for the distribution of steam; principles and working of different condensers; the arrangement and action of detail parts of the engine; different kinds of fuel and their advantages, their condition, and their economical use and preservation; thorough knowledge of the methods of calculating the performances of the engine, under different circumstances of speed and consumption of coal; nominal, indicated, and effective horse-power; use of the indicator diagram as a basis of calculation; preservation of engine and boiler; explosions; different forms of distilling apparatus; all the pumps used in the German Navy.

VI.—FORTIFICATION.

1. *Field fortification.*—Nomenclature, erection and disposition of redoubts and field-works, and methods of attack and defense.
2. *Permanent fortification.*—Main conditions for the erection of permanent fortifications, and a brief survey of the construction of existing forts; outlines of the attack and defense of fortresses.
3. *Coast fortification.*—Necessary conditions for shore-batteries, in reference to choice of site, details of construction, and armament; laying harbor obstructions; use of submarine torpedoes; attack and defense of coast-batteries.

VII.—DRAWING.

Preparation of a sketch of coast-line, a survey, and some line-drawings from subjects in ordnance, ship-building, and engineering.

· VIII.—MATHEMATICS.

1. Review of arithmetic and algebra; higher equations and equations with several unknown quantities; fundamental principles of the differential calculus, and its application to the development of functions in series, and the evaluation of indeterminate forms; maxima and minima of functions; interpolation, and the analysis of equations.

2. Geometry and stereometry; analytic geometry of two and three dimensions; theory of co-ordinates; equations of surfaces; conic sections; higher plane curves; rectification, quadrature, and cubature; solids of revolution; Simpson's rules.

3. *Trigonometry*—The application of spherical trigonometry and the use of tables, with special reference to the problems of navigation.

4. *Pure mechanics.*—Determination of centers of gravity; principles of dynamics: law of falling bodies; path of projectiles; the pendulum; moments of turning, moments of inertia.

IX.—PHYSICS AND CHEMISTRY.

Higher knowledge of the principles of chemistry, and the important substances and processes in use in naval arts, or in fabrication of ordnance or machinery; gravitation and phenomena connected therewith; molecular energy, hardness, elasticity, capillarity, crystallization; statics and dynamics of solids and fluids; undulatory theory; reflection and refraction of light, interference, polarization, and chemical action of light; use of the spectroscope; photography; telescope. Theory of heat; measurement and effect of heat; expansion, liquefaction, evaporation, conduction of heat, capacity for heat, heat-unit, and equivalent of work; electricity, magnetism; correlation of forces; measurement and application of forces. Theory of the compass, disturbing influences, &c. Physical geography and meteorology, with practice in the use of meteorological instruments; methods of graphic representation of meteorological phenomena, law of storms, terrestrial magnetism.

The programme of the officers' professional examination (*See-Offizier-Berufsprüfung*) and the grouping of studies are as follows:

FIRST CLASS: COEFFICIENT, 3.

I.—NAVIGATION.

Nine papers are given, of one hour each, of which six are practical and three descriptive. Of the last, one is on instruments, one on deviation, and one on coast surveying.

II.—GUNNERY.

Six papers, of an hour each, of which one is on powder, three on the structure, fabrication, and inspection of guns and ammunition, and two on the use and working of guns.

III.—STEAM-ENGINEERING.

Four papers, of which two are descriptive, and two involving the theory of the steam-engine, and calculations. Three hours are allowed for this subject.

SECOND CLASS: COEFFICIENT, 2.

IV.—SHIP-BUILDING AND NAVAL ARCHITECTURE.

The same number of papers and hours as in the preceding subject, divided equally between a mathematical and a descriptive treatment of the subject.

V.—LAND TACTICS.

One hour and a half.

VI.—PHYSICS.

Four hours are allowed, and six papers are given, on the following subjects, respectively: Chemistry, mechanics, light and heat, electricty, physical geography, and meteorology.

THIRD CLASS: COEFFICIENT, 1.

VII.—FORTIFICATION.

Two papers of an hour each, one on field and permanent fortification, and one on coast defense.

VIII.—DRAWING.

The examination in drawing consists of the inspection of drawings made during the course of instruction, which must include at least the following: One hydrographical survey, one topographical survey (using the plane table); three drawings of objects pertaining to ordnance, one of which is an isometric projection; one calculated trajectory; one working drawing for ship-building; and three on detail parts of engines.

IX.—MATHEMATICS.

Four hours allowed. Six papers are given, of which two are in pure mathematics, two in geometry and stereometry, one in mechanics, and one in the application of mathematics to navigation.

Upon the close of the examination, the board transmits to the faculty of the Naval School its proceedings, and a summary of the results of the examination. The faculty thereupon makes up its general report or petition (*Gesuchsliste*) for the award of certificates of proficiency as officers, to those sub-lieutenants who have passed, which it transmits to the Admiralty, together with the proceedings of the board, and a list arranged according to conduct. In the transactions of the board must appear recommendations in regard to officers who failed at their examinations, stating whether they should be granted a second trial. A report to the Emperor is drawn up by the Admiralty on the basis of these reports from the board and from the Naval School, with the previous certificates and reports from the captain of the midshipmen's training ship; and the Minister requests the award of certificates in accordance therewith, which is forthwith granted. Sub-lieutenants who show extraordinary proficiency, receive the commendation of the Emperor. The seniority of the officers having been determined by the report, commissions as sub-lieutenants, and later, as lieutenants, are granted by the Emperor as vacancies occur; but in any case the passing of the examination and a total of five years' sea-service are indispensable qualifications for a commission.

CHAPTER XXVIII.

The students of the Naval Academy are of two classes, viz, officers ordered to the Academy for study in the regular course, and officers entered as transient students (*Hospitanten*). For admission to the latter class, that is, for the purpose of attending a partial course or certain special lectures, it is only necessary for an officer to have permission from the superior officer under whose orders he is for the time being. The conditions of entrance as a regular student are, however, much more exacting. In the first place, only those officers are ordered who show by their general character and conduct, by their professional zeal, and by their mental attainments that they will make the most of the privilege of attendance at the Academy to advance the interests of the service. They must then pass an examination of a peculiar character. Certain subjects are drawn up and published in November of each year by the Admiralty in the departments of military history, seamanship and naval tactics, navigation, gunnery, marine engineering, and ship-building. The candidate selects any three of these subjects, and is required to work up a paper in each subject. The papers that he prepares must be sent in to the Direction of the Naval Academy and School by the 1st of June in the following year, with a statement of the works of reference and other means of assistance employed. Applications for orders to the Academy on behalf of any individual are sent, before July, by his commanding officer to the Admiralty through the Commander-in-Chief of the station; and they must contain a report upon the qualifications of the candidate, and an explicit statement of the points referred to above, including character, ability, &c. At the same time the director of the Academy presents to the committee on studies a report of the opinion of the committee of the faculty appointed to examine the papers of candidates; and the papers themselves are also turned over to the committee on studies. The report of the examining committee states in terms whether the candidates, in the papers they have presented, have shown sufficient experience, proficiency, talent for observation, sound judgment, and practical skill to appear qualified to receive with advantage a further education. The reports, accompanied by the opinion of the committee on studies, are then sent to the Admiralty, with which rests the final decision upon the application.

The course at the Academy covers three years or classes, the session lasting from October to May, in each year. The general arrangement of the studies is such that the first or lowest class (*Cœtus*) takes up the auxiliary sciences forming a groundwork and preparation for scientific study in general, and for professional study in particular; the second class continues this fundamental training and takes up professional studies; and the third completes the professional course, and has a full course in those branches of natural and social science which concern most nearly the duties of a naval officer.

The subjects in the course of instruction of each class are as follows:

FIRST CLASS.

1. Logic, ethics, and the elements of psychology.
2. A short review of elementary mathematics, and the fundamental principles of analytical geometry and higher calculus.
3. Organic and inorganic chemistry.
4. The whole range of pure physics, treated both experimentally and mathematically.
5. Naval organization and naval tactics.
6. Tactics of land forces, considered strategically, and with special reference to landing parties.
7. Permanent fortification, especially in coast-defense.
8. * Military administrative law and international law.

SECOND CLASS.

1. Thorough course in higher mathematics, especially in its application to geometry, mechanics, and the calculation of probabilities.
2. History of naval wars.
3. Ordnance and gunnery.
4. Steam-engineering.
5. Naval architecture.
6. Nautical astronomy.
7. Coast survey.
8. Electricity, particularly in its application to torpedoes.
9. † Laws of war; maritime law.
10. † Sanitary administration, especially in relation to the conditions of life on shipboard.
11. † General survey of the history of civilization.

THIRD CLASS.

1. History of naval wars.
2. Ordnance and torpedoes.
3. Steam-engineering.
4. Naval architecture.
5. Nautical astronomy, the construction of charts, and the principles of geodesy.
6. Observation with instruments, including the preparation and use of all the instruments employed in navigation, and in geodetic surveys.
7. Physical geography.
8. ‡ History of civilization.
9. ‡ Construction of harbors.
10. ‡ Natural history of the sea.
11. ‡ Political economy.

---

* All the above studies are obligatory with the exception of the last, which the student may take or not, as he pleases.

† These three subjects are electives, but the student is required to take one of them.

‡ Of these four elective subjects each student is required to take two. All the other subjects are obligatory.

In addition to the course described, opportunity is given to the students to extend and perfect their knowledge of foreign languages. Instruction in this branch includes English, French, and Danish, and, when practicable, Russian and Spanish also. All this is independent of the regular course, but each student is required during all the years of attendance to take instruction in at least one foreign language.

During the summer months, from May till the end of September, the students of the Academy return to active service, being ordered either to a cruising-ship, or to the gunnery-ship, or attached for the time to the coast survey, or to the torpedo division. During the session practical instruction is confined to certain excursions made in connection with the course of study. These include, in the first class, topographical surveys by students who have had no such practice before; and, in the second class, hydrographic surveys in a small vessel attached to the station. There is also a special course of instruction in observing with fixed instruments at the observatory, and practice in the use of nautical, astronomical, and geodetic instruments, the latter in connection with triangulations and geodetic surveys.

Instruction is given in the usual academic manner, by lectures and questioning, supplemented by laboratory work, and by the use of drawings, models, and apparatus, of which there is a full collection at the Academy. The method of instruction here, as in all German institutions, aims directly at the thorough understanding of the fundamental principles of the subjects studied. To clear up obscurities, and to keep alive an active interest on the part of the students, the closest attention is directed to special cases and circumstances as they arise in the service, and to the demands of actual practice. The freest interchange of opinions between teachers and students is encouraged, and every method is adopted to stimulate thoughtful and intelligent effort, by informal expositions and discussions, by the preparation of written papers, and by frequent interrogations of and by the students.

Besides the questions, problems, &c., given out from time to time during the session, short examinations are held quarterly in each subject. Papers are set by the professors, and each examination lasts an hour, the students working under the supervision of the instructor. The results of the examination are tabulated, and sent, together with the report of the director, to the committee on studies. At the final examination for the year, another merit-roll is made out, giving the general results for the year; and a report of the Direction on the work done by students is sent in to the committee on studies. Upon the result of these examinations depends the return of the student for the next year's course. At the close of the three years, a diploma or certificate is given by the committee on studies, to each graduate, and the form of the certificate depends upon the character of the student's work, as shown from time to time in the quarterly and annual examinations.

Officers ordered as students to the Academy are entirely detached from their division (*Marine-Theile*). They are under the orders of the director and direction officer, but they are also obliged to report separately to the Commander-in-Chief, the military members of the committee on studies, and the commanding officer of the first division of seamen. The temporary students, or *Hospitanten*, report only to the director and direction officer.

The regulations governing attendance at lectures and exercises are exceedingly precise and strict, considering the age of the officer-students. Each has his place in the lecture-room, which he must always occupy. The most punctual attendance is required, and no officer can absent himself from an academic exercise, except in case of illness, leave of absence, or special dispensation; or, in the case of hospitants, details for special duty. In case of sickness the fact must be reported without delay. Applications for leave or for other dispensations are made in writing to the director, and are only granted in cases of urgent necessity.

S. Ex. 51——13

The school of engineering and pilotage is at Kiel. It has its own officers of government, but, as in the case of the Naval School, they are under the general orders of the Commander-in-Chief of the Baltic station. At the head of the school is a line-officer, as director, who is appointed by the Emperor, and who has the same authority over the personnel of the institution as the commander of a naval division. The director is therefore responsible for the military discipline of the school, and has general supervision of the studies and course of instruction. He has an asssistant, called the direction officer, acting in the capacity of a chief of staff. The latter is a line-officer ordered by the Ministry to this duty, on the recommendation of the Commander-in-Chief of the station; and besides his ordinary duties, he has charge of the director's office and the library, and he may be detailed for the supervision of practical exercises.

The instructors are either military or civil officers of the Navy, ordered on the application of the Commander-in-Chief, or special teachers, or military or civil instructors, under provisions similar to those of the Naval School. The director engages the special teachers, paying them regular fees, subject to the approval of the Admiralty. For the management and care of instruments, models, &c., a lieutenant is ordered as house-superintendent (*Haus-Verwalter*). The officers ordered as instructors perform the duty of supervision of the students, for which duties, as well as for instruction, the house superintendent may be specially detailed.

Students are quartered and messed, as far as possible, in the school itself, and in the barracks. The rest board at one of the naval messes of the station, or at private boarding-houses, but in company, as far as is practicable.

The school consists of three divisions, (1) the school of engineering, (2) the school of pilotage, and (3) the class of paymaster-applicants. The first and third are the only divisions that can be considered as forming a part of the system of education of officers. The object of the pilotage school is to prepare for the prescribed examinations such boatswain's-mates, and quartermasters as wish to become navigating boatswains. It does not therefore properly come within the scope of the present subject.

The object of the school of engineering is to give machinists (*Maschinisten*) a scientific training, and to prepare them for the examinations for promotion to mechanical engineers. It includes four classes, as follows (the lowest class being the first):

1. Machinist's-mates' class.
2. Second machinists' class.
3. First machinists' class.
4. Engineer class.

There are also several parallel courses, each of which is limited to twenty-five students.

In the machinist's-mates' class applicants for appointment as machinists (*Maschinisten-Applikanten*) are prepared for the first examination. In the second machinists' class the machinists and upper machinist's-mates are prepared for the examination of machinist of the watch. In the first machinists' class machinists of the watch are prepared for the leading machinists' examination. This course is also open to those machinists and upper machinist's-mates, provisionally examined at sea, who have shown remarkable proficiency in the machinist's-mates' examination. As a general rule, each student must attend the second machinists' class and pass the examination before he can be received into the next class. In the engineer class upper machinists, and properly qualified machinists near the head of their grade, are prepared for the engineer examination; on passing which they receive commissions in place of their warrants.

The course for all four classes generally begins on the 1st of October. At that time is held the test examination for upper machinists. There is a vacation of two weeks at Christmas, and the last fortnight in March is occupied by the examination, at the close of which the students are transferred to their respective divisions. Attendance upon the engineer and first and second machinists' classes is indispensable in order to receive permission to come up for the examinations at the close of these courses.

The dockyard divisions send up those upper machinists, machinists, upper machinist's-mates, and applicants who are qualified to attend the school, by September 1 in each year, and they are transferred at the time fixed for the beginning of the course. At its close they return again to their divisions. Those machinists or upper machinists who have not been classed as *good* at the preceding examination, are required to pass anew at the beginning of their course by presenting an essay on a prescribed subject, and performing certain other exercises. The director decides as to the fitness of the candidates, and those who fail are immediately sent back to the dockyards.

The course of instruction is arranged in accordance with a programme approved each year by the Admiralty. The only addition allowed to the prescribed course is in a general permission to give extra instruction in modern languages not specified in the programme, which are useful in maritime intercourse, and for which there may be teachers at the school.

For the examinations examining boards are convened by the Commander-in-Chief, at the request of the director. The oral examination is conducted by the board. For the written examination each member of the board prepares a paper on all the subjects; and from these the president of the board makes a selection. The instructors of the school cannot examine at all in their own branches of instruction, except informally.

The scale of marks, and method of obtaining final results, are nearly similar to those used by the board for the examination of line officers, and already described. A final mark, equivalent to 5, "satisfactory" (55 per cent. of the maximum), is required in order to pass; but if the candidate has less than 77 per cent. in the majority of subjects he is characterized as having "barely passed." There are also certain subjects in each of which 55 per cent. is required.

At the final meeting (*Schluss Konferenz*) of the board a report (*Prüfungs-Protokoll*) is drawn up, containing the names of the candidates in order of merit, and recommendations as to re-examination, &c. Certificates are granted at the same time, stating the character and aptitude for the service of the candidates and their general qualifications for promotion. In cases of re-examination the necessary preparation may be made at the school, if the director permits, and if it is attended with no inconvenience; otherwise, the unsuccessful candidate must depend on himself.

The course of instruction is chiefly devoted to steam-engineering, mathematics, and modern languages. Considerable time is given to physics and chemistry, but the course is not of a very high character. The details of instruction are as follows:

I.—MACHINIST'S-MATES' CLASS.

(20 lessons a week.)

1.—STEAM-ENGINEERING: 3 lessons.

General study of steam and its properties as far as is necessary for the management of a ship's engine, but omitting all calculations; different kinds of engines, high and low pressure, compound engines, &c., and their details, and the peculiarities of various types; different types of boilers; propellers and fitting them; boat engines and boilers, and their details; care and preservation of engines, and the prompt repair of injuries; auxiliary engines and machinery used on shipboard, including that for raising the anchor and for turning turrets, steam-steering apparatus, ventilating apparatus, fire-extinguishers, &c.; distilling apparatus, its details, use, and preservation; diving apparatus; German and foreign measures, weights, and coins.*

2.—MATHEMATICS: 4 lessons.

Arithmetic; algebra through equations of first degree with one unknown quantity; elementary plane geometry.

3.—MECHANICS: 2 lessons.

Elementary propositions relating to motion, velocity, acceleration. The combination, division, and equilibrium of forces; laws of inertia, reciprocal action, gravity, mechanical work.

---

* Classified under this head in the authorized school programme.

4.—PHYSICS: 1 lesson.

Physical properties of bodies, expansion, volume, form, impenetrability and porosity, attraction, gravity, weight, divisibility, cohesion, density, hardness, elasticity; laws of pressure of gases, and instruments relating thereto; laws of hydrostatic pressure, hydraulic press, Segner's water-wheel, turbine, &c.

5.—CHEMISTRY: 1 lesson.

Elementary principles and processes, filtration, distillation, evaporation, crystallization; reaction and reagents; chemical combinations and elements; salts, metals, and metalloids; affinity and atomic weight; oxygen, hydrogen, nitrogen, and carbon.

6.—MECHANICAL DRAWING: 5 lessons.

7.—GERMAN: 2 lessons.

Preparation of an essay free from rhetorical and orthographical errors; word formation and classification; punctuation.

8.—ENGLISH: 2 lessons.

Reading and translation of easy authors.

The examination of machinist-applicants who have completed the course is held as usual by a board, composed of the director of the school as president; a lieutenant-commander (*Kapitän-Lieutenant*) for German and English; a mechanical engineer, for mechanics and drawing; and a constructing engineer, for engineering and mathematics, physics, and chemistry.

The examination in English is oral, and is purely optional; in all the other subjects there are required examinations, both oral and written. In the first class of subjects, with a coefficient of three, are engineering, mathematics, and mechanics; in the second class, physics, chemistry, drawing, and German; and in the third class, English. In each of the first-class subjects, 55 per cent. is required in order to pass. Sixteen papers are given in all; of which one is in German; two each in mechanics, physics, and chemistry; four in mathematics; and five in steam-engineering, two of which are on weights and measures.

After passing the examination, the machinist-applicants are qualified for the grade of upper machinist's-mates, and are prepared to enter the second machinists' class.

II.—SECOND MACHINISTS' CLASS.

(22 lessons a week.)

1.—STEAM ENGINEERING: 4 lessons.

History of the development of the steam-engine, and especially of the marine engine; the origin and extension of steam navigation; distilling apparatus; diving apparatus in all its details, its preservation, use and repair; different systems of pumps for armored ships, details and use, and instructions for critical occasions; arrangement and effective use of a ship's machinery in all its parts; arrangement of different kinds of boilers; super-heaters, flues, chimneys, uptakes, propellers; dynamometer, indicator, interpretation of indicator diagrams.

2.—MATHEMATICS: 3 lessons.

Arithmetic; algebra, through equations of the first degree, with several unknown quantities; geometry and mensuration; trigonometric functions.

3.—MECHANICS: 2 lessons.

Laws of statical moments; determination of the center of gravity of lines, surfaces, and solids; Simpson's and Guldinus's rules, and their application to the computation of the volume of solids, by first determining centers of gravity; laws of friction.

4.—PHYSICS: 2 lessons.

Laws of heat; thermometer, pyrometer, diffusion of heat by radiation, transmission, absorption, and conduction; expansion from heat; determination of hygrometric state; specific heat; sources of heat; mechanical equivalent of heat; heat developed by chemical combinations; cosmic and terrestrial heat; vital heat; measurement of heat.

5.—CHEMISTRY: 2 lessons.

Properties of the metalloids; potassium, sodium, calcium; iron and its fabrication; cast and wrought iron, slag, cemented steel, Bessemer steel, cast steel.

6.—MECHANICAL DRAWING: 3 lessons.

7.—GERMAN: 2 lessons.

Preparation of essays; rhetorical rules; grammatical analysis.

8.—ENGLISH: 3 lessons.

Examination optional.

9.—OFFICIAL DUTIES: 1 lesson.

Organization of the Navy; general acquaintance with the army organization; general regulations for warrant officers (*Deck Offiziere*); instructions for machinists afloat.

At the examination the regulations as to the composition of the board, classification of subjects, and method of examination are the same as those for the previous class. Of the eighteen papers given, one is in German; two each in mechanics, physics, chemistry, and official duties; three in engineering; and six in mathematics, of which two are in arithmetic and algebra, and two each in geometry and trigonometry.

III.—FIRST MACHINISTS' CLASS.

(22 lessons a week.)

I.—STEAM ENGINEERING: 4 lessons.

The mechanical theory of heat; general theory of the steam-engine; calculations entering into actual practice, as horse-power, consumption of coal and of water for ascertained distances and given periods of time, according to different indications of the engine; principles of construction of the various systems of engines, especially their reversing gear, condensers, and pumps; different kinds of boilers, with their various stays and braces; propellers of all kinds; method of proceeding in preparing for action as well as in case of damages or injuries; proper preparation of the engine and boilers when going into commission and precautions for their preservation when going out of commission; blowing out of the boiler; boiler-record (*Kessel-protokoll*); properties of engine materials; good and bad qualities of different kinds of hard coal and the care of coal.

2.—MATHEMATICS: 4 lessons.

Algebra through quadratics; plane geometry; deduction and application of formulas of plane trigonometry; stereometry, including calculation of surfaces and contents of solid bodies.

3.—MECHANICS: 2 lessons.

Mechanical powers, lever, pulley, screw, inclined plane; principle of virtual velocities; law of the pendulum, of falling bodies; Keppler's laws; centrifugal force; *vis viva*.

4.—PHYSICS: 1 lesson.

Laws of equilibrium and motion of fluids; metacenter; specific gravity; aërometer; surface-tension; capillarity, osmose, discharge of fluids. Laws of equilibrium and motion of gases; barometer; Dalton's and Mariotte's laws; diffusion; Giffard's injector. The laws of heat in their application to meteorological phenomena; winds and currents, climatic influences, magnetism.

5.—CHEMISTRY: 1 lesson.

Mining, and the treatment of ores; properties of the metals used in engine building; principal combinations of these metals; technical processes.

6.—MECHANICAL DRAWING: 4 lessons.

Preparation of working drawings of the parts of the engine, from which repairs might be made.

7.—GERMAN: 2 lessons.

Preparation of theses; higher rhetoric.

8.—ENGLISH: 3 lessons.

Oral examination only.

9.—OFFICIAL DUTIES: 1 lesson.

Military dispatches and papers, such as the draught of reports and opinions; preparation of requests, telegrams, &c.; military courts, regulations in regard to punishment; service regulations for leading machinists; regulations for the care and disposition of stores and materials of a ship in commission.

The board of examiners is composed of the director of the school as president; a lieutenant-commander for German, English, and official duties; a constructing engineer for engineering, mechanics, and drawing; and two instructors at the Naval School, one of them for mathematics, the other for physics and chemistry. In the classification of subjects, engineering, mathematics, mechanics, and physics are in the first class; chemistry, drawing, and German in the second; and English and official duties in the third. The number of papers set is eighteen, of which one is in German; two each in mechanics, physics, chemistry, and official duties; three in engineering, two of them descriptive, and one of calculations; and six in mathematics, two in arithmetic and algebra, two in geometry, and the others in trigonometry and stereometry. In drawing the board may either pass upon the work already performed or allow a certain time in the examination for a new drawing.

IV.—ENGINEER CLASS.

(14 lessons a week.)

1.—STEAM-ENGINEERING: 2 lessons.

Advanced course in the theory of heat, with special reference to the properties of steam; principles governing the action of condensers; complete course in calculations of work, &c., performed by an engine; theory of Zeuner's diagrams for construction and control of the valve-gear; contrivances for expansion, and their governing principles; economical engines and their use.

2.—MATHEMATICS: 2 lessons.

Arithmetic; algebra, to cubic equations; review of the whole of plane geometry and of plane trigonometry; mensuration of surfaces and solids.

3.—MECHANICS: 2 lessons.

Short review of statics and kinematics; laws of equilibrium; theory of machines.

•4.—PHYSICS: 1 lesson.

Wave motion; acoustics, optics, electricity, and magnetism.

5.—CHEMISTRY: 1 lesson.

Preliminary study of organic chemistry; atomic and molecular weight; chemical formulas; density of steam; the heaviest alcohols, ether, sebaceous acids, fats, glycerine; hydrocarbons (petroleum, paraffine, &c., sugar, starch, cellulose); gunpowder and other explosives.

6.—GERMAN: 1 lesson.

Essays and reports.

7.—ENGLISH: 2 lessons.

Reading and translation of technical works; practice in writing and in translation from German into English; practice in conversation.

8.—FRENCH: 2 lessons.

Grammar; translation of easy French works; translation into French; conversation.

9.—OFFICIAL DUTIES: 1 lesson.

Naval and military organization; dispatches and reports; official relations between executive officers and engineers (*Dienstverhältnisse der Offiziere und Maschinen-Ingenieure*).

Of the above subjects, engineering only belongs to the first class, German and English to the second, and French to the third. There is no examination in mathematics, chemistry, or official duties; and the remaining branches, mechanics and physics, are joined in the examination with engineering. In engineering and German 55 per cent. is required in each subject in order to pass.

The object of the examination at the close of the course of the engineer class is to test the proficiency of candidates for promotion to the grade of assistant mechanical engineer (*Maschinen-unter-Ingenieur*), the lowest commissioned rank in this branch of the service. The board is composed of the director of the school, the director of engine construction at the dockyard, or his deputy, as examiner in engineering, and the instructor in languages at the Naval Academy as examiner in German, English, and French.

The Commander-in-Chief of the station is required to attend the oral examinations and to sign the report embodying the result. In the examination in engineering three extensive papers are given, two of which are descriptive in their character, and the third dealing exclusively with calculations. The oral examination is chiefly upon the construction and use of economical engines. In German an essay is prepared. In English and French the examination is both oral and written, and calls for a high standard of proficiency in reading, translation, and conversation, especially in English.

PAYMASTER-APPLICANTS' CLASS.

The object of this class or school is to give to applicants for appointment as naval paymasters, the necessary preliminary training for their profession. For convenience of organization it is made a part of the machinists' and pilotage school, and it has the same direction and government. The instructors are naval officers, intendants, and paymasters. They are detailed for this service, in addition to their other duties, by the Commander-in-Chief at Kiel, on recommendation of the director of the school; and they receive extra pay according to the work they perform.

The paymaster-applicants are ordered to attendance at the school by the Commander-in-Chief at Kiel. If they belong to the division of the North Sea, their orders are issued upon an understanding between the commanding officers at Kiel and Wilhelmshafen. They may perform regular duty at Kiel while pursuing their studies in the school, and they have accordingly to maintain official relations not only with the school director, but also with the commander of the naval battalion, and the intendant or the commander of the naval division, according to their assignment.

The course lasts from October 1 to March 31, and the studies are so arranged as not to interfere with the regular duties of the students. The studies comprised in the course are as follows:

1. Laws of exchange and commercial law: 2 hours a week.
2. Currencies, merchandise, and mercantile affairs: 2 hours.
3. Official duties: 1 hour.
4. English: 4 hours.
5. French: 2 hours.
6. Other modern languages or stenography: 1 hour.

All the duties are obligatory except the last, which is purely optional; but the programme is subject to such minor modifications as the director and instructors may judge necessary.

At the close of the course, examinations, both oral and written, are held in all the subjects. Certificates of graduation are given to the students, signed by the director and all the instructors, stating the acquirements, conduct, and application of the students. In case of failure, the student may be recommended for a second trial of the course; and the final decision as to this recommendation rests with the Commander-in-Chief of the station.

PART IV.

ITALY.

CHAPTER XXX.

THE NAVAL SCHOOL (*Regia scuola di marina*).

The Royal Naval School of Italy is composed of two divisions, the first at Naples and the second at Genoa. The course lasts four years, of which the first two are passed in the first division, and the last two in the second. The Naples school may therefore be considered as preparatory to that of Genoa.

The existence of two separate schools is explained by the fact that upon the formation of the Kingdom of Italy, in 1861, the government found itself in possession of two educational establishments, one of which had formerly belonged to the Sardinian Government, the other to the Kingdom of the Two Sicilies. Neither was a school of a very high order, but it was thought best to build upon the foundations already existing rather than to attempt immediately the formation of a new establishment. The schools, therefore, continue to occupy their antiquated and ill-adapted buildings—at Naples, the old palace of the consulate, and at Genoa, the former convent of Santa Teresa—but their courses have been so arranged that they form respectively the preliminary and advanced stages of a single system of education.

This arrangement is, of course, attended with inconveniences, and is particularly objectionable on the score of expense, as it nearly doubles the necessary force of officers, instructors, and employés. Attempts have accordingly been made from time to time to unite the schools, or to establish a new school, with an academy for advanced instruction, at Spezia, Leghorn, or elsewhere. When such a plan is adopted it will doubtless lead to essential changes in the details of the present system, but as yet none of the attempts at reorganization have succeeded.

The examination for admission takes place on the 15th of June of each year, before a commission appointed by the Minister of Marine. Applications from those who desire to become candidates are sent some time previously to the commandant of the school at Naples. Candidates must be natives of Italy, and they must be not less than thirteen nor more than seventeen years of age at the date of admission. The physical qualifications are similar to those of other countries, and are ascertained by the usual medical examination.

The mental examination is both written and oral, the latter being conducted in public. The subjects of the examination are arithmetic, algebra, geometry, history, geography, Italian, and French. The examination in arithmetic embraces the whole subject, including common and decimal fractions, interest, proportion, and roots. In algebra it extends through quadratic equations, summation of series, exponential equations, and logarithms. In geometry it includes the whole of plane geometry, and measurement of superficial areas and volumes of sections of the sphere, cone, &c. History is confined to sacred history and the history of Greece and Rome. In Italian, candidates must write an essay and answer questions in grammar. In French, the examination includes reading, translation, and grammar.

The examination is divided into two parts, the scientific and literary; the first includes arithmetic, algebra, and geometry ; the second, literature, history, geography, French, and handwriting. The work is marked on a scale of 10, and the mark in each branch is multiplied by the coefficient of the branch, the sum of the products giving the final result. In each of the scientific subjects the coefficient is 3, in handwriting 1, and in all the others 2. In order to pass, candidates are required to obtain 60 per cent. in each of the mathematical subjects and in Italian, and 60 per cent. of the maximum aggregate in the remainder. If the number of successful candidates is greater than the number of vacancies, the examination becomes competitive in its character, as only the highest on the list are accepted. But candidates who are rejected either from the want of vacancies or from failure to pass, are allowed to compete the next year if they are still within the prescribed limits of age.

Students at the naval school are required to pay the government a fee of 900 lire, or $180, per annum, payable quarterly in advance. A certain sum, however, is annually appropriated by the Ministry for scholarships and half scholarships, equal in amount to the required fees. These are given each year to the student who passed highest in his class at the examination of the preceding year, counting the examination for admission as the first. The full scholarship is only given to : sons of naval officers or of civil functionaries connected with the naval administration. The outfit required by regulation is provided at the student's expense and is not covered by the fees; students are also required to provide their own text-books and instruments.

There is an annual practice-cruise, beginning always on the 15th of July and ending about the 1st of November. It is followed immediately by the session of eight months, lasting until the 1st of July. The first fortnight in July is always devoted to the annual examination. The first practice-cruise follows immediately the examination for admission, and therefore precedes the first session of theoretical instruction. In this way the first three years are passed. The fourth year, called the complementary course, is divided into two parts : the first, of eight months, from November to June 20, is the final course of the school at Genoa; the second, of six months, is spent on board a practice-ship, but includes rather more theoretical instruction than is generally the case in practice-cruises.

The *personnel* of the schools is divided into the corps of administration and the corps of instruction. The corps of administration includes at each school (Naples and Genoa) the commandant, two lieutenants as inspecting officers, a chaplain, a paymaster or storekeeper, and four naval officers to assist in executive duties (*aiutante*). The corps of instruction consists at Naples of a director of studies, eleven civil professors, a mechanical engineer, a writing-master and three tutors; at Genoa, of a director of studies, nine civil professors, five officers detailed for duties of instruction, and three tutors.

The professorships are distributed as follows:

At Naples :
> Algebra, trigonometry, navigation.
> Analytical and descriptive geometry.
> Calculus.
> Physics and chemistry.
> Italian literature (2).
> Geography.
> French (2).
> English.
> Drawing.

At Genoa:
> Pure mechanics.
> Applied mechanics and theory of ship construction.
> Nautical astronomy.
> Nautical surveying.
> Italian literature.
> History.
> Geography (political).
> French.
> English.
> Naval construction.
> Fortification and military art.
> Gunnery. } Officers detailed for this duty.
> Hydrographic drawing.
> Naval tactics.

There are also at each school two sword-masters, a dancing-master, a petty officer to give instruction in knotting and splicing, and instructors in great-gun drill, in small-arm drill, and in gymnastics. The total number of persons directly engaged in instruction and government is about sixty. All are approved by the Ministry except the tutors and subordinate masters, who are named by the commandant and approved by the Commander-in-Chief of the department. The command of the practice-ship is usually given to the commandant of one school or the other. In addition to the regular officers of the ship, an instructor in navigation is regularly detailed by the Commander-in-Chief of the department. One of the officers of the ship is the instructor in gunnery.

At each school there are two boards of government—the council of administration and the superior council—corresponding to the two divisions of the staff of the establishment. Each of these is presided over by the commandant of the school. The council of administration is composed of the commandant, the senior inspecting officer, the chaplain, the senior executive adjutant, and the paymaster. It is charged with all questions of administration and prepares the estimates; it audits accounts, and is responsible for all receipts and disbursements. The accounting officer, though a member of the council, has only a consulting voice; he also performs the duty of secretary.

The council reserves 200 lire a year from the fees of each student, which are credited to him on the books, and expended for the repair and replenishment of his wardrobe. All this is done without the intervention of the student concerned. The cast-off clothing and effects of students are sold and the proceeds turned into the treasury of the school, to be disposed of as the council sees fit. Any balance of the fees reserved remaining to the credit of a student is paid over to him at the close of his course.

The superior council is composed of the commandant, the director of studies, the senior professor, the professor of history or of literature, and one other professor, the last being selected by the commandant. The superior council is authorized to direct and carry out the instruction in conformity with the general outline prescribed by the Ministry; to select text-books; to name the members of the staff of professors; to conduct the annual examinations; to report upon the result of the examinations; to award prizes, and to make an annual report in detail upon the progress of the school. It has also authority to make recommendations to the ministry in regard to the following subjects: Changes in the course, charges for instruction, cases of negligence or misconduct, discharge of undeserving students. The council has meetings regularly once a quarter, and at other times as the commandant may direct or the members desire. The essential feature of the system is the entire separation of the details of instruction, (*i. e.*, theoretical instruction), and administration, except, of course, as they are united in the person of the commandant. With this exception, none of the officers in the administrative branch have anything to do with instruction, nor have the instructors any part in the military administration.

The commandant of the school is responsible to the Commander-in-Chief of the department, and holds to that officer much the same relation that the captain of the Borda holds to the *Préfet Maritime* at Brest. He makes a formal inspection of the school every week, and reports the result of his observations in person or in writing to the Commander-in-Chief.

Applications for officers to be ordered to duty at the school are made by name by the commandant to the Commander-in-Chief, who forwards them to the Ministry with his indorsement. Lieutenants are thus ordered as inspecting officers, and junior officers as executive officers (*aiutante*), for the daily routine of the school.

The two inspecting officers remain at the school one year, and may be renominated if applied for; their duty is to superintend the inner work-ing of the school, but more particularly to watch over the discipline and conduct of the cadets and to supervise their practical and military in-struction; they are on duty during alternate weeks. During his tour of duty, the inspecting officer receives every morning the reports of the professors and of the janitor, and makes a *résumé* of whatever has hap-pened, which he presents to the commandant; he is present at prayers, at the inspection of rooms, at meals, and at all exercises, including swimming; and he may inflict punishments, subject to the approval of the commandant.

Subject to the authority of the superior council, of which he is a mem-ber, the director of studies has oversight of everything relating to the theoretical instruction of the cadets and carries out the regulations of the superior council; and the orders of the commandant are transmitted by him to the professors.

A few days before the opening of the school he receives from the pro-fessors a programme of the studies intrusted to each one, with modifi-cations suggested by the previous year's course of study, and with a pro-posed list of text-books. These propositions are discussed separately and collectively in the superior council, to which the director of studies makes a report on the programme. He is also charged with the exe-cution of the details of the course of instruction, the supervision of the methods pursued by instructors, and the progress made by the cadets. He receives from the instructors monthly reports, which form in part the basis of the monthly reports on the school; and he prepares the materials for the annual report, transmitted by the superior council, on the condition and working of the school.

The professors and tutors are answerable to the commandant, but receive orders relative to discipline from the inspecting officers and those relating to instruction from the director of studies.

Monthly marks are given by the professors for the work done by the students as shown in recitations and exercises; these marks, like all others, are on a scale of 10. Monthly reports are sent to the inspect-ing officer, containing the marks, together with statements of conduct (in recitation-rooms), diligence, and progress of the students. A gen-eral merit-roll is made up from the marks given, multiplied by the pre-scribed coefficients, and this roll is published each month at the school. A special report of the conduct, progress, and diligence of each student is sent to his parents.

The following list shows the arrangement of the studies in the course of theoretical instruction, with the coefficient of each branch:

First year at Naples:

	Coefficient.
Algebra	3
Plane and spherical trigonometry; navigation	3
Analytical geometry; descriptive geometry	3

Coefficient.

Italian literature... 2
Geography... 2
French.. 2
Drawing... 1

Second year at Naples:

Calculus.. 3
Experimental physics; elementary chemistry... 3
Political geography... 2
Italian literature.. 2
French.. 2
English... 2
Drawing... 1

Third year at Genoa:

Elementary mechanics.. 3
Nautical astronomy.. 3
Nautical surveying.. 3
History... 2
Political geography... 2
Italian literature.. 2
French language and literature.. 2
English... 2

Complementary course at Genoa:

Applied mathematics; theory of ship construction.................................... 3
Naval construction.. 2
Naval tactics... 2
Fortification and military art.. 2
Gunnery... 2
Modern history.. 2
Italian literature.. 2
English... 2

Complementary course; final practice cruise:

Naval maneuvers... 2
Steam-engineering... 1
Practical navigation.. 2
Description and use of naval ordnance... 1
Regulations for service on shipboard.. 1
Maritime laws and responsibilities.. 1
Hydrographic surveys.. 1

During all the school sessions practical instruction is given every
week in the following branches, alternate days being taken:

Rigging ship, exercise with sails and spars... 2
Ship-building in the dockyard... 1
Great guns.. 1
Small-arms.. 1
Fencing, gymnastics, swimming.

All but the last are subjects of the annual examination. First year
students have also exercises in penmanship. In addition to the above
exercises, which take place on week days, students have regularly on
Sundays lessons in fencing and dancing, and occasional exercises in
great guns, small-arms, and gymnastics.

In the two upper classes a weekly lecture of a familiar character is given by the professor of history on subjects connected with social and political matters, and the relations of officers to society and to the State. These lectures are largely attended by the officers of the station. *

Each school is provided with a library, physical and chemical laboratories, an observatory, and a collection of astronomical and geodetical instruments. The library is used by students as well as officers, and is under the direction of a professor. The professional apparatus appertaining directly to the schools is not so extensive as the necessities of the case would seem to warrant, though the immediate neighborhood of the dockyards supplies these deficiencies to a limited extent. A mast and bowsprit, fully rigged and sparred, are set up in one of the court-yards, and are used for exercises in seamanship.

The course outlined above requires some further observations. As usual, a large part of the time at disposal is given to mathematics. The examination for admission is of a sufficiently high character to do away with the necessity of much elementary work. Accordingly, the first year's course in algebra is extensive, going thoroughly into the theory of equations. The students in this year also go through plane and spherical trigonometry and a simple course in plane sailing.. The analytical geometry, plane and solid, is a reasonably high course, while that in descriptive geometry is comparatively simple. In the second year, mathematical instruction is confined to the differential and integral calculus. The course is extensive, going largely into differential equations, but it is too exclusively theoretical, and contains much that is of little practical value. Judged by the standards of instruction of the present day, it would be considered somewhat antiquated. The third year contains a full course in analytical mechanics, including statics, dynamics, and kinematics. This year also includes the whole of navigation and nautical astronomy, and a complete course in hydrography and topography. The course in applied mathematics of the fourth year consists of an exhaustive treatment of what is generally called naval architecture.

The course in physics and chemistry of the second year calls for no special comment; the former includes heat, light, sound, electricity, and magnetism; the latter is a simple course of a descriptive character in inorganic chemistry.

The important professional subjects are chiefly reserved for the fourth year; these include, in addition to the mathematical treatment of naval architecture, already referred to, the subjects of ship-building; naval tactics under sail and steam; fortification and the military art, including field and permanent fortification, organization of the Italian and foreign armies, and military tactics; gunnery, including guns of all modern systems, projectiles, carriages, powder, fabrication of guns and ammunition, and torpedoes.

Under the name of Italian literature several subjects are included. The course for the first year consists of rhetoric and exercises in various kinds of composition. In the second year the rhetoric and composition are continued with the addition of a course in general literary criticism, treating of different branches of literature, and including history, travels, fiction, and poetry. The third year's course takes up in detail the leading Italian authors from the early development of the language to the present century—from Cavalcanti to Manzoni and d'Azeglio—and includes a special study of Dante. In the fourth year the subject proper of Italian literature is dropped, and international law takes its place. This last subject is treated in its most modern aspect, in the light of the most recent cases. In this last year there is also some practice in writing compositions on literary subjects.

Instruction in French is given during the first three years, in English during the last three; in both, instruction is thorough as far as it goes, though not so successful as in Germany. It includes grammar, writing letters and exercises, conversation, and translating from and into Italian. In English one of the text-books is Irving's Life of Columbus.

The course in history which has a place in the two final years is very thorough. It begins with mediæval history, extending to the crusades. In the last year the whole of European history, especially Italian, is gone over from the thirteenth century down to the present time. The later courses in geography consist largely of historical study, and form the complement of the historical course. Under the name of political geography the course includes the study of comparative politics, and the territorial and diplomatic relations of states. The first year's course in geography is merely descriptive.

The theoretical course of instruction receives its practical application in the annual practice-cruises. These cruises are five in number, the first four of three and a half months each, and the last of six months, making in all nearly two years of training on shipboard. The training is not confined to practical exercises, though the latter occupy the important place. Regular instruction is given in navigation, gunnery, and seamanship during all the cruises; in naval tactics during the last three; in ship-building during the last two; and in official regulations and steam-engineering in the complementary course. All this represents an essential and very important part of the instruction given in the Italian scheme of education. Its most marked defect is the postponement of all instruction in steam-engineering to the close of the course, and the inadequate provision of time for carrying it on by relegating it to a part of the final practice-cruise.

Examinations are held every year at the end of the academic session. The board of examiners at Naples consists of professors designated by the superior council. At Genoa, the board is a mixed commission, composed of the following persons:

The commandant as president.

One superior officer, junior to the commandant, and one lieutenant; both designated by the commandant.

One professor, charged with instruction in the branch which is the subject of examination.

One other professor, named by the superior council.

The boards receive from the commandants of the two schools a report on the ability, application, and conduct of the students during the past year, which is read aloud as each candidate comes up. The examination is chiefly oral, and may therefore be materially different for different students.

Two subjects selected from the programme in each branch of study are given to each student, and he is allowed to take his choice. Any one of the examiners may add such questions as he sees fit on any part of the established programme. After the examination of each candidate, the board decides by ballot whether he can pass or not, and he is then marked; 60 per cent. being fixed as the passing standard for the general result.

The order of the subjects for examination is, first, scientific or professional subjects; second, literary; and, third, practical subjects. If the candidate gets below 60 per cent. in any one of the scientific subjects, or if his average in literary subjects is below 60 per cent., his examination is closed and he has failed to pass. If he receives less than 60 per cent. as his average in the practical subjects, he fails likewise. As in the German system, the only coefficients for different subjects are 3, 2, and 1. In general the scientific subjects have the largest coefficient, the professional and literary the next, while the smallest coefficient is given to drawing and certain practical exercises; but, in any case, the differences in relative weight are not excessive.

Deficient students may be turned back once, but never twice. A second failure inevitably results in dismissal. Bad conduct is also a cause of dismissal. If a student develops, during his stay at the school, some physical defect which unfits him for active service, but otherwise makes satisfactory progress, he may at the end of the third year **be appointed** to some other branch of the naval service.

On passing the final examination at the close of the complementary course, students are admitted to the Navy with the rank of midshipman (*guardia marina*). Further examinations are held for promotion from the grade of midshipman to that of sub-lieutenant, and from sub-lieutenant to lieutenant. Candidates for promotion to sub-lieutenant (*sotto tenente di vascello*) are examined in practical navigation, seamanship, naval tactics, official duties, including regulations, correspondence, reports, charge of men, &c., ordnance and gunnery, and steam-engineering; the latter only a general examination. The last examination, for promotion to lieutenant (*luogo tenente di vascello*), includes seamanship, naval tactics, steam-engineering, ship-building, ordnance, and gunnery. These examinations are progressive in character; the last being exceedingly searching in its requirements of the prescribed subjects.

Students of high merit at the schools are selected as chiefs of divisions (*contro-maestri*) and chiefs of classes (*quartier-maestri*), each class being composed of students of a given year. There are, therefore, in all, two chiefs of divisions and five chiefs of classes. They have some slight privileges and wear distinctive badges. At all formations, by class or division, they give orders, call the roll, and report absentees to the adjutant on duty. They have military authority over their respective divisions and classes, and they are responsible for order at studies, exercises, and recreation. Every morning they make a report of offenses coming under their cognizance during the last twenty-four hours. The senior chief of division, in the absence of superior officers, has authority to order the arrest of a student; chiefs of division are in charge at . breakfast and dinner, superintend the distribution of provisions, and instruct the steward as to withholding certain dishes from students under punishment.

There are play-rooms and a billiard hall, the last only for the upper division. The students are divided into five or six parties, and use the billiard hall on alternate evenings.

Students are allowed to receive visitors only on certain days of the week in the reception-room. The only visitors they are allowed to see are their parents, or near relatives, or a designated agent of the parents. Permission to leave the school inclosure is only granted to the higher classes every Sunday, and to the three lower classes on alternate Sundays. The leave extends from 1.30 to 8 p. m. in the fall and winter, and to 9 p. m. in the spring and summer. There are seven holidays during the year, chiefly religious festivals, upon which days the leave begins at 10 a. m. This permission is only given for the purpose of visiting the parents or guardians of the students; if they do not reside in the city, they designate a suitable person, who must be the head of a family, to take their place. In either case, the person whom the student is authorized to visit must accompany him to and from the school; the students are under no circumstances allowed to go alone through the streets. If parents living at a distance come to town for a short time, in which no liberty-day happens to fall, their son may take one day's liberty during their visit, but he makes up for it at the next regular holiday. Leave of absence to visit their families is given to meritorious students during the short period intervening between the annual examination and the practice-cruise.

The restraints imposed in general by discipline are more severe at the Italian Naval School than at similar institutions in most other states. As in France, the students have no standing as officers; they are designated simply pupils (*allievi*); and the regulations for their government are based on an extreme form of this theory. This must be kept in mind, as the only explanation of a system which subjects a body of lads, whose average age at the start is fifteen or sixteen years, not only to extreme care and minute attention, but to discipline and restrictions that seem only suitable for young children. The scale of punishments, however, does not wholly bear out this view; it ranges from the extreme of pettiness to the extreme of severity; from deprivation of dessert at dinner to thirty days' solitary confinement on bread and water. In regard to the latter penalty, it is to be presumed that it is rarely inflicted; but even so, a young man whose offenses were such as to warrant this punishment could hardly be fit to continue at the school or in the service.

It is probable that in the near future considerable modifications may be made in the system of naval education in Italy. The national navy is yet in its infancy, having been less than twenty years in existence. During this time it has made immense progress, and there are growing signs of a spirit of reform, which must soon be felt in naval education.*

---

* Since the above was written, information has been received of an important change in the organization of the Naval School by the addition of a fifth year to the course. As it now stands, the first three years of the course are passed at Naples, and the last two at Genoa. The number of pupils is as follows:

Naples:	Pupils.
First year	38
Second year	16
Third year	11
Genoa:	
Fourth year	22
Fifth year	28

A parliamentary commission has also recently made a report, recommending the union of the schools, and the establishment of the new school at Leghorn, the selection being partly based on the fact that Leghorn is not one of the great naval ports.

CHAPTER XXXI.

THE GUNNERY SCHOOL (*Regia nave-scuola di artigleria navale*).

The gunnery school was established by a royal decree of April 2, 1873. It is placed on board the screw frigate Maria Adelaide, permanently stationed at Spezia. Attached to the school are a gun-boat for target practice in motion, and a small steamer for auxiliary purposes. The establishment is under the immediate control of the Ministry of Marine in all matters relating to instruction, exercises, and the special gunnery service; in other respects it is subject to the authority of the Commander-in-Chief of the first maritime department.

The *personnel* of the school includes the following officers, instructors, students, &c.:

Captain	1
Commander	1
Lieutenants	6
Sub-lieutenants	8
Midshipmen (the number regulated each year by the Ministry).	
Commissaries and surgeons.	
Seamen (12 as servants)	60
Seamen as gunnery pupils	360
Petty officers, pilotage class	11
Gunners	13
Gunnery corporals	24
Seamen gunners	28
Machinists	3
Stokers	12
Clerks	10
Artificers	12
Draughtsmen as required from time to time.	

The captain and commander of the ship, with one of the lieutenants as recorder, form a board (*consiglio*), which regulates the course, and introduces all the necessary modifications.

The ordinary executive duty of the ship is performed by the senior lieutenant, acting under the commander, while the latter gives his personal attention to the school instruction and to the school exercises. The lieutenant who acts as recorder of the board has special charge of the instruction of the officers' class. The work of the other lieutenants is divided between routine duties and duties of instruction, in the first as officers of the watch, and in the second as officers of divisions. The whole *personnel* of the ship, exclusive of commissioned officers, is arranged in six divisions, with a lieutenant in charge of each. Of the eight sub-lieutenants, four assist in instruction and four in routine duties.

The gunners and gunnery corporals are selected as the best in their grade in the final course of each year; they act as instructors and sub-instructors of the class of seamen. The seamen gunners are also picked men, the best in the ordinary course, and during their stay at the school they act as assistant instructors. All of these commissioned, warrant, and petty officers who serve as instructors at the school have special advantages in the way of promotion and of detail for duty.

Two courses of instruction are carried out on board the Maria Adelaide; the ordinary course, and the final course of application. The first is for the purpose of obtaining a certain number of men eligible as seamen gunners. The second course is to perfect the practical training in naval gunnery of officers, midshipmen, and seamen gunners, not only to enable them to pass the required examination for promotion, but to fit them for duty as instructors at the school itself.

I.—THE ORDINARY COURSE.

Men pursuing the ordinary course are known as gunnery pupils. They are selected from among the seamen of the three divisions of the fleet, those being preferred who apply voluntarily, and who can read and write. They must be active men, of quick intelligence, and robust constitution, with unimpaired eyesight, and not less than 4 feet 9 inches in height. They are also required to have completed three years of active service. If any men admitted as gunnery pupils appear to the captain of the ship unfit for gunners, he has authority to send them back to their divisions; with this view, 10 per cent. more than the required number are attached to the ship at the beginning of the course.

The course of instruction lasts eight months, and is both practical and theoretical. The first embraces those parts of the 1st and 2d volumes of the "Military Instructions for the Royal Navy" included in their programme, and also target practice with great guns, rifles, and revolvers. The second embraces the programme contained in the 3d volume of the Military Instructions, and is limited to general descriptions and information.

The course comprises three periods, as follows:

FIRST PERIOD.

Practice.—School of the soldier without arms; manual of the rifle, loading and firing; school of the company and of the battalion; great-gun drill; sabre exercise.

Theory.—Nomenclature and general descriptions of guns, carriages, equipments, and small-arms.

SECOND PERIOD.

Practice.—Exercise at will with great guns, sponging and loading; school of the battery; target practice at anchor; school of the platoon and company; skirmish drills; small-arm target practice.

Theory.—General observations in regard to drills; description and use of projectiles, charges, and fuses; pointing with director.

THIRD PERIOD.

Practice.—General exercise of combat; target firing in motion; school of the magazine, of the piece, division, and battery; boat guns and landing parties; target practice with revolvers; general review.

Theory.—Pyrotechny, and the service construction, &c., of magazines; deviations in the flight of projectiles; boat-gun carriages; general review.

Practical and theoretical instruction is given by divisions. One division of the six is always engaged in duties connected with the ship's routine, during which the guard receives instruction in the duties of sentinels from the sub-lieutenant of the division. The other five divisions have instruction for 5½ hours daily, except on holidays, and on Thursday, which is occupied in cleaning ship. In the afternoon hour in summer the divisions are exercised in rowing and swimming.

Each series of students at the beginning of the second period must make five hits with the following pieces: 3 with 6.25-inch rifle and 2 with 32-pounder smooth-bore. In the school of the platoon, each man must make five bull's-eyes with the rifle.

Table of firings.

Projectiles.	Period.	Manner.	Number of shots to each man.	Weapon.
Empty shell	2	At anchor	1 to every 8 men	10-inch M. L. R.
Do	3	Ship in motion	1 to every 4 men	8-inch M. L. R.
Do	2	At anchor	2	6.25-inch R.
Solid shot	2	do	2	32-pounder S. B.
4 empty shells; 1 loaded shell; 1 grape.	3	Gunboat in motion	6	4-inch R. or 32-pdr. S. B.
Empty shell	3	One from launch, 2 from shore.	3	3-inch boat-gun.
	2		5 with arms extended; 5 with a rest; 5 kneeling; 5 lying down.	Rifle.
	3		6	Revolver
	3		1	Rocket.

Prizes are given for excellence in firing.

The commandant makes weekly reports of the condition of the school to the Ministry. At the end of each period, lieutenants in command of divisions make reports of the progress of each man in their divisions, giving him separate marks on his practical and theoretical work and a final mark based on the separate marks; these marks, like those at the Naval School, are on a scale of 10. At the close of the course, lieutenants of divisions send in a final report, taking the arithmetical mean of the marks for the three periods. Men receiving a final mark between 8 and 10 are said to be approved for the first class (*approvati per la 1ª classa*); those whose final mark is between 6 and 8 are approved for the second class. Men whose final mark is below 6 are considered not qualified for advancement, and are required to go over the course anew.

Successful pupils receive a certificate of fitness according to their class and are returned to their respective divisions, when they are immediately appointed seamen-gunners. About 300 a year are sent out in this way into the service; the 30 best men are retained at the school and put through the final course. In case of bad conduct, the graduates are returned in the same way to their divisions, but their advancement is withheld until they have received a good-conduct certificate from a commanding officer.

II.—THE FINAL COURSE OF APPLICATION.

Non-commissioned officers.—The following non-commissioned officers form the class in the final course.

I. The 30 gunnery pupils selected as the best at the close of the final course.

II. All the seamen-gunners not employed in the different divisions.

. III. The gunnery corporals sent up from the divisions who have not yet passed the final course in their grade; the number being fixed each year by the Ministry of Marine.

No one can be admitted to this course who has less than a year to complete of his term of enlistment, unless he has engaged to renew it; nor can any subordinate officers be admitted who have already passed through the course in their grade and have passed their examination for promotion. The course lasts three months and follows immediately the close of the ordinary course.

The seamen-gunners complete their practical instruction, and pursue a theoretical course to fit them for examination for the grade of gunnery corporal. The gunnery corporals are practiced in the use of guns and small-arms, in the school of command, and are fitted by their theoretical course for the examinations for promotion.

The captain assigns one or more instructors to duty with each class according to the necessities of the case, and two sword masters are added to the staff of the school during this period for instruction in fencing.

Table of firings performed by all students in the final course.

Projectiles.	Manner.	Number of shots per man.	Weapon.
Empty shell	In motion	1 to every 5 men	10-in. and 8-in. M. L. R.
Do	At anchor	1	6.25-in. R.
Do	In motion	1	4-in. R. or 32-pdr. S. B.
		5	Rifle.
		6	Revolver.
		1	Rocket.

At the close of the final course all those who apply are examined for ● promotion to the higher grades. The examination is conducted by the board of instruction, which is augmented for the purpose by the officers who have taken part in the instruction. Certificates of fitness are given to the successful candidates, and from them the captain selects instructors for the next ordinary course. The rest are sent back to their divisions.

Commissioned officers.—Sub-lieutenants and midshipmen are sent in turn to pursue the final course at the gunnery school. The course of instruction is similar to the final course for non-commissioned officers in its principal details, notably in the firings. The course in naval ordnance is conducted by the lieutenant designated for the purpose (the recorder of the board of instruction), and it deals chiefly with the recent improvements in the science. The students attend the experiments taking place in the polygon on shore, and they take part in the practice-drills, and are specially exercised in the school of command. A course in fortification and military art is carried on by a lieutenant or by an officer of engineers.

At the close of the course, detailed reports of the progress of each student are made by the instructors to the captain, and by the captain to the Minister of Marine. The general system is largely founded on that of the Excellent.

CHAPTER XXXII.

THE TORPEDO SCHOOL.

The torpedo school, in its present form, is of very recent date. At first the study and working of torpedoes, as in England, were confined to the gunnery officers; but in 1874 a separate torpedo corps was established, with grades of petty officers corresponding to those of the gunnery corps, as follows:

GUNNERY CORPS.	TORPEDO CORPS.
Chief gunner.	
Second chief gunner.	Second chief torpedo officer.
Gunnery corporal.	Torpedo corporal.
Seaman gunner, first class.	Torpedo seaman, first class.
Seaman gunner, second class.	Torpedo seaman. second class.

The number of petty officers in the torpedo corps is limited to 60 in the highest grade, 100 in the second, and 180 in each of the two lower grades. There is no grade in this corps corresponding to that of chief gunner, but the highest grade of torpedo officers may compete for promotion to the grade of chief gunner; the number selected from the two corps being in the same proportion as the number of individuals in the second grade. To obtain the promotion, however, torpedo officers must pass through the final course at the gunnery school, and obtain the certificate of fitness.

The torpedo school is established on board the screw corvette Caracciolo, and its object is to fit men for the various grades of the torpedo corps, and also to give officers such training as may be necessary. The organization is similar to that of the gunnery school-ship. The officers are a captain, 4 lieutenants, 4 sub-lieutenants, 4 midshipmen, and the usual staff officers. A large number of skilled artificers is attached to the ship in addition to its ordinary complement of machinists, firemen, and seamen.

As in the gunnery school, there are two distinct courses—the ordinary course and the final course. The first prepares seamen for the lowest grade of torpedo seamen. During the course they are known as torpedo pupils (*allievi torpedinieri*). The final course is a shorter and higher course for junior officers and for the higher grades of the torpedo corps and also to give preparation for the examinations for promotion. The torpedo pupils are taken from the school for apprentices (*mozzi*), from enlisted men, and from men obtained by the maritime inscription. The regulations governing reports and certificates are similar to those in the gunnery school.

221

CHAPTER XXXIII.

SCHOOL FOR ENGINEER MECHANICIANS (*Scuola degli allievi macchinisti.*)

This school was established in 1862 at Genoa. In technical matters and matters pertaining to instruction, it is under the control of the director of naval construction, while matters of discipline are regulated by the commander of the division of seamen of the first maritime department. The immediate head of the establishment is a director, selected by the Ministry of Marine either from the corps of executive officers, of constructing engineers, or of mechanicians. The duty of the director is to carry out the established programme, to maintain discipline, to assign the students to the workshops, and to call meetings of the examining boards and of professors.

Candidates for admission to the school must be Italian subjects, of not less than 14 nor more than 17 years of age; they must pass the usual medical examination, and a mental examination in reading, writing, composition, and arithmetic. They must also have served an apprenticeship to a founder, boiler-maker, metal-worker, or machinist, and must give evidence of their skill by work done at the naval arsenal in presence of the examining board.

The examination for admission is held annually, and begins on the 15th of September; it is held successively at Genoa, Naples, and Venice, before a board composed of an engineer officer of high rank as president, a lieutenant, a chief mechanician, and two professors of the school. It is a test-examination, but it may become competitive when the number of those who pass the required standard is greater than the number of vacancies. Of candidates having the same mark, preference is given (1) to the orphan sons of persons in the military service; (2) to orphans generally; (3) to sons of persons in military life not orphans.

The school session opens on the 15th of October. The successful candidates must report within fifteen days from the time of notification, otherwise they are considered to have forfeited their right to enter, unless they can show that it was impossible for them to join before. At their admission, students are required to engage that they will remain in the service until the age of twenty-nine years; but their parents may withdraw them at any time before they have reached the age of 18 by indemnifying the government for all expenses incurred on their account.

222

Students who fail at the annual examination are allowed a second trial, but never a third. In case of a second failure, they are finally dismissed. Students whose conduct is habitually bad, and who do not reform under punishment, are expelled from the corps of mechanicians, and placed in that of stokers or coal-heavers, where they are obliged to perform the duties of this position during their engagement.

The course lasts four years, and is arranged according to the following programme:

Studies.	No. of lessons per week.	No. of hours per week.	Coefficient.
FIRST YEAR.			
Arithmetic and algebra	6	12	6
Line drawing	5	10	5
Handwriting	5	5	1
SECOND YEAR.			
Elementary plane and solid geometry	6	12	6
Rudiments of physics and chemistry......................	3	6	4
Drawing	5	11	5
THIRD YEAR.			
Elementary mechanics....................................	2	4	4
Descriptive geometry and kinematics	4	8	6
Drawing	5	15	5
FOURTH YEAR.			
Steam-engines..	4	8	8
Descriptive geometry and kinematics.....................	2	4	6
Drawing (mechanical)....................................	5	15	5
Political geography	1	2	1

In addition to the above, students have practical work in the shops for 5 hours a day during the whole course. The coefficient for shop-work is 10. In the first three years, on the day in each week not occupied with drawing, students have a drill at great-guns or small-arms, or an exercise in composition.

The theoretical course occupies nine months of the year, but the shop-work is continued throughout the year. During the quarter in which studies are suspended students may have one month's leave, unless there is some special reason to the contrary.

Monthly marks on the usual scale of 10 are given by the instructors in all branches, including shop-work, and reports based on these marks are transmitted to the Ministry of Marine. The monthly marks do not, however, appear to count in the final standing, which is determined by the examinations alone.

The examinations are held at the end of the year, in each subject, before a board appointed for the purpose. The board for the three lower classes is composed of the director of the school as president, of the professor of the subject on which the examination turns, of another professor, of a chief mechanician, and of a constructing engineer. The board for examining the highest class is composed of the director of naval construction as president, the professor whose branch is the subject of examination, the chief of the section of machines, a constructing engineer, and a lieutenant. The examinations are oral, and last half an hour for each student in each subject, except in shop-work, which lasts not less than two hours. Two master-mechanics are added to the board for this examination, but they have only a consulting voice.

In order to pass to a higher class, students must obtain a mark of 6 on their general average, as well as in the separate branches of shop-work (all four years), arithmetic and algebra (1), geometry (2), mechanics (3), and steam-engines (4). The sum of the products obtained by multiplying the mark in each branch by the coefficient of the branch gives the final mark. A prize of 50 lire ($10) is given to the students who pass first in the different classes. Seven professors are attached to the school to carry on the course of instruction. These may be appointed from civil life, or may be taken from the corps of constructing engineers or of mechanicians; in the latter case they receive a considerable accession to their pay while on duty at the school. The professors meet at the end of each academic year to arrange programmes of study and distribution of time, to propose changes in the course, and to make reports on the working of the school. The financial management is in the hands of a council of administration composed of the officer in command of the first division as president, the director of the school, a mechanician and a staff officer.

Mechanician pupils receive pay at the rate of 60 centesimi (12 cents) per diem and a ration; those of the two higher classes have an addition of ten cents a day for their work in the shops. The last sum may be slightly raised or diminished according to the quality of the work. An allowance of 100 lire ($20) is made to each pupil at the beginning of the course for outfit, which is doubled if the pupil has reached his seventeenth year. The pupils are regularly attached to the section of mechanicians and stokers of the first division of seamen.

Executive duty in the school is performed by a mechanician, assisted by four petty officers, two from the marine infantry and two from the seamen's division. The detailed regulations concerning leave, clothing, and discipline are similar to those of the Naval School, and the council of administration has general supervision of all matters connected with the school.

APPENDIX.

APPENDIX.

NOTE A.

EXTRA PAY ALLOWED TO OFFICERS BORNE FOR SPECIAL DUTIES.

NAVIGATING OFFICERS.

	Per diem.
	£. s. d.
Commander, for navigating duties, when appointed to a flagship {	2 6
to	5 0
Lieutenant	2 6
Lieutenant of five years' seniority	3 0
Lieutenant who has passed for first-class ships	4 0
Sub-lieutenant	.2 6

GUNNERY OFFICERS.

First-class certificate (Greenwich and Excellent)	3 6
Second-class certificate (Greenwich and Excellent)	2 6
Third-class certificate (Greenwich and Excellent)	1 6

TORPEDO OFFICERS.

First-class certificate (Greenwich and Vernon)	3 6
Second-class certificate (Greenwich and Vernon)	2 6

INTERPRETERS.

First-class certificate (R. N. C.)	2 6
Second-class certificate (R. N. C.)	1 6

NOTE B.

COURSE OF STUDIES, HER MAJESTY'S SHIP BRITANNIA.

Subjects.	First form.	Second form.	Third form.	Fourth form.
Arithmetic (Hamblin Smith)	Part I, omit chapter 13; Part II, chapters 14 to 20, omitting chapter 18.	Part II, chapters 21 to 31, omitting chapters 22, 23, 26, 28.		Chapters 30, 31; chapter 29 is omitted; recapitulation.
Algebra (Hamblin Smith)	Chapters 1 to 12.	Chapters 13 to 19, omitting chapter 18.	Chapters 20 to 28, omitting chapter 22.	
Euclid (Todhunter)	Book I, proposition 7 may be omitted, the 8th being proved as in the notes.	Book II, omitting proposition 8; Book III, to proposition 20.	Book III, and Book IV to proposition 10.	Book V, definitions. Book VI, selected propositions; problems.
Plane trigonometry; theoretical (Hamblin Smith).	Chapters 1, 3 4, 5, 6, 7	Chapters 4, 8, 9, 10, omitting French measure.	Chapters 2, 3 iii, 12, 13	Chapters 14 to 19; chapter 20, only 219, 228; recapitulation.
Plane trigonometry; practical (Johnson).	Logarithmic expressions; use of tables; solution of plane triangles.	Easy problems in solution of triangles.	More difficult problems.	Recapitulation.
Spherical trigonometry; theoretical (Todhunter).		Rules 1, 2, 3	Chapters 1, 2, 3	Proofs of all rules used in solutions of triangles. Adaptation of the same to logarithmic computations.
Spherical trigonometry; practical (Johnson).		Plane, parallel, traverse, and Mercator's Sailings; correcting courses; all necessary definitions.	Solution of all triangles, not using rules 4 and 5; easy problems.	Application of rules to solution of astronomical problems.
Navigation and nautical astronomy (Jeans).			Day's work; latitude by meridian altitude; longitude by meridian altitude; conversion of arc into time; necessary definitions, proofs, and corrections.	Variation by amplitude; longitude by sun chronometer; practical navigation paper; explanations of all necessary rules and definitions.
Dictation	Twice a week	Once a week		
Essay	Naval history	A play of Shakspeare.	An essay of Macaulay.	A subject of history.
Instruments		Parts of sextant	Use and adjustment of sextant, azimuth-compass, taking angles and bearings.	Latitude by meridian altitude from sea horizon; longitude, error and rate of chronometer by artificial horizon; variation by azimuth-compass.
Charts		Construction of charts on a scale of 1.2 inches to a degree of longitude.	Construction of charts on a scale of 1.2 inches to a degree of longitude; laying off courses; placing ships.	Theodolite, thermometer, and barometer; construction and use of charts on any scale.
Latin	Cornelius Nepos; Arnold's Exercises; Latin Primer.	Extracts from Cicero; Arnold's Exercises; Latin Primer.	Extracts from Cicero; Arnold's Exercises; Latin Primer.	Extracts from Livy; Arnold's Exercises; Latin Primer.
Physical geography (Johnson) Astronomy (Norman Lockyer)	Elementary			
French	Grammar: substantive, adjective, pronoun; verbs active, neuter, and passive, all regular; conversation.	Elementary. Verbs reflexive and pronominal; all irregular verbs; conversation.	Syntax to beginning of verbs; grammatical analysis; conversation.	Syntax, verbs to end; logical analysis; conversation.

Drawing............	Rectangular objects in parallel perspective.	Rectangular models; advanced parallel perspective; oblique perspective commenced.	Models, including circles, oblique perspective.	Free-hand drawing from outside nature; perspective, with all the terms used.
Seamanship............	Parts of ship, spars and sails, standing rigging, knotting and splicing, bends and hitches; compass, first part.	Signals, rigging spars, tackles, running rigging, strapping blocks, log and lead, compass; questions.	Use of signals, night and day; making up, bending, reefing, and furling; making and shortening sail.	Setting up rigging, staying masts, anchor (model), rule of the road, palm and needle; recapitulation of all subjects taught.
Natural philosophy (Ganot) ..	....	....	Mechanics; hydro-mechanics; pneumatics; acoustics.	Light; heat; magnetism; electricity.

All cadets are taught and exercised in rowing; the third and fourth forms only are taught the handling of boats under sail and oars.

NOTE C.

(Page 58.)

PUNISHMENTS.—HER MAJESTY'S SHIP BRITANNIA.

SECOND CLASS FOR CONDUCT.

Wear a white stripe on left arm.

Turn out one hour earlier than others, and stand on middle deck.

Stand apart at Sunday morning muster.

One hour's drill every afternoon; leave stopped; march out one hour with corporal.

Stand on middle deck one hour after evening prayers; sit at second-class table in mess-room at meals; not allowed soup, beer, or second course; pocket-money to be stopped.

Sit in front at church and at prayers in mess-room.

Limit of punishment.—From seven to fourteen days.

THIRD CLASS FOR CONDUCT.

Wear a white stripe on each arm.

Get up at 6 a. m. in winter and 5 a. m. in summer; at 6.30 winter and 5.30 summer, fall in and drill until prayer-time.

Stand apart from other cadets at all musters.

One-and-a-quarter hour's drill every afternoon; leave stopped; march out one hour with corporal. .

Stand on middle deck half-an-hour after evening prayers.

Alternate days in cell, on bread and water; other days take their meals at cockpit mess; not allowed soup, beer, or second course.

Pocket-money to be stopped.

Sit on stool on half deck when not in cell; sit in front in church and at prayers in mess-room.

Limit of punishment.—Six days.

Commanders' punishments.

No.	Nature.	Not to exceed—
1	Imposition of copy of regulation broken	
2	{ Turn out and fall in one hour before the others............*.................. } { Called at 5.25, dismissed at 6.25................................. }	4 days.
3	Extra drill for one hour*	3 days.
4	Stand on middle deck for one hour after prayers (evening).....................	3 days.
5	Defaulters' tablet†	1 week.
	(The above punishments may be inflicted by lieutenants for one day.)	
6	Confined to cricket field;	1 month.
7	Extra drill one hour, leave stopped, go ashore one hour with corporal, and No. 4.	4 days.

* Extra drill of one hour is divided into exercises of ten minutes each, in the following order: Rifles, poles, clubs.

† No. 5. Diet—dry bread for breakfast and tea; not allowed soup, beer, or second course.

‡ Cadets confined to the cricket field are to report themselves to the gymnasium sergeant.

Mess offenses will receive mess punishments.

Permission to use the blue boats and sailing cutters will be stopped for offenses committed when away in them.

Offenses after the third time to be considered habitual, and the punishment to be doubled, or the case reported to the captain.

Pocket money to be stopped if four offenses of any description are recorded during the week.

OFFENSES TO BE REPORTED TO THE CAPTAIN.

Improper use of lights, immoral language or conduct, including falsehood or subterfuge, insubordination or disrespect to superiors, improper possession of others' property, all grievous and repeated offenses.

NOTE.—A cadet reported more than three times in the week, or more than ten times in the month, to a lieutenant, becomes a habitual offender, and remains as such until clear of the defaulters' book for one week. He is not allowed any privileges, and for any offense is reported to the commander.

NOTE D.
CADETS' MESS—ROUTINE OF DIET.

Days.	Breakfast.	Dinner.	Tea.
Sunday.......	Tea; eggs; bread and butter.	Soup; lamb, mutton, or veal, and bacon; two vegetables.	Tea; cold meat; bread and butter.
Monday......	Coffee; eggs and bacon; bread and butter.	Beef, roast and stewed; two vegetables.	Tea; cold meat; bread and butter.
Tuesday	Cocoa; currie; bread and butter.	Soup; roast mutton or lamb; two vegetables.	Tea; water cress;* cold meat; bread and butter.
Wednesday ..	Coffee; eggs and bacon; bread and butter.	Roast beef; two vegetables; two sweets.	Tea; cold meat; bread and butter.
Thursday	Cocoa; sausages;† bread and butter.	Soup; boiled mutton, caper sauce; two vegetables.	Tea; cold meat; bread and butter.
Friday	Coffee; eggs and bacon; bread and butter.	Roast mutton or lamb;‡ two vegetables.	Tea; water cress;* cold meat; bread, jam and cream.
Saturday.....	Cocoa; fish or bacon; bread and butter.	Roast beef; two vegetables; two sweets.	Tea; cold meat; bread and butter.

* If procurable.   † Except in summer.   ‡ In season.

Half pound bun loaf and quarter pint of milk on afternoons of Monday, Tuesday, Thursday, and Friday.

NOTE E.
EXAMINATION PAPERS: H. M. S. BRITANNIA.
July, 1878.
FIRST TERM.
ALGEBRA.

(Time allowed, 3 hours.)

1. Resolve into factors the expressions $x^3 + 1$ and $x^2 - x - 30$.
2. From $5x^5 + 4ax^4 - 3a^2x^3 + 5a^3x^2 + a^5$ subtract $3x^5 - 7a^2x^3 + 12a^3x^2 - 11a^4x - 7a^5$.
3. Multiply $\dfrac{x^2}{4} + \dfrac{2xy}{3} - \dfrac{y^2}{5}$ by $\dfrac{2x^2}{3} - \dfrac{xy}{2} + \dfrac{y^2}{4}$.

4. Divide $10a^3 + 11a^2b - 15a^2c - 19abc + 3ab^2 + 15bc^2 - 5b^2c$ by $5a^2 + 3ab - 5bc$.

5. Simplify the expressions—

$$(a)\ \frac{a^2 + b^2}{a^2 - b^2} + \frac{a}{a+b} - \frac{b}{a-b}.$$

$$(b)\ \cfrac{1}{2a + 3b - \cfrac{1}{a + \cfrac{1}{\ldots}}}.$$

6. Find the H. C. F. of the expressions $6x^4 - 2x^3 + 7x^2 - x + 2$ and $6x^4 - 12x^3 + 21x^2 - 6x + 9$.

7. Solve the equations—

$$(a)\ (x+3)(x-1) = (x+2)^2 - 9.$$

$$(b)\ \frac{5(2-x)}{6} - \frac{7+x}{2} - \frac{3}{10} = \frac{2x+9}{5} + 3.$$

8. A certain number, when increased by 72, is three times as large as the original number. Find the original number.

9. A ship's company of 260 men was composed of seamen, marines, and stokers; the stokers were 10 more in number than the marines, but 90 less than the seamen. How many were there in all?

ARITHMETIC.

(Time allowed, 3 hours.)

1. What is meant by the Highest Common Factor, and what by the Lowest Common Multiple of two or more numbers?
Find the H. C. F. and L. C. M. of 266, 399, and 456.

2. How many Mexican dollars worth 3s. 10¾d. each are worth 35,530 Spanish dollars at 3s. 11½d. each?

3. Simplify

$$(a.)\ \frac{2\frac{1}{3} - 1\frac{1}{4} + 9\frac{1}{11}}{4\frac{1}{2} - 2\frac{1}{4} + 13\frac{1}{11}},$$

$$(b.)\ \frac{7\ (1\frac{1}{4} \text{ of } \frac{3}{4})}{\frac{1}{5}\ (3\frac{1}{2} \text{ of } 7)} \div \frac{3}{4}.$$

4. Reduce $\frac{2}{3}$ of $7\frac{1}{4}$ of $16\frac{1}{4}$ yards to the fraction of a furlong; and $\frac{1}{5}$ of 1 oz. 13 dwt. to the decimal of $1\frac{1}{4}$ of 5 dwt. 15 grs.

5. Find the value of .375 of a guinea + .54 of 8s. 3d. + 1.027 of 2l. 15s., and express the result as the decimal of 5l.

6. Find the square roots of 1.014049 and $175\frac{1}{16}$.

7. Find, by practice, the value of 6 tons 7 cwt. 2 qrs. 17 lb. at 3l. 10s. 7d. per cwt.

8. A mine is worth 3,700l., and a man who owns $\frac{7}{16}$ of it sells .1351 of his share. What money does he receive for it?

9. If 7 men working 10¾ hours a day can earn 4l. 15s. 3d. in 5¼ days, what sum will 28 men earn in 15¾ days if they work 5.3 hours a day?

LATIN.

(Time allowed, 3 hours.)

1. Translate into English, parsing the words in *italics:*

(a.) Æthiopes, pardorum leonumque *pellibus amicti*, arcus habent prælongos: *sagittas* vero *breves:* his pro ferro lapides acuti præfixi sunt. Hastas præterea habent, his *præfixa sunt* cornua cervorum: habent etiam clavas nodosas. Corporis dimidium, in pugnam prodeuntes, creta dealbatum habent, dimidium minio pictum. Alii caput tectum habent pelle equina, de capite equi detracta, cum auribus et juba. Pro scutis gruum pellibus corpora tegunt.

(*b.*) Interim advenit Persicus exercitus; sed quum pontem solutum vidissent, magnopere timebant ne ab Ionibus desererentur. Erat tunc apud Darium vir Ægyptius, omnium hominum maxima voce præditus. Hunc Darius jussit, in ripa stantem, vocare Histiæum Milesium. Quod ubi fecit, Histiæus statim, navibus omnibus ad trajiciendum exercitum paratis, pontem junxit. Ita Persæ e Scytharum manibus effugerunt.

(*c.*) Ad hæc Iones responderunt: "Nos Ionia misit ut mare custodiamus; non ut naves nostras tradentes Cypriis, ipsi cum Persis pedestri acie confligamus. Nos igitur, qua parte locati sumus, in ea utilem præstare operam conabimur: vos autem, memores qualia Persis parentes passi ab illis sitis, fortes viros esse oportet." Post hæc, quum Persæ in Salaminiorum advenissent campum, aciem instruxerunt reges Cypriorum: ita quidem ut ceteros Cyprios hostium ceteris militibus oppouerent, Persis autem fortissimos e Salaminiis selectos. Contra Artybium vero, ducem Persarum, lubens stetit Onesilus.

(*d.*) Post digressum Persarum ex hac regione, commota tremuit Delos: quod nec ante id tempus, ut aiunt Delii, nec post, ad meam usque ætatem, factum est. Et hoc quidem prodigium edidit Deus, ut imminentia hominibus mala significaret. Constat autem regnantibus Dario, Xerxe, Artaxerxe, plura mala afflixisse Græciam quam per viginti alias generationes quæ ante Darium exstiterint. Itaque non sine causa commota est Delos. Et in vaticinio ita scriptum est.

Et Delum, quamvis sit adhuc immota, movebo.

Hæc autem tria nomina hoc significant Græco sermone: Darius coërcitorem, Xerxes bellatorem, Artaxerxes magnum bellatorem.

2. (*a.*) Give the principal parts of the verbs to which belong *jussit, trajiciendum, junxit, tradentes, stetit, scriptum est*.

(*b.*) Where was Delos, and for what was it famous?

(N. B.—The aid of dictionaries is permitted for the rest of the paper.)

3. Translate into English:

Otho, occiso Galbâ, invasit imperium, materno genere nobilior, quam paterno, neutro tamen obscuro: in privatâ vitâ mollis, et Neronis familiaris: in imperio documentum sui non potuit ostendere. Nam cum iisdem temporibus, quibus Otho Galbam occiderat, etiam Vitellius factus esset a Germanicianis exercitibus imperator, bello contra eum suscepto, cum apud Bebriacum in Italiâ levi prælio victus esset, ingentes tamen copias ad bellum haberet, sponte semetipsum occidit, petentibus militibus, ne tam cito de belli desperaret eventu, cum tanti se non esse dixisset, ut propter eum civile bellum commoveretur. Voluntariâ morte obiit, trigesimo et octavo ætatis anno, uonagesimo et quinto imperii die.

4. Translate into Latin:

(*a.*) The whole of the country was desolated by a great famine.

(*b.*) The bees are storing up honey that they may be able to live through the winter.

(*c.*) The messenger declared that, having recruited his strength, he was willing to finish his journey.

(*d.*) At Dodona, a town of Epirus, the very doves are said to have delivered oracles from the trees.

(*e.*) We are permitted to fight on horseback.

FRENCH.

(Time allowed, 3 hours.)

1. Translate into French:

The master of the parish.

As a country schoolmaster was one day entering his school-room he was met by a certain nobleman, who asked him his name and vocation. Having declared his name, he added: "And I am master of this parish." "Master of this parish!" observed the peer; "how can that be?" "I am master of the children of the parish," said the man; "the children are masters of their mothers; the mothers are the rulers of the fathers, and consequently I am master of the whole parish."

II. Translate into English:

Un toast littéraire.

Le poète anglais Campbell fut invité à dîner par son éditeur. Tous les autres convives étaient des libraires. Au dessert, l'hôte invita le célèbre auteur à porter un toast littéraire. Campbell se leva et dit solennellement: "à Napoléon"! Les convives furent fort surpris, car les sentiments libéraux de l'éminent écrivain étaient connus, et de plus on savait qu'il était grand ennemi de Bonaparte. *"Mais, monsieur," dit l'éditeur, "je vous ai demandé un toast littéraire."* "Et c'est un toast littéraire que je vous donne," répondit Campbell; "je propose la santé de l'empereur Napoléon, parce qu'il a rendu un immense service à la littérature, en faisant fusiller un libraire." (Il faisait allusion au pauvre libraire allemand Palm, condamné par un conseil de guerre français et exécuté sur l'ordre de Napoléon, pour la publication d'une brochure politique.)

III. GRAMMAR:

1. Parse the words printed *in italics.*
2. Give in a table—
    (*a*) the present participle,
    (*b*) the past participle, masculine and feminine,
    (*c*) the present indicative, second person singular and plural,
    (*d*) the subjunctive present, first person singular and plural—of *dîner, être, lever, apercevoir, pendre.*
3. Give the plural of *mal, bal, balle; eau, beau, bleu, fou, clou; voix, fils, gaz.*
4. Give the feminine forms corresponding to *ture, grec, blanc, malin, marin, espagnol, cruel, acteur, majeur, voleur.*

SECOND TERM.

ALGEBRA.

(Time allowed, 3 hours.)

1. Simplify the fraction—

$$\frac{\frac{b}{a^2}+\frac{1}{b}+\frac{1}{a}}{b-a} \div \frac{\frac{1}{a^3}-\frac{1}{b^3}}{\frac{1}{a^2}-\frac{1}{b^2}}.$$

2. Resolve into factors the expression: $x^3 - 4ax^2 - 5a^2x$.

3. If $2s = a + b + c$, prove that—
$$s^2 + (s-a)^2 + (s-b)^2 + (s-c)^2 = a^2 + b^2 + c^2.$$

4. Find the L. C. M. of $x^3 - 4x^2 + 4x - 3$, and $x^3 - x^2 - 7x + 3$.

5. Extract the square root of $4x^4 - \frac{8}{3}x^3 + \frac{112}{9}x^2 - 4x + 9$.

6. Solve the equations—

(*a*) $\dfrac{2x-13}{5} = \dfrac{15-x}{2} - \dfrac{x+3}{6}$.

(*b*) $\left.\begin{array}{l}\dfrac{2x+y}{4} = \dfrac{11-y}{3} + 1 \\[2mm] \dfrac{x+4y}{5} = 4 - \dfrac{9-x}{2}\end{array}\right\}$

7. In a certain examination three candidates competed. The second obtained half many marks again as the third, but 350 less than the first, while they had in all 1,950 marks between them. How many marks did each obtain?

8. A person has $9\frac{3}{4}$ hours at his disposal. How far may he travel on a coach at 9 miles an hour, so as to return home in time, walking back at the rate of 4 miles an hour?

9. Solve the equations—

(a) $(2x-1)(x+7)=2(x+2)(7-x)+10.$

(b) $\dfrac{x}{9-x}-\dfrac{9-x}{x}=\dfrac{9}{20}.$

ARITHMETIC.

(Time allowed, 3 hours.)

1. Divide .14 by 7, 140 by .07, and .014 by 7000; add the results together, and express the sum as a vulgar fraction.

2. Express $\dfrac{(2+\frac{1}{3})\div(3+\frac{1}{7})}{(\frac{1}{4}-\frac{1}{5})\times(4-3\frac{2}{3})}$ of $1l.$ as the fraction of $1\frac{1}{2}$ guineas.

3. Add together .60625 of $1l.$, .142857 of 14s. $10\frac{1}{2}d.$, and $2\frac{1}{11}$ of $\frac{3}{17}$ of $3l.$ 5s. 1d., and express the sum as the decimal of $17l.$ 10s.

4. Extract the square roots of 1.002001 and 6.249.

5. Find the cost of carpeting a room whose dimensions are 19 ft. 6 in. long and 16 ft. 3 in. wide, with carpet $\frac{3}{4}$ yard wide at 4s. 6d. a yard.

6. How much will $375l.$ 17s. 6d. amount to in $15\frac{1}{2}$ years at $3\frac{1}{4}$ per cent. simple interest?

7. One tap can fill a cistern in 30 minutes, and another can fill it in 40 minutes, and the discharge tap can empty it in 25 minutes. Suppose the cistern to be empty and all the taps to be open, in what time will the cistern be full?

8. If 3 men working 11 hours a day can reap a field of 20 acres in 11 days, in how many days can 9 men working 12 hours a day reap a field of 100 acres?

9. If flour be worth $10.13 per sack of 280 lbs. in America, and the cost of conveyance be $1 per sack, at what price per pound must it be sold in England so that the gain may be $12\frac{1}{2}$ per cent?

Exchange $477 = $100l.$

NAVIGATION.

(Time allowed, 3 hours.)

1. Define, with diagrams, course, distance, middle latitude, diff. long.

2. A ship sails S. 430 miles from a place in lat. 33° 50′ S., long 11° 50′ E.; find her lat. and long.

3. How must a ship sail from lat. 35° 50′ N., long. 36° 50′ W., to reach a port in lat. 35° 50′ N., long. 13° 25′ W., and what is her distance from it?

4. A ship sails NE. by E. from a port in lat. 38° 50′ S., long. 93° 35′ W., until her diff. lat. is 285 miles; find the distance she has sailed, and her lat. and long.

5. Find (by middle latitude sailing) the course and distance from Cape Race to Cape Finisterre:

Cape Race.	Cape Finisterre.
Lat. 46° 10′ N.	Lat. 42° 54′ N.
Long. 53° 3′ W.	Long. 9° 16′ W.

6. Find the compass course and distance from A to B:

Lat. A 5° 50′ S.	Lat. B 18° 10′ N.
Long. A 168° 35′ E.	Long. B 173° 20′ W.

Variation 12° 30′ E.; deviation 8° 50′ E.

7. Find the true course made by a ship, the compass course being E. by S. $\frac{1}{4}$ S., variation $2\frac{1}{4}$ points W., deviation 9° 25′ E., leeway $1\frac{1}{4}$ points, wind S. by E.

8. Give definitions of variation and deviation of the compass, and find how a ship

must be steered to reach an island bearing SW. ¾ W. (true) if the variation be 8° 50' W., deviation 5° 45' W. Also, if the wind be SSE., and the leeway be 1 point.

9. A ship sails from a port in lat. 6° 50' N., long. 88° 55' E. as follows: SW. ¼ W. 120', N. by E. ¼ E. 85', W. by N. ¼ N. 105', S. 55°, W. 42'. Find the lat. and long. arrived at.

FRENCH.

(Time allowed, 3 hours.)

I. Translate into French:

Havelock's general order to his troops after the battle of Cawnpore.

Soldiers! Your general is satisfied and more than satisfied with you. He has never seen steadier troops. But your labors are only beginning. Between the 7th and the 16th instant you have, under the Indian sun of July, marched 126 miles and fought four actions. But your comrades at Lucknow are in peril; Agra is besieged; Delhi still the focus of mutiny and rebellion. Three cities have to be saved; two strong places to be blockaded. Your general is confident that he can effect all these things and restore this part of India to tranquility, if you only second him with your efforts, and if your discipline is equal to your valor.

II. Translate into English (dictionary allowed):

C'était en Angleterre, à l'époque de la Révolution française. Le Duc de Bedford avait offert au Duc de G——, émigré, un splendide repas, une de ces fêtes quasi-royales que les grands seigneurs anglais mettent leur honneur à donner à des souverains, leur bon goût à offrir à des exilés. Au dessert, on apporta une certaine bouteille d'un vin de Coustance merveilleux, sans pareil, sans âge, sans prix. C'était de l'or liquide, dans un cristal sacré; un trésor fondu qu'on vous admettait à déguster; un rayon de soleil qu'on faisait descendre dans votre verre: c'était le nectar suprême, le dernier mot de Bacchus. Le Duc de Bedford voulut verser lui-même à son hôte cette liqueur des dieux. Le Duc de G—— prit le verre, goûta le prétendu vin, et le déclara excellent. Le Duc de Bedford voulut en boire à son tour; mais à peine a-t-il porté le verre à ses lèvres qu'il s'écria avec un horrible dégoût: "Ah! qu'est-ce que c'est que ça?" On accourt vers lui, on examine la bouteille, ou interroge le parfum: c'était de l'huile de castor! . . . . Le Duc de G—— avait avalé cette détestable drogue sans sourciller. *Ce trait sublime fit grand honneur à la noblesse de France; on conçut une haute idée d'un pays où la politesse allait jusqu'à l'héroïsme.*—MADAME EMILE DE GIRARDIN.

III. GRAMMAR:

1. Parse the words of the first and last sentences (printed in *italics*).

2. Give in a table—

    (a) the present participle,

    (b) the past participle, masculine and feminine,

    (c) the present subjunctive, 2d person singular and plural,

    (d) the future, 1st person singular and plural—

    of *offrir, mettre, vouloir, boire, s'écrier.*

Also the whole imperative of *s'en aller* (negatively).

3. Give the feminine forms corresponding to *doux, triste, blanc, duc, bon, merveilleux, pareil, liquide, hôte.*

Also the feminine of *tailleur, bailleur, tuteur, acheteur, docteur, empereur, ambassadeur, majeur, sauveur; gras, ras, gros, clos, exclu, reclus, favori, poli, citoyen, secret, muet.*

4. Of what gender are *Angleterre, âge, prix, or, verre, lèvres, huile, honneur, héroïsme?* State the rules.

Tour is of both genders. What is its meaning when feminine? Give a list of words which are both masculine and feminine; state also their meanings.

(Time allowed, three hours.)

1. Translate into English, parsing the words in *italics:*
   (*a.*) His de rebus Cæsar certior factus, et infirmitatem Gallorum *veritus*, quod sunt in consiliis capiendis mobiles, et novis plerumque rebus student, nihil his committendum existimavit. Est autem hoc Gallicæ consuetudinis, uti et *viatores*, etiam invitos, consistere cogant; et, quod quisque eorum de quaque re audierit aut cognoverit, quærant; et mercatores in oppidis vulgus circumsistat, quibusque ex regionibus veniant, quasque ibi res cognoverint, pronuntiare cogant. His rumoribus atque auditionibus permoti de summis sæpe rebus consilia ineunt, *quorum* eos e vestigio *pœnitere* necesse est, quum incertis rumoribus serviant, et plerique ad voluntatem eorum ficta *respondeant*.

   (*b.*) Cæsar, paucos dies in eorum finibus moratus, omnibus vicis ædificiisque incensis frumentisque succisis, se in fines Ubiorum recepit; atque iis auxilium suum pollicitus, si ab Suevis premerentur, hæc ab iis cognovit: Suevos, posteaquam per exploratores pontem fieri comperissent, more suo concilio habito, nuntios in omnes partes dimisisse, uti de oppidis demigrarent, liberos, uxores, suaque omnia in silvas apponerent, atque omnes qui arma ferre possent unum in locum convenirent: hunc esse delectum medium fere regionum earum quas Suevi obtinerent: hic Romanorum adventum exspectare atque ibi decertare constituisse. Quod ubi Cæsar comperit, omnibus his rebus confectis, quarum rerum causa transducere exercitum constituerat, ut Germanis metum injiceret, ut Sigambros ulcisceretur, ut Ubios obsidione liberaret, diebus omnino decem et octo trans Rhenum consumptis, satis et ad laudem et ad utilitatem profectum arbitratus, se in Galliam recepit pontemque rescidit.

   (*c.*) Hunc ad egrediendum nequaquam idoneum arbitratus locum, dum reliquæ naves eo convenirent, ad horam nonam in anchoris exspectavit. Interim legatis tribunisque militum convocatis, et quæ ex Volusseno cognosset, et quæ fieri vellet, ostendit, monuitque—ut rei militaris ratio, maxime ut maritimæ res postularent, ut quæ celerem atque instabilem motum haberent—ad nutum et ad tempus omnes res ab iis administrarentur. His dimissis, et ventum et æstum uno tempore nactus secundum, dato signo et sublatis anchoris, circiter millia passuum septem ab eo loco progressus aperto ac plano litore naves constituit.

2. (*a.*) Give the rules which govern the following constructions: *in consiliis capiendis* (1*a*); *omnibus vicis......succisis* (1*b*); *millia passuum septem......progressus* (1*c*).

   (*b.*) Decline, in both the singular and plural number: *sui, dies, reliquus, tempus, celer.*

   (N. B.—Dictionaries are permitted for the rest of the paper.)

3. Translate into English:
   Cæsar, omni exercitu ad utramque partem munitionum disposito, ut si usus veniat, suum quisque locum teneat et noverit, equitatum ex castris educi, et prœlium committi jubet. Erat ex omnibus castris, quæ summum undique jugum tenebant, despectus, atque omnium militum intenti animi pugnæ proventum exspectabant. Galli inter equites raros sagittarios expeditosque levis armaturæ interjecerant, qui suis cedentibus auxilio succurrerent, et nostrorum equitum impetus sustinerent. Ab his complures, de improviso vulnerati, prœlio excedebant. Cum suos pugna superiores esse Galli confiderent, et nostros multitudine premi viderent, ex omnibus partibus et ii, qui munitionibus continebantur, et hi, qui ad auxilium convenerant, clamore et ululatu suorum animos confirmabant.

4. Translate into Latin:

(a.) Now that the clouds were dispersed, the sun burst forth with increased splendor.

(b.) To some of the tribes the name Macrobii was given, because they were supposed to live somewhat longer than others.

(c.) Is not iron a metal more valuable to mankind than gold?

(d.) Perseus came to the assistance of Andromeda, who had been bound to a rock.

(e.) Are we permitted to take away our horses, our arms, and our baggage?

THIRD TERM.

ARITHMETIC AND ALGEBRA.

(Time allowed, 3 hours.)

1. Simplify—

$$\frac{22.4}{.25} + \frac{250}{.8} + \frac{1.2}{.0075}.$$

2. At what rate per cent. will 1,250*l.*, lent at simple interest, amount to 1,412*l.* 10*s.* in 3½ years?

3. If 120 cwt. of sugar at 38 shillings per cwt. be mixed with 90 cwt. at 42 shillings per cwt., at what price per lb. must the mixture be sold to realize a profit of 10 per cent.?

4. Find the L. C. M. of $6(x^2 - 3x - 28)$, $9(x^2 + 7x + 12)$, and $8(x^2 - 4x - 21)$.

5. Extract the square root of $4x^{-1} + 12x^{-\frac{3}{4}} + 29x^{-\frac{1}{2}} + 30x^{-\frac{1}{4}} + 25$.

6. Simplify the expressions—

(a) $\dfrac{2\sqrt{5} + 3\sqrt{7}}{3\sqrt{5} - 2\sqrt{7}}$.

(b) $\sqrt{29 + 12\sqrt{5}}$.

7. Solve the equations—

(a) $\dfrac{7\frac{1}{2} - \frac{1}{6}x}{3} + \dfrac{12 - x}{8} = \dfrac{\frac{x}{3} + 7}{30} + 2\frac{1}{4}$.

(b) $\dfrac{x + 8}{7} - \dfrac{12 - y}{3} = 2x - 11$

$\dfrac{2y - x}{6} + \dfrac{3(y - 4)}{5} = 14 - y$ }

8. At what time will the hands of a watch be at right angles to one another between four and five o'clock.

9. Solve the equations—

(a) $\sqrt{x + 9} - \sqrt{x - 6} = \sqrt{2x - 5}$.

(b) $\dfrac{x + 3}{3x - 1} - \dfrac{4x + 1}{3} = 2(x - 3)$.

(c) $x^3 + y^3 = 65$ }
$x^2 - xy + y^2 = 13$ }

10. Two travelers, one walking half an hour more than the other, start together to walk 2½ miles; if the first reaches his destination one hour before the other, find their rates of walking.

11. When are three quantities said to be in continued proportion?

If a, b, c be in continued proportion, prove that—

$$3a + 7b : 3b + 7c :: 5a - 7b : 5b - 7.$$

CHART DRAWING.

(Time allowed, 3 hours.)

N. B.—The meridional parts, as taken out of the tables, and all the work must be sent up.

1. Construct a chart, and, laying down the bearings and courses, find the latitude and longitude in, from the following:

A headland *A* (lat. 54° 46′ S., long. 141° 44′ W.) and an island *B* (lat. 54° 27′ S., long. 140° 57′ W.) bore respectively W. by N. and NNE.

Afterwards sailed as follows:

True courses.	Distance.
ENE.	83′
SW. ¼ W.	95′
SSE.	63′
N. by E. ¼ E.	152′
NW. by W.	·68′
S. by E.	76′

To be drawn on a scale of 1.2 inches to a degree of longitude, and to extend from lat. 53° S. to lat. 57° S., and from long. 138° W. to 142° W.

2. Required the distance and compass bearing of the ship in her last position from a port in lat. 54° 52′ S., long. 138° 12′ W. Variation 18° E.; deviation 5° E.

3. What is the true bearing of a rock which bears from the ship in her last position E. by S. ¼ S. ? Variation 18° E.; deviation 9° W. Place the rock in your chart, its distance being 18 miles.

NAVIGATION.

(Time allowed, 3 hours.)

1. Define: Course, Distance, Diff. Lat., and Departure. Illustrate your definitions by diagrams.

2. Prove that departure = diff. long. × cos. mid. lat.

3. A ship entering a harbor on a NNE. course, with a deviation of 8° 40′ E., has to proceed until a beacon bears NW. by W. *magnetic*, and then has to alter course to the NW., on which course the deviation will be 10° 30′ W. Find the *compass* bearing of the beacon immediately before and after the change in the course.

4. Find the time at New York (long. 73° 59′ W.) and also at Hong Kong (long. 113° 40′ E.) when it is 9ʰ 30ᵐ a. m. at Gibraltar (long. 5° 22′ W.).

5. Required the compass course and distance from A to B.

Lat. A. 56° 30′ N.	Long. A. 5° 15′ E.
Lat. B. 56° 30′ N.	Long. B. 3° 36′ W.

Variation 25° W. Deviation 10° W.

6. A ship sails from a port in lat. 17° 50′ S., long. 163° 30′ W., as follows: E. by N. ¼ N. 117 and S. by W. 83 miles. Find the bearing and distance of the port of departure.

7. Define, with diagrams: Right Ascension, Azimuth, and Hour Angle of a heavenly body. Also, Dip, Refraction, and Semidiameter.

8. May 27, 1866, at noon, in long. 128° 30′ W., the obs. mer. alt. of the sun's L.L. was 62° 10′ 50″ (zenith N. of the sun), the index error was + 1′ 52″, height of the eye above the sea 23 feet: required the latitude.

9. January 10, 1878, at noon, a point of land in lat. 46° 40′ N., long. 53° 3′ W. bore by compass N. ¼ W. (ship's head W. by S.), distant 9 miles; afterwards sailed as by the following log account: required the lat. and long. on January 11 at noon.

H.	K.	$\frac{1}{10}$ths.	Standard compass courses.	Leeway.	Winds.	Deviation.	Remarks.
1	5	5	W. by S.	$\frac{2}{3}$	S. by W.	9° 10′ W.	P. M.
2	5	4					
3	5	0					
4	4	8					
5	4	5					
6	5	0					
7	5	5					
8	6	2	{ SE by M. $\frac{1}{2}$ E. }	{ 1$\frac{1}{4}$ }	S.	8° 0′ E.	
9	6	5					
10	6	8					
11	6	2					
12	6	0					Variation of compass, 26° 30′ W.
1	10	2	N. by E. $\frac{1}{4}$ E.	0	SW.	2° 30′ E.	A. M.
2	10	0					
3	10	5					
4	10	8					
5	5	5	W. $\frac{3}{4}$ N.	$\frac{1}{2}$	SSW.	10° 50′ W.	
6	5	8					
7	6	4					
8	6	8					
9	7	2	NW. by W.	$\frac{1}{4}$	SW.	7° 20′ W.	
10	7	0					
11	7	5					
12	7	8					

LATIN.

(Time allowed, 3 hours.)

1. Translate into English, parsing the words in *italics*:

(a.) Nec enim, dum eram vobiscum, animum meum videbatis, sed eum esse in hoc corpore ex iis rebus quas gerebam intelligebatis. Eundem igitur esse *creditote*, etiam si nullum videbitis. Nec vero clarorum virorum post mortem honores permanerent, si nihil eorum ipsorum animi efficerent, quo *diutius* memoriam sui teneremus. Mihi quidem numquam persuaderi potuit, animos, dum in corporibus essent mortalibus, vivere; quum exissent ex iis, emori: nec vero tum animum esse insipientem, quum ex insipienti corpore evasisset; sed quum omni admixtione corporis libertus purus et integer esse cœpisset, tum esse *sapientem*. Atque etiam, quum hominis *natura* morte dissolvitur, ceterarum rerum perspicuum est quo *quæque* discedat; abeunt enim illuc omnia, unde orta sunt: animus autem solus nec quum adest nec quum discedit apparet.

(b.) Sophocles ad summam senectutem tragœdias fecit: quod propter studium quum rem negligere familiarem videretur, a filiis in judicium vocatus est, ut, quemadmodum nostro more malo rem gerentibus patribus bonis interdici solet, sic illum quasi desipientem a re familiari removerent judices. Tum senex dicitur eam fabulam quam in manibus habebat et proxime scripserat, Œdipum Coloneum, recitasse judicibus, quæsisseque num illud carmen desipientis videretur. Quo recitato, sententiis judicum est liberatus.

(c.) Inventi autem multi sunt qui non modo pecuniam sed vitam etiam profundere pro patria parati essent, iidem gloriæ jacturam ne minimam quidem facere vellent, ne republica quidem postulante: ut Callicratidas, qui quum Lacedæmoniorum dux fuisset Peloponnesiaco bello, multaque fecisset egregie, vertit ad extremum omnia, quum consilio non paruit eorum qui classem ab Arginusis removendam, nec cum Atheniensibus dimicandum putabant. Quibus ille respondit, Lacedæmonios, classe illa amissa, aliam parare posse, se fugere sine suo dedecore non posse. Atque hæc quidem

Lacedæmoniis plaga mediocris: illa pestifera, qua quum Cleombrotus, invidiam timens, temere cum Epaminonda conflixisset, Lacedæmoniorum opes corruerunt,

2. (*a.*) "Invidiam timens" (passage 1 *c*). To what battle is reference here made?

(*b.*) Write out in full the imperative mood active of the verb *credo*, the perfect tense subjunctive mood of *possum*, and the future perfect tense active voice of *video*.

(N. B.—The aid of a dictionary is permitted for the rest of the paper.)

3. Translate into English:

Cæsar, omni exercitu ad utramque partem munitionum disposito, ut, si usus veniat, suum quisque locum teneat et noverit, equitatum ex castris educi, et prœlium committi jubet. Erat ex omnibus castris, quæ summum undique jugum tenebant, despectus, atque omnium militum intenti animi pugnæ proventum exspectabant. Galli inter equites raros sagittarios expeditosque levis armaturæ interjecerant, qui suis cedentibus auxilio succurrerent, et nostrorum equitum impetus sustinerent. Ab his complures, de improviso vulnerati, prœlio excedebant. Cum suos pugna superiores esse Galli confiderent, et nostros multitudine premi viderent, ex omnibus partibus et ii, qui munitionibus continebantur, et hi, qui ad auxilium convenerant, clamore et ululatu suorum animos confirmabant.

4. Translate the following passages into Latin:

(*a.*) Who can doubt that the greatest general owes many of his victories to good luck?

(*b.*) The Nile supports the crocodile, a monster not less terrible to man on land than in the river.

(*c.*) I think that re-enforcements should be sent the consul, that he may attack the enemy with greater confidence.

(*d.*) Some are wanting in courage, others in honesty.

(*e.*) Apollo was supposed to carry in his right hand a bow and arrows, in his left hand a lyre.

FINAL EXAMINATION.

ARITHMETIC AND ALGEBRA.

(Time allowed, 3 hours.)

1. A vulgar fraction has for its numerator 209, and its nearest approximate value to three places of decimals is .511; what is its denominator?

2. A bankrupt owes 1,765*l.*; his property realizes 540*l.*, and he has debts owing to him of 927*l.* 10*s.*, of which he recovers 12*s.* in the pound. What dividend can he pay?

3. Express in its simplest form

$$\left(1 - \frac{2xy}{x^2 + y^2}\right) \div \left(\frac{x^3 - y^3}{x - y} - 3xy\right).$$

4. Reduce to its lowest terms the fraction—

$$\frac{x^3 - 2ax^2 - 5a^2x - 12a^3}{x^3 - 7ax^2 + 13a^2x - 4a^3}.$$

5. (*a.*) Divide $3\sqrt{5} + 2\sqrt{3}$ by $4\sqrt{3} - 3\sqrt{5}$.

(*b.*) Extract the fourth root of $193 + 132\sqrt{2}$.

6. Multiply $x^{-2} + 2x^{-\frac{3}{2}}y^{-\frac{1}{2}} + y^{-2}$ by $x^{\frac{3}{2}} - x^{\frac{1}{2}}y + y^{\frac{3}{2}}$.

7. Solve the equations—

(*a.*) $5(4x + 1) - \frac{2}{3}(8x + 7) = \frac{3}{2}(9 - 2x) + 27.$

(*b.*) $\sqrt{x} - \sqrt{a - \sqrt{ax + x^2}} = \sqrt{a}.$

8. The rent of a farm consisting partly of arable land at 1*l.* 10*s.* per acre, and partly of pasture land at 1*l.* 5*s.* per acre, was 375*l.* Had the number of acres of arable been one-fifth greater, and of pasture land one-third less than it was, the rent would have been 370*l.* Find the number of acres in the farm.

9. Solve the equations—

$$(a.) \frac{x+4}{x-2} - \frac{2x-7}{x+2} = 1\tfrac{1}{6}.$$

$$(b.) \begin{array}{r} xy + x + y = 23 \\ x^2y + xy^2 = 120 \end{array} \}$$

10. A contractor spent 200*l.* every week in the wages of his workmen; being compelled to raise the wages of each by 5*s.* 4*d.* per week, he found that he was able to employ 25 men less than before for the same sum. How many men did he employ?

11. When are three quantities said to be in continued proportion?

If *a*, *b*, *c* be in continued proportion, prove that $(a^2 + b^2)(b^2 + c^2) = (ab + bc)^2$.

12. (*a.*) The sum of the first and fourth term of an arithmetical series is 2, and the sum of the second, third, and sixth terms is −11; find the sum of 8 terms of the series.

(*b.*) Investigate an expression for the sum of *n* terms of a geometric series. Sum the series $-\tfrac{2}{3} + \tfrac{1}{4} - \tfrac{1}{70} + \&c.$, to 7 terms.

GEOMETRY.

(Time allowed, 3 hours.)

1. If a straight line fall on two parallel straight lines it makes the alternate angles equal to one another, and the exterior angle equal to the interior and opposite angle on the same side; and also the two interior angles on the same side together equal to two right angles.

Show that the difference between the alternate angles made by any straight line falling upon two given straight lines is invariable. Under what conditions will this difference be zero?

2. Equal triangles on the same base, and on the same side of it, are between the same parallels.

The sides *AB* and *AC* of a triangle are bisected in *D* and *E* respectively, and *BE*, *CD* are produced until *EF* = *EB*, and *GD* = *DC*, show that the line *GF* passes through *A*.

3. If a straight line be divided into two equal parts and also into two unequal parts, the rectangle contained by the unequal parts, together with the square on the line between the points of section, is equal to the square on half the line.

Prove that the area of a square is greater than the area of a rectangle of the same perimeter.

4. Describe a circle about a given triangle.

If a circle be described about a triangle *ABC*, and perpendiculars be let fall from the angular points *A*, *B*, *C* on the opposite sides, and produced to meet the circle in *D*, *E*, *F*, respectively, show that the arcs *EF*, *FD*, *DE* are bisected in the points *A*, *B*, *C*.

5. If the vertical angle of a triangle be bisected by a straight line which also cuts the base, the segments of the base shall have the same ratio which the other sides of the triangle have to one another; and if the segments of the base have the same ratio which the other sides of the triangle have to one another, the straight line drawn from the vertex to the point of section shall bisect the vertical angle.

6. The sides about the equal angles of triangles which are equiangular to one another are proportionals, and those which are opposite to the equal angles are homologous sides; that is, are the antecedents or the consequents of the ratios.

7. Explain the terms: Similar rectilineal figures, duplicate ratio.

Similar polygons may be divided into the same number of similar triangles, having the same ratio to one another that the polygons have; and the polygons are to one another in the duplicate ratio of their homologous sides.

TRIGONOMETRY.

(Time allowed, 3 hours.)

1. Define the tangent and versine of an angle, and express the tangent in terms of the versine.

Between what limits does the value of the versine lie, and for what angle is it greatest?

2. What are the units chiefly made use of in measuring angles?

If the unit of angular measurement be the angle of an equilateral triangle, how many degrees are there in the angle represented by .6?

3. Prove geometrically that—

$$(a)\ \cos{(90^\circ + A)} = -\sin{A}.$$
$$(b)\ \cos{(A + B)} = \cos{A}\cos{B} - \sin{A}\sin{B}.$$

From the second of these equations deduce the value of $\sin{(A - B)}$.

4. Show that if $x = 45^\circ$ $\sin{(x + a)} = \cos{(x - a)}$.

5. Prove the identities:

$$(a)\ \cos{3A} = 4\cos^3{A} - 3\cos{A}.$$

$$(b)\ \frac{1 + \tan{\left(\dfrac{\pi}{4} - \theta\right)}}{1 - \tan{\left(\dfrac{\pi}{4} - \theta\right)}} = \cot{\theta}.$$

$$(c)\ \frac{\cos{A} - \cos{B}}{\cos{A} + \cos{B}} = -\frac{\tan{\dfrac{A - B}{2}}}{\cot{\dfrac{A + B}{2}}}$$

6. Investigate a formula, adapted to logarithmic calculation, for finding the angles of a plane triangle, the sides of which are given.

In the triangle ABC, $a = 97.6$, $b = 101.4$, and $c = 119.6$, find the angle A.

7. The area of a quadrilateral field $ABCD$ is 3,495 square yards. The sides AB, BC and diagonal AC are 75, 81, and 64 yards respectively, the remaining sides AD, DC are equal to one another, and the angle D is $107^\circ 33'$. Find the length of the equal sides AD, DC.

8. Define the terms: sphere, spherical triangle.

Show that the angles of a spherical triangle are together greater than two and less than six right angles.

9. Prove that in any spherical triangle ABC

$$\cos{A} = \frac{\cos{a} - \cos{b}\cos{c}}{\sin{b}\sin{c}}.$$

What does this formula become (1) when the angle A is a right angle, (2) when the side (a) is a quadrant, and (3) when applied to the Polar Triangle?

10. In the spherical triangle ABC, given $a = 86^\circ 59'$, $b = 106^\circ 17'$, and $A = 90^\circ$, find the remaining parts.

CHART DRAWING.

(Time allowed, 3 hours.)

N.]B.—The meridional parts, as taken out of the Tables, and all the work must be sent up.

1. Construct a chart, and, laying down the bearings and courses, find the latitude and longitude in, from the following:

Two headlands *A* (lat. 54° 41′ N., long. 19° 40′ W.) and *B* (lat. 54° 44′ N., long. 20° 25′ W.), bore respectively NNE. and NW. by W.

Afterwards sailed as under:

True courses.	Distances.
SSE.	55′
SW.	78′
ESE.	91′
SSW. ¼ W.	45′
NW. by N.	144′
E. by S.	93′

To be drawn on a scale of 1.22 inches to a degree of longitude, and to extend from latitude 51° to latitude 55° N., longitude 18° W. to 22° W.

2. Find the compass bearing and distance of the ship in her last position from a port in lat. 52° 26′ N. and long. 21° 15′ W.; variation, 30° W.; deviation, 3° E.

3. What is the true bearing of a rock which bears from the ship in her last position SW. ¼ W.; variation, 30° W.; deviation, 9° W.

Place the rock on your chart, supposing its distance 14 miles.

THEORETICAL NAVIGATION.

(Time allowed, 3 hours.)

1. Define the terms: Equator, Parallels of Latitude, Course, Distance, and Middle Latitude.

Prove the formula made use of in Middle Latitude Sailing.

2. A ship in lat. 27° 30′ S. sails west 550 miles and then due south; subsequently she alters course to east, and after sailing 420 miles reaches the meridian from which she started. How many miles has she sailed in the southerly direction?

3. Define the term Celestial Meridian. What do we require to know with regard to the sun in finding latitude by meridian altitude?

On a certain day, at two places on opposite sides of the equator, the sun's meridian zenith distance was double the latitude of the place. Having given the declination 10° N., find the respective latitudes.

4. Explain the terms: Rational Horizon, Circles of Altitude, Prime Vertical.

At what times will the sun be on the prime vertical of a place in latitude 15° 30′ S. when its declination has the following values: 0°, 15° 30′ S. and 5° S. respectively?

5. What are the causes of twilight?

How long does twilight last at a place on the equator when the sun's declination is 12° 30′ N.?

6. Having given the distance between the sun and moon 86° 48′, the declination of the sun 19° 10′ 30″ S., of the moon 10° 8′ 45″ N., and the right ascension of the sun 15ʰ 32ᵐ 47ˢ, find the right ascension of the moon, which lies to the westward of the sun.

7. What is meant by the Amplitude of a heavenly body?

Find the compass bearing of the sun when it rises at 6ʰ 54ᵐ a. m. (apparent time) at a place in lat. 30° 10′ N., Var. 16° W., Dev. 7° E.

8. A ship sails from lat. 51° 26′ S., long. 17° 25′ W., to lat. 39° 22′ S., long. 21° 55′ E. Find the highest latitude reached.

PRACTICAL NAVIGATION.

(Time allowed, 3 hours.)

October 27, 1866, at noon, a point of land in lat. 25° 10′ S., long. 46° 30′ E., bore by compass N. by W. ¼ W. (ship's head being SSE.) distant 12 miles; afterwards sailed as by the following log account; work up the reckoning to noon, October 28.

H.	K.	₁₀ths.	Standard compass courses.	Leeway.	Winds.	Force.	Bar. Ther.	Devia-tion.	Remarks.
1	5	8	SSE.	¼	E. by N.			5° 20' E.	P. M.
2	5	4							
3	5	5							
4	6	0							
5	9	5	W. ¾ N.	0	E..			8° 50' W.	
6	9	8							
7	10	2							6ʰ 25ᵐ, variation by amp.
8	9	5							
9	8	8							
10	8	5							
11	4	8	S. ¼ W.	¾	ESE.			1° 10' W.	
12	4	5							
1	4	8							A. M.
2	5	0							
3	4	5							
4	4	5							
5	4	0							
6	5	5	NE. ¾ N.	¼	E. by S.			6° 40' E.	
7	5	8							
8	6	0							8ʰ 30ᵐ obs. for long.
9	6	2							
10	6	0							
11	5	8							
12	5	5							Noon, obs. for lat.

Course and distance made good.	Latitude.	Longitude.	Variation allowed.	True bearing and distance. Mauritius.
	D. R.	D. R.		
Current.	Obs.	Obs.	18° W.	Lat. 20° 10' S. Long. 57° 30' E.

October 27, at 6ʰ 25ᵐ p. m., the sun set by compass W. 12° 30' N. (Deviation as for ship's head), find the Variation.

October 28, at 8ʰ 30ᵐ a. m., the following sights were taken to find the longitude:

Times by chron.	Alt. sun's L.L.	
7ʰ 42ᵐ 33ˢ	38° 55' 40''	Index error — 35''
43ᵐ 0ˢ	59' 50''	Height of eye 24 feet.
43ᵐ 25ˢ	39° 4' 20''	

September 26, at G. M. noon, the chronometer was fast on G.M.T. 2ʰ 27ᵐ 38.5ˢ, gaining daily 6.5 seconds.

October 28, at noon, the obs. mer. alt. of the sun's L.L. was 76° 48' 10'' (zenith S.), index error and height of eye as above, find the latitude.

PHYSICS.

(Time allowed, 3 hours.)

1. Explain carefully how a Centigrade thermometer is made and graduated. What temperature will be represented by the same number in the Centigrade and Fahrenheit thermometers?

2. We require to measure accurately the changes in the length of a rod of metal due to changes of temperature. Explain how this can be done.

3. If ice at 0°C. is placed in a vessel to which heat is continuously applied trace the changes in size, state, and temperature which the heat will produce.

4. Water may be frozen in consequence of the evaporation from its surface. Explain under what circumstances this may be made to take place.

5. How has the velocity of light been determined?

6. What is a photometer? Describe some form of photometer.

7. A candle is placed in front of a concave mirror whose radius is 2 feet. Where must we stand in order to see the images when the distance of the candle from the mirror is first 4 feet, then 1 foot 6 inches, and finally 6 inches? Say in each of the three cases whether the images are larger or smaller than the candle, and whether they are erect or inverted.

8. Describe the *eye* as an optical instrument. What defects of the eye are corrected by eye-glasses? When should the eye-glass be convex and when concave?

9. What is the fundamental law of magnetism? What differences as regards magnetic properties are there between soft iron and steel?

10. How may the intensity of the earth's magnetism in two different places be compared?

11. Describe the electrophorus, and show how by its means we may charge a Leyden jar.

12. What properties of an electric current are utilized in telegraphy?

LATIN.

(Time allowed, 3 hours.)

1. Translate into English the following passages, parsing the words in *italics:*

(a.) Et conversus ad simulacrum Jovis, " Andi, Juppiter, hæc scelera," inquit ; " audite jus Fasque. Peregrinos consules et peregrinum senatum in tuo, Juppiter, augurato templo, captus ipse atque oppressus, visurus es? Hæcine fœdera Tullus, Romanus rex, cum Albanis, patribus vestris, Latini, hæc L. Tarquinius vobiscum postea fecit? Non venit in mentem pugna apud Regillum lacum? Adeo et cladium veterum vestrarum et *beneficiorum nostrorum* erga vos *obliti estis?*" Quum consulis vocem subsecuta patrum indignatio esset, proditur memoriæ, adversus crebram implorationem deum quos testes fœderum *sæpius invocabant* consules, vocem Annii spernentis numina Jovis Romani auditam.

(b.) Vicit tamen pars, quæ in præsentia videri potuit majoris animi quam consilii ; sed eventus docuit, fortes fortunam juvare. Bellum ex auctoritate patrum populus adversus Vestinos jussit. Provincia ea Bruto, Saminium Camillo sorte evenit. Exercitus utroque ducti, et cura tuendorum finium hostes prohibiti conjungere arma. Ceterum alterum consulem L. Furium, cui major moles rerum imposita erat, morbo gravi implicitum fortuna bello subtraxit ; jussusque dictatorem dicere rei gerendæ causa longe clarissimum bello ea tempestate dixit, L. Papirium Cursorem, a quo Q. Fabius Maximus Rullianus magister equitum est dictus, par nobile rebus in eo magistratu gestis, discordia tamen, qua prope ad ultimum dimicationis ventum est, nobilius.

(c.) Fabius contione extemplo advocata obtestatus milites est, ut, qua virtute rem publicam ab infestissimis hostibus defendissent, eadem se, cujus ductu auspicioque vicissent, ab impotenti crudelitate dictatoris tutarentur ; venire amentem invidia, iratum virtuti alienæ felicitatique ; furere, quod se absente respublica egregie gesta esset ; malle, si mutare fortunam posset, apud Samnites quam Romanos victoriam esse ; imperium dictitare spretum, tanquam non'eadem mente pugnari vetuerit, qua pugnatum doleat.

2. (a.) Give the derivations of the words : *simulacrum, numen, prorincia, auspicium, egregius.*

(b.) Give some particulars of the battle at the lake Regillus.

(N. N.—Dictionaries are permitted for the remainder of the paper.)

3. Translate into English:

Cum in omnibus locis, consumpta jam reliqua parte noctis, pugnaretur, semperque hostibus spes victoriæ redintegraretur, eo magis quod denstos pluteos turrium videbant, nec facile adire apertos ad auxiliandum animum advertebant, semperque ipsi recentes defessis succederent, omnemque Galliæ salutem in illo vestigio temporis positam arbitrarentur, accidit inspectantibus nobis, quod, dignum memoria visum, prætermittendum non existimavimus. Quidam ante portam oppidi Gallus, qui per manus sevi ac picis traditas glebas in ignem e regione turris projiciebat, scorpione ab latere dextro trajectus exanimatusque concidit. Hunc ex proximis unus jacentem transgressus, eodem illo munere fungebatur; eadem ratione ictu scorpionis exanimato altero, successit tertius, et tertio quartus; nec prius ille est a propugnatoribus vacuus relictus locus, quam, restincto aggere atque omni ex parte submotis hostibus, finis est pugnandi factus.

4. Translate into Latin:

(*a.*) The love of money is so great that not even the fear of death can altogether quell it.

(*b.*) After that battle they sent Pausanius to Cyprus, to drive out the barbarians from that island.

(*c.*) Perseus showed his enemies the head of the Medusa, by the sight of which all were changed into stones.

(*d.*) Cicero warned the conspirators to conceal nothing from the judges.

(*e.*) We pity our fellow-citizens, who have lost all in the conflagration.

FRENCH.

(Time allowed, 3 hours.)

I. Translate into French:

The British army.

The first corps raised in England in accordance with our present system, and in fact the first germ of an English standing army, was the Coldstream Guards, raised by General Monk at Coldstream. In the course of a few years several others were added, and by 1665 the British infantry consisted of four regiments besides the Guards. Before the close of the century, a grenadier company, furnished with hand grenades, had been added to each regiment; bayonets had been introduced; several regiments of fusiliers, originally intended to protect artillery, had been raised; and the principle of a standing army of considerable numbers fairly established. Light horse were introduced in 1745, and lancers in the reign of George III. It is within the last few years, however, that the greatest changes have taken place in the British army. But the advancement and elevation of the soldier himself only render him more capable of appreciating the traditions of his corps.

II. Translate into English:

UN ÉPISODE DEVANT SÉBASTOPOL.

Je m'étais arrêté, et je regardais avec une émotion profonde tous ces hommes qui, un instant auparavant, étaient pleins d'audace et de courage; tous étaient immobiles. Cependant, sur l'un des brancards les plus rapprochés de moi se soulevait faiblement une capote, et le bras du blessé cherchait à atteindre le brancard que l'on avait placé à côté du sien; un instant après, deux mains se touchaient. Celui qui le premier avait cherché cette étreinte fraternelle, rejeta tout-à-coup la capote dont on l'avait couvert, et aux premières lueurs du jour je le vis lever la tête, essayer de se soulever, puis retomber.—Je me penchai sur lui; le pauvre soldat était mort.—Cette main étendue, qui voulait presser une autre main, avait été le dernier adieu du mourant à un frère d'armes.—CINQ MOIS AU CAMP DEVANT SÉBASTOPOL, PAR LE BARON DE BAZANCOURT.

III. GRAMMAR:

1. Write down the present participle and past participle of each of the following verbs: *Savoir, voir, mouvoir, pouvoir; rire, dire, écrire; manger, placer, régler, mener.* Give also the present subjunctive 2d person singular and plural of each.

Give the future, first person singular, of *savoir, voir, pouvoir.*

2. Give the feminine form of each of the following adjectives: *Sec, grec; malin, marin; aigu, perdu; net, complet; vif, royal, formel.*

‡ 3. Translate and write in full: "16 July, 1878; 200 men; 220 men; 2,220 men; 80 pounds; 84 pounds; 2,000 pounds; thousands of pounds." Explain the rule about *vingt, cent, and mille.*

4. "*Cherchait à atteindre*"; "*je le vis lever* la tête, *essayer* de se *soulever.*" Why are these verbs in the infinitive? State the rule, or rules.

5. "*Le brancard que l'on avait placé*"; "*celui qui avait cherché*"; "*le pauvre soldat était mort*". With what part of the sentence do the participles *placé, cherché, mort,* agree? State the rule of agreement of the participle conjugated with *avoir.*

NOTE F.

Apportionment of naval instructors, midshipmen, and naval cadets among sea-going ships and otherwise, January 1, 1879.

Ship or station.	Description.	Midshipmen.	Naval cadets.	Total.	Naval instructors.
Monarch	Iron, screw turret-ship, armor-plated	11	1	12	1
Alexandra	Iron, double screw, armor-plated	18	1	19	1
Temeraire	do	10	1	11	1
Shannon	Iron, screw, armor-plated	8	2	10	1
Bellerophon	do	11	3	14	1
Audacious	Iron, double screw, armor-plated	10	.....	10	1
Invincible	do	10	2	12	1
Iron Duke	do	8	5	13	1
Triumph	Iron, screw, armor-plated	.....	2	2	1
Minotaur	do	13	4	17	1
Achilles	do	10	3	13	1
Agincourt	do	8	1	9	1
Defence	do	6	4	10	1
Pallas	Screw corvette, armor-plated	7	.....	7	1
Shah	Iron, screw frigate, cased in wood	8	.....	8	1
Raleigh	do	10	.....	10	1
Boadicea	Iron, screw corvette, cased in wood	9	2	11	1
Euryalus	do	4	3	7	1
Volage	do	4	.....	4	1
Active	do	3	.....	3	1
Rover	Iron, screw corvette	4	.....	4	1
Charybdis	Screw corvette	4	.....	4	1
Wolverene	do	7	1	8	1
Diamond	do	2	.....	2	1
Garnet	do	2	.....	2	1
Opal	do	4	.....	4	1
Ruby	do	4	.....	4	1
Sapphire	do	2	.....	2	1
Tourmaline	do	4	2	6	1
Turquoise	do	2	.....	2	1
Blanche	do	3	.....	3	1
Danaë	do	1	.....	1	....
Spartan	do	1	.....	1	....
Admiralty		.....	.....	.....	1
Britannia*		.....	171	171	9
Royal Naval College		.....	.....	.....	6
Training-ships for boys		.....	.....	.....	5
Chaplains in other ships		.....	.....	.....	13
Unattached		16	9	25	7
Total		224	217	441	71

* Exclusive of the class that entered in the preceding November, whose admission to the Britannia dated from January 15, 1879.

NOTE G.

EXAMINATION PAPERS: ROYAL NAVAL COLLEGE, JUNE, 1878.

EXAMINATION FOR RANK OF LIEUTENANT, R. N.

I.—ALGEBRA.

(Time allowed, 3 hours.)

1. (*a*) Simplify the fraction—

$$\cfrac{1}{x+\cfrac{1}{2+\cfrac{x-3}{2-x}}}$$

(*b*) Resolve the expression $4a^2c^2 - (a^2 + c^2 - b^2)^2$ into four factors.

2. Prove the rule for finding the L.C.M. of two algebraical expressions. Find the L.C.M. of

$6a^4 - a^3b - 3a^2b^2 + 3ab^3 - b^4$ and $9a^4 - 3a^3b - 2a^2b^2 + 3ab^3 - b^4$.

3. Simplify $\dfrac{x^{3n}}{x^n - 1} - \dfrac{x^{2n}}{x^n + 1} + \dfrac{1}{x^n + 1} - \dfrac{1}{x^n - 1}.$.

4. (*a*) Explain the terms: Surds, surds of the same order.

Express as surds of the same order $\sqrt{5}$ and $\sqrt[3]{7}$.

(*b*) Simplify the expression $\sqrt[4]{\dfrac{113}{16} + \dfrac{9}{\sqrt{2}}}$.

5. Solve the equations—

(*a*) $\dfrac{8x + 5}{14} + \dfrac{7x - 3}{6x + 2} = \dfrac{16x + 15}{28} + \dfrac{2\frac{1}{4}}{7}$.

(*b*) $\sqrt{a^2 + bx} + \sqrt{a^2 - bx} = 2c$.

(*c*) $\left. \begin{array}{l} x + ay + a^2z = a^3 \\ x + by + b^2z = b^3 \\ x + cy + c^2z = c^3 \end{array} \right\}$

6. A railway train 66 yards long, traveling at the rate of 50 miles an hour, met another train traveling at the rate of 22 miles an hour, which it passed in 5 seconds. Find the length of the second train.

7. Solve the equations—

(*a*) $\dfrac{3x - 11}{x - 2} + \dfrac{2(x + 2)}{15} = \dfrac{2}{5}(x + 1)$.

(*b*) $\left. \begin{array}{l} x^2 - y^2 = 32 \\ \dfrac{1}{y^2} - \dfrac{1}{x^2} = \dfrac{2}{9} \end{array} \right\}$

8. If a and β are the roots of the equation $ax^2 + bx + c = 0$, find the value of $a^4 + a^2\beta^2 + \beta^4$.

9. Explain the terms: Ratio, Ratio of less inequality.

If $a : b$ be a ratio of less inequality, and x a positive quantity, show that the ratio $a - x : b - x$ is less than the ratio $a : b$.

10. (*a*) Prove that the ratio of the sum of the latter half of $2n$ terms of any arithmetical series is to the sum of $3n$ terms of the same series as 1 to 3.

(*b*) Sum the series $\frac{1}{3} - \frac{1}{2} + \frac{3}{4} -$ &c. to 7 terms.

11. Having given two numbers and the difference of their logarithms, show how the base of the system may be determined.

In what system does the logarithm of 40 exceed that of 5 by 3?

II.—GEOMETRY.

(Time allowed, 3 hours.)

1. Describe an equilateral triangle upon a given finite straight line.

If a second equilateral triangle be described on the other side of the given line, what figure will the two triangles form?

2. If two triangles have two angles of the one equal to two angles of the other, each to each, and one side equal to one side, namely, either the sides adjacent to the equal angles, or sides which are opposite to equal angles in each, then shall the other sides be equal, each to each, and also the third angle of the one equal to the third angle of the other.

3. The complements of the parallelograms, which are about the diameter of any parallelogram, are equal to one another.

4. On the sides AB, BC, CD of a parallelogram are described equilateral triangles ABE, CDF without the parallelogram and BCG within it; prove that EG, FG are equal to the two diagonals respectively.

5. Describe a square equal to a given rectilineal figure.

6. Define a circle.

If two circles touch one another internally, the straight line which joins their centers, being produced, shall pass through the point of contact.

7. Define the segment of a circle.

On a given straight line describe a segment of a circle containing an angle equal to a given rectilineal angle.

8. In a given circle, inscribe a triangle equiangular to a given triangle.

If the inscribed triangle be equilateral, show that the distance of any point in the circumference of the circle from the most remote angle of the triangle is equal to the sum of the distances from the other angles.

9. Enunciate the axioms of Euclid's Fifth Book.

Find a mean proportional between two given straight lines.

10. In equal circles, angles, whether at the centers or circumferences, have the same ratio which the arcs on which they stand have to one another; so also have the sectors.

III.—TRIGONOMETRY.

(Time allowed, 3 hours.)

1. What are the methods of measuring angles commonly made use of?

If the radius of a circle be 20 feet, find to four places of decimals the length of the arc subtending an angle of 7° at the center of the circle.

2. Trace the changes in the sign and magnitude of the fraction $\dfrac{\sin 3\theta}{\cos 2\theta}$ as θ varies from 0 to $\dfrac{\pi}{2}$.

3. Prove that $\cos (A + B) = \cos A \cos B - \sin A \sin B$.
Apply this formula to find the value of $\cos 135°$.

4. Solve the equations—

$$(a)\ \ \mathrm{Tan}\ ^2\theta - 3 \cot\ ^2\theta = \frac{1}{2} \sec\ ^2\theta.$$

$$(b)\ \ \mathrm{Cos}^{-1} \frac{\sqrt{3}}{2} + \sec^{-1} x = \frac{\pi}{4}.$$

5. Prove the identities—

$(a)\ \mathrm{Sin}\ A\ (1 + \tan A) + \cos A\ (1 + \cot A) = \sec A + \csc A.$

$(b)\ \dfrac{\mathrm{Cos}\ A + \sin C - \sin B}{\mathrm{Cos}\ B + \sin C - \sin A} = \dfrac{1 + \tan \dfrac{A}{2}}{1 + \tan \dfrac{B}{2}}$ when $A + B + C = 90°$.

6. Investigate a formula giving the value of the sine of half an angle of a plane triangle in terms of the sides.

7. The area of a quadrilateral figure ABCD is 9,688 square yards. AB is 110 yards, BC 91 yards, AC 125 yards, and CD 82 yards. Find the side AD.

8. If in the plane triangle ABC the angle A be three times the angle B, prove that—

$$\mathrm{Sin\,B} = \frac{1}{2}\sqrt{\frac{3b - a}{b}}.$$

9. Find an expression for the value of the cosine of an angle of a spherical triangle in terms of functions of the sides, and thence the cosine of a side in terms of functions of the angles.

10. If a, b, c, be the sides of a spherical triangle, and if the arc δ be drawn from the angle A to bisect the side a, show that—

$$\cos\frac{a}{2}\cos\delta = \cos\frac{b + c}{2}\cos\frac{b - c}{2}.$$

11. In the spherical triangle ABC, given $a = 90°$, $b = 71°\,39'$, and $A = 104°\,15'\,30''$, find the remaining parts.

IV.—MECHANICS AND HYDROSTATICS.

(Time allowed, 3 hours.)

1. Assuming the Parallelogram of Forces, so far as the direction of the Resultant is concerned, prove it for the magnitude of the Resultant.

Show that if the angle at which two forces are inclined to each other be increased, their resultant is diminished.

2. If two forces be inclined to one another at an angle of 150°, find the ratio between them when the resultant is equal to the smaller force.

3. Explain the term: Center of Parallel Forces.

Prove that if three forces, acting on a rigid body, balance each other, the lines in which they act must either be parallel or must pass through a point.

4. A uniform lever of the first kind is 12 feet in length. A weight of 50 lbs. is suspended from each extremity in turn, and it is found that weights of 20 lbs. and 110 lbs. are required at the other extremities to preserve equilibrium. Determine the weight of the lever and the position of the fulcrum.

5. The ninth part of the area of a triangle is cut off by a line parallel to the base. Find the center of gravity of the remaining area.

6. Investigate the conditions of equilibrium for the single movable pulley when the parts of the string are not parallel.

A cord fastened at A passes under a movable pulley bearing a weight, P; it then passes over a fixed pulley at B, and under a second movable pulley bearing a weight, Q, and is fastened to a peg at C. Find the tension of the cord when the angle at one of the movable pulleys is double that at the other, P being greater than Q.

7. Define velocity, distinguishing between uniform and variable velocity.

If 2 seconds be the unit of time and an acre be represented by 40, what is the measure of a velocity of 30 miles an hour?

8. A body projected up a smooth plane, with a velocity of 320 feet per second, returned to the foot of the plane in 30 seconds. How high did it ascend, and what was the inclination of the plane to the horizon?

9. If a circle be placed in a vertical plane, determine the cord passing through the lowest point, down which a body must fall so that it may acquire the greatest horizontal velocity at the bottom.

10. Distinguish between the terms: "Fluid pressure at a point" and "Fluid pressure on a point."

An area of $1\frac{1}{2}$ square feet is subject to a uniform fluid pressure of 2,568 lbs. Determine the measure of the pressure at any point when the unit of length is $\frac{1}{4}$ inch.

11. Describe the Hydrostatic Balance, and show how it may be used to compare the specific gravities of a solid and a fluid, the solid being of less specific gravity than the fluid.

12. What must be the weight of a mass of silver (sp. gr. 10.5) which, when weighed in a fluid of specific gravity 4.5, appears to weigh the same as a mass of lead (sp. gr. 11.4) the weight of which in vacuo is 15 lbs.?

V.—PHYSICS.

(Time allowed, 3 hours.)

1. Give a diagram of a common pump, and show how the height of the barometer influences the limit of its action.

2. Give an account of the phenomena of capillary action, and describe the different ways in which it manifests itself in a water and a mercury barometer.

3. Describe Nicholson's Hydrometer.

A liquid is known to consist of water and alcohol; explain how a Nicholson's Hydrometer could be used to determine the proportions in which they are mixed.

4. What is meant by magnetic induction? Explain the statement that repulsion is a surer test of magnetization than attraction, and show that it is possible for one end of an iron bar under certain circumstances neither to attract nor repel the red pole of a magnet brought near to it.

5. What are meant by the lines of force in a magnetic field? In what directions do the lines of force of the terrestrial magnetism pass through the magnetic equator?

6. Distinguish between the temperature of a body and the quantity of heat which it contains.

If m lbs. of one body and n lbs. of another rise respectively p degrees and q degrees in temperature on the addition of the same quantity of heat to each, compare their specific heats.

7. What is meant by the hygrometric state of the air?

Explain the significance of the indications of the wet and dry bulb thermometers.

8. A ray of light passes from one medium to another; explain by what law the change in its direction is governed.

A ray of light falls on one surface of a triangular glass prism, is totally reflected at the second, and emerges from the third face. Draw a diagram showing the course of the ray.

9. Distinguish between a virtual and a real image formed by a lens.

A large convex lens is used as a reading glass in the ordinary way; explain with a diagram the formation of the magnified image.

VI.—STEAM ENGINE.

(Time allowed, 3 hours.)

1. Water of 93° F. is under the atmospheric pressure. How many units of heat will be required to raise one pound of it to the boiling point, and how many more units to evaporate nine-tenths of a pound?

If the pressure to which it is subjected is 2 atmospheres, of which the corresponding temperature is 249° F., how much heat will be required to make it boil, and how much more to evaporate the whole of it?

2. What is meant by an *indicated horse-power*?

If the volume of a pound of steam under a pressure of 30 lbs. per sq. in. is 13½ cubic feet, how many foot-pounds of external work is done during its formation from water under this pressure; and how many pounds of steam will be required per hour to develop one indicated horse-power?

3. Give a short description of coal, stating what its great calorific value is due to.

4. Describe the indicator, and how it is used.

5. How much salt does ordinary sea-water contain? At what density should the water in the boiler be kept? Describe how the density of the boiler water is determined, and how it is kept from becoming too great.

6. Draw a section of a marine boiler, and insert in, or point out on, the drawing, the following parts :

> Combustion chamber,
> Fire bridge,
> A stay tube,
> An ordinary stay,
> Damper door (half open), and
> The smoke-consuming apparatus.

7. Describe the action of the slide valve in admitting and cutting off the steam, explaining what is meant by the terms *lead* and *cushioning*.

8. Describe either a jet condenser *or* a surface condenser, showing how the vacuum is maintained, and mention the temperature and pressure when in good working order.

9. Describe the arrangements by which the piston works steam tight in the cylinder and the piston rod in the cylinder end.

10. Describe the principal action of the screw propeller on the water, to which the onward motion of the ship is due.

At how many revolutions per minute must a screw propeller of 20 feet pitch be driven that the speed of the ship may be 12 knots an hour, supposing the slip to be 8 per cent ?

VII.—FRENCH.

(Time allowed, 2 hours.)

(*Dictionaries are permitted, except for the grammatical questions.*)

I. Translate into French :

Learning, on its revival, was held in high estimation by the English princes and nobles. The four successive sovereigns, Henry, Edward, Mary, and Elizabeth, may be admitted into the class of authors. Queen Catherine Parr translated a book ; Lady Jane Grey, considering her age, her sex, and her station, may be regarded as a prodigy of literature. Queen Elizabeth wrote and translated several books, and was familiarly acquainted with the Greek as well as Latin tongue. It is pretended that she made an extemporary reply in Greek to the University of Cambridge, who had addressed her in that language. It is certain that she answered in Latin without premeditation, and in a very spirited manner, to the Polish ambassador, who had been wanting in respect to her. When she had finished, she turned about to her courtiers and said, "S'death, my Lords," (for she was much addicted to swearing,) "I have been obliged to scour up my old Latin, that hath long lain rusting."—HUME.

II. Translate into English :

Science et travail.

"Le monde appartient à l'énergie," disait Alexis de Tocqueville, " il n'y a jamais d'époque dans la vie où l'on puisse se reposer; l'effort au dehors de soi, et plus encore au dedans de soi, est aussi nécessaire et même bien plus nécessaire à mesure qu'on vieillit que dans la jeunesse. Je compare l'homme en ce monde à un voyageur qui marche sans cesse vers une région de plus en plus froide, et qui est obligé de remuer davantage à mesure qu'il va plus loin. La grande maladie de l'âme, c'est le froid. Et pour combattre ce mal redoutable, il faut, non seulement entretenir le mouvement vif de son esprit par le travail, mais encore par le contact de ses semblables et des affaires de ce monde." L'exemple personnel de l'auteur de ces paroles vient à leur appui et le confirme. Au milieu de ses grands travaux il perdit la vue, puis la santé, mais jamais il ne perdit l'amour de la vérité. Lorsqu'il fut réduit à un tel état de faiblesse qu'il fallait qu'une garde-malade le portât dans ses bras de chambre en chambre, comme un frêle enfant, son courage ne l'abandonna pas.—FLAMMARION.

III. GRAMMAR:

1. Give the present subj., 2d pers. sing. and plural, of the verbs *appartient, puisse, vieillit, obligé, va, combattre, faut.* Give also the present and past participle of each of those verbs.

2. Form negative sentences with the following: " Le monde appartient à l'énergie"; "C'est le froid"; and state the rule for the formation of negative sentences.

3. "Un voyageur." Give the feminine form of that word; also of. *tuteur, auteur, voleur, buveur, mineur, vengeur.*

4. "Qui est obligé": with which part of the sentence does the participle *obligé* agree? State the rules.

VIII.—PRACTICAL NAVIGATION.

(Time allowed, 1 hour.)

1. Required the compass course and distance from A to B:

Lat. A...... 41° 56' } N.   *Var. 1¾ pts. W.   Long. A...... 43° 27' } W.
B...... 29° 52' }   B...... 18° 56' }

2. May 29, 1878, at noon, a point of land in lat. 51° 26' S. and long. 47° 28' E. bore by compass NW. by W., distant 17 miles (ship's head being SE.),* afterwards sailed as by the following log account; find the latitude and longitude in on May 30th at noon:

H.	K.	₁₀ths.	Course.	Wind.	Leeway.	Remarks.
1	7	5	SE.	NW. by N.		
2	8	0				
3	8	5				
4	9	0				
5	11	5	N. ¼ W.	W.	¼	
6	12	0				
7	13	5				Variation of compass, 2¾ pts. W.
8	14	0				
9	9	0				
10	10	5				
11	11	5	E. by N. ¾ N.	SE.	¼	
12	7	0				
1	5	5				
2	3	0				
3	9	0				
4	7	5	W.	NW. by N.	¼	
5	6	0				
6	12	0				
7	14	5				
8	8	5				
9	7	5	SE. by E. ¼ E.	NE.	1	
10	6	0				
11	7	0				
12	3	0				

3. What stars of the first and second magnitudes will pass the meridian of a place in long. 117° E., between the hours of 11ʰ p. m. October 8 and 1ʰ 20ᵐ a. m. October 9?

4. September 27, 1878, in long. 101° E. the observed meridian altitude of the sun's L.L. was 47° 26' 50" (zenith S. of the sun), the index correction was + 2' 40", and the height of the eye above the sea was 15 feet; required the latitude.

5. April 20, 1878, in long. 56° 30' W., the observed meridian altitude of the moon's L.L. below Pole was 18° 26' 30", the index correction was — 2' 40", and the height of the eye above the sea was 18 feet; required the latitude.

6. April 7, 1878, in lat. 24° S., long. 39° 26' W., when a chronometer showed 3ʰ 44ᵐ 1ˢ, the observed altitude of the sun's L.L. (near the meridian) was 58° 54' 20", the height of eye above the sea 23 feet, the index correction was + 0' 50", and the error of chronometer fast on G.M.T. 1ʰ 17ᵐ 49ˢ; required the latitude.

---

*See table of deviations.

7. September 18, 1878, in lat. by account 35° S., the following double altitude of Altair was observed:

Time by chron.	Obs. alt. L.L.	Bearing.
7ʰ 8ᵐ 14ˢ	12° 51' 20''	E. by N. ¼ N.
11ʰ 3ᵐ 25ˢ	45° 42' 13''	N. by E.

The run of the ship in the interval was N. ¼ W. 15 miles, the index correction + 2' 10'', and the height of the eye 22 feet; required the true latitude at the last observation.

8. January 2, 1878, at 10ʰ 5ᵐ p. m. (mean time nearly), in lat. 48° 2' N. and long. 133° W., when a chronometer showed 4ʰ 14ᵐ 13ˢ, the observed altitude of the planet Mars, W. of mer., was 22° 51' 30'', the index correction was — 0' 20'', and the height of the eye above the sea was 28 feet; required the longitude.

On November 26, 1877, at noon, the chron. was slow on G.M.T. 2ʰ 45ᵐ 15ˢ, and its daily rate was 3.4 seconds gaining.

9. July 7, 1878, at 3ʰ 30ᵐ p. m., in lat. 43° 46' N. and long. 141° W., the following lunar was taken:

Obs. alt, sun's L.L.	Obs. alt. moon's L.L.	Obs. dist. N.L.
42° 57' 7''	18° 58' 57''	99° 10' 55''
Index error — 4' 20''	+ 2' 20''	+ 3' 30''

The height of the eye above the sea was 21 feet; required the longitude.

10. October 30, 1878, about 5ʰ 25ᵐ p. m., in latitude 31° 20' N., longitude 107° W., the sun set by compass W. 16° 20' S., the ship's head being west; required the variation of the compass.

11. December 2, 1878, about 8ʰ 25ᵐ a. m. mean time, in lat. 22° 21' S., long. 43° 12 ' W., the sun bore by compass S. 57° 20' E., when a chronometer showed 9ʰ 16ᵐ 12ˢ. The error of the chronometer slow on G.M.T. was 1ʰ 59ᵐ 26ˢ, and the ship's head WSW.; required the variation of the compass.

Deviation of the compass (caused by the local attraction of the ship) for given positions of the ship's head.

Direction of ship's head.	Deviation of compass.	Direction of ship's head.	Deviation of compass.
N.	2° 45' E.	S.	3° 0' W.
N. by E.	4 57	S. by W.	4 20
NNE.	7 30	SSW.	5 0
NE. by N.	9 0	SW. by S.	6 7
NE.	10 0	SW.	7 0
NE by E.	10 55	SW. by W.	7 27
ENE.	10 40	WSW.	7 50
E. by N.	9 55	W. by S.	8 20
E.	8 50	W.	8 50
E. by S.	7 15	W. by N.	8 10
ESE.	5 35	WNW.	6 50
SE. by E.	3 40	NW. by W.	5 40
SE.	1 50	NW.	4 50
SE. by S.	0 20 E.	NW. by N.	3 20
SSE.	0 56 W.	NNW.	1 40 W.
S. by E.	2 20	N. by W.	1 10 E.

Answers to navigation.

1. S. 40° E.; 1390 m.
2. Lat. in 50° 52' S.; long. in 48° 11' E.
3. From β Ceti to α Arietis.
4. Lat. 43° 53' S.
5. Lat. 82° 7' S.
6. Lat. 23° 48' S.
7. Lat. 35° 1' S.
8. Long. 133° 8' W.
9. Long. 141° 18' W.
10. Var. 8° 42' E.
11. Var. 14° 52' W.

IX.—THEORY OF NAVIGATION AND NAUTICAL ASTRONOMY.

(Time allowed, 3 hours.)

1. Define, giving diagrams, the terms: Axis of the Earth, Departure, True Zenith, Ecliptic, Latitude and Longitude of a Heavenly Body.

Show that if the declination of a star be equal to its longitude, its latitude must be equal to its Right Ascension.

2. Investigate a formula connecting Diff. Long., Mer. Diff. Lat. and Course.

3. Explain the terms: First Point of Aries, Sidereal Time.

The time of transit of the first point of Aries, given in the Nautical Almanac for July 4, is $17^h 7^m 44.1^s$; find the right ascension of the mean sun at Greenwich mean noon on July 5.

4. What should be the length of the knot on the log-line to correspond to a 33-second glass?

If the distance entered in the log was 135 miles, the actual distance sailed being 118 miles, find the length of knot used.

5. Show how to find the latitude by altitude of a heavenly body near the meridian below pole, having given the approximate latitude.

Give approximately the limits of latitude in which this observation is possible in the case of the moon.

6. (a.) Define the term Polar Angle, used in a double altitude observation.

In a double altitude of a planet, the elapsed time, as measured by a chronometer keeping mean time, was $3^h 56^m 30^s$, and the decrease of the planet's right ascension in the interval was 31.5 seconds. Required the polar angle.

(b.) Prove the rule for finding the index error of a sextant by measuring the sun's diameter on and off the arc.

7. At a certain place the sun's altitude at 6 o'clock was 11° 50', and his azimuth N. 85° 30' W. Required the declination and latitude of the place.

8. A ship sailed from A to B, a place to the westward, on the arc of a great circle, the distance being 6,150 miles. Find the latitude and longitude of B, having given the latitude of A, 57° 58' S., longitude of A, 17? 39' E., and latitude of the vertex 61° 54'.

9. Calculate the augmentation of the moon's semi-diameter, when the apparent altitude of the center is 30° 46', the horizontal parallax being 56' 10'', and the horizontal semi-diameter 16' 5''.

10. What is the sun's declination when the day is 3 hours longer in latitude 55° than in latitude 15°?

X.—WINDS AND CURRENTS.

(Time allowed, 2 hours.)

1. Explain in detail how the northern hemisphere is divided into distinct basins by the several ranges of mountains upon its surface.

2. Give an account of the winds which prevail (1) on the coast of Brazil, (2) on the west coast of Australia.

3. Give an account of the monsoons of the China seas.

The interior of China is but little known, but it is considered probable that the monsoons prevail over a great part of the Chinese continent. What reasons may be adduced in support of this theory?

4. What are the characteristics of the West Indian hurricane?

It has been observed that the path of these storms coincides for some distance with the course of the Gulf Stream. How has this been explained?

5. Describe the Cape St. Roque current. What remarks have been made by Maury and by Horsburgh, respectively, relative to this current?

6. What currents have been noticed in the Grecian Archipelago? What bearing has

the existence of such currents upon the theory of ocean currents based upon the effects of excessive evaporation ?

7. Give an account of the system of currents of the South Pacific Ocean.

8. Explain the agency of currents in causing (1) the fogs of Newfoundland, (2) the heavy sea found off the Cape of Good Hope.

XI.—NAUTICAL SURVEYING.

(Time allowed, 3 hours.)

1. Construct a Mercator's chart on a scale of 1.15 ins. = to a degree of longitude, extending from lat. 54° to 58° N., and long. 32° to 37° E.

A ship sailed from lat. 56° 25′ N., long. 32° 10′ E., as follows:

Mag. courses.	Dist.	
SW. by S	60′	
SE. by E. ¼ E	115′	
NNE. ¼ E	120′	Variation 33° W.
NW. ¼ W	55′	
SW.	95′	

Required lat. and long. arrived at.

2. Chron^r slow on mean time, Montevideo (long. 56° 10′ W.), 1^h 14^m 14^s, daily rate losing 1^s.8—fourteen days afterwards at sea.

Summit of Tristan d'Acunha in lat. 37° 17′ S., long. 12° 36′ W., bore N. 38° E. (*true*), distant 5 miles, when chron^r was slow on mean time ship 4^h 09^m 0^s.

How much had the rate changed?

3. *A*, *B*, *C* are in line on an E. by N. bearing. *B* in the center is equidistant from *A* and *C* 1.5 miles. Variation 10° easterly.

At ship.
A 48° 30′ *B* 50° 15′ *C*.

Protract on scale of 1.5 ins. = a mile, and fix ship's position.

4. At *Y*, *X* was elevated 7° 12′; at *X*, *Z* was elevated 2° 12′.

Horizontal distance between *Y* and *X*=1,350 feet.

Horizontal distance between *X* and *Z*= 980 feet.

Required the height in feet of *Z* above *Y*.

5. By means of ruler and compasses construct a compass having a diameter of 6 inches, marked to half points, and the variation being one point westerly, show the direction of the magnetic meridian.

EXAMINATION PAPERS: BEAUFORT TESTIMONIAL.

I.—MATHEMATICS.

(Time allowed, 3 hours.)

1. If *y* be the harmonic mean between *x* and *z*, and *x* and *z* respectively the arithmetical and geometrical means between *a* and *b*, prove that

$$y = \frac{2(a+b)}{\left\{ \left(\frac{a}{b}\right)^{\frac{1}{4}} + \left(\frac{b}{a}\right)^{\frac{1}{4}} \right\}^{4}}.$$

2. Find an expression for the number of combinations of *n* things taken *r* at a time.

There are 10 white and 6 red balls in a bag; in how many different ways may 6 balls be drawn out, so that there may be at least 2 red balls each time ?

3. The sides about the vertical angle of a triangle are 25 and 16, and the line bisecting the vertical angle is 12. Find the base.

4. Investigate an expression for the radius of a circle inscribed in a triangle.

If α, β, γ be the angles which the sides of a triangle subtend at the center of the inscribed circle, show that

$$4 \sin \alpha \sin \beta \sin \gamma = \sin A + \sin B + \sin C.$$

S. Ex. 51——17

5. ABC, ABD are two spherical triangles such that AC + CB = AD + DB; if O be the middle point of AB, show that

$$\frac{\cos CO}{\cos DO} = \frac{\cos \dfrac{BC - AC}{2}}{\cos \dfrac{AD - BD}{2}}$$

6. Investigate the conditions of equilibrium when a weight W is supported on a rough plane inclined to the horizon at an angle α, by a power P inclined to the plane at an angle β, μ being the coefficient of friction.

7. One ball impinges on another ball at rest; find the condition in order that after impact their directions of motion may be at right angles, e being the coefficient of elasticity.

8. A cylinder is filled with equal volumes of n different fluids which do not mix; the density of the uppermost is ρ, that of the second 2ρ, and so on, that of the lowest being $n\rho$; show that the whole pressures on the different portions of the curved surface of the cylinder are in the ratio

$$1^2 : 2^2 : 3^2 \ldots : n^2.$$

II.—NAUTICAL SURVEYING.

(Time allowed, 1½ hours.)

In a running survey of part of a river, landed on the east bank from gunboat lying at anchor.

At Dead Tree sextant station. Height of eye above river 5 feet.

Lat. 4° 40′ N., long. 108° 30′ E. (by observation):

Gunboat's main truck...N. 56° 30′ W.
Gunboat's main truck elevated above mark on hull level with eye...........1° 43′ 6
Pagoda, 143° 30′. Gunboat's mainmast.

After which gunboat weighed and proceeded as follows:

Mag. course.	Dist. by pat. log.	Time on course.
S. 45° W.	4′ .9	1ʰ 30ᵐ
S. 80° E.	3′ .15	1ʰ
S. 17° W.	5′ .06	1ʰ 42ᵐ
S. 48° E.	2′ .92	0ʰ 45ᵐ

Then anchored a second time, observing on the south bank.

At Lime Kiln sextant station. Height of eye above river 5 feet.

Lat 4° 31′ N., long. 108° 32′ E. (by observation): ·

Gunboat's main truck at second anchorage in line with Pagoda..............N. 45° E.
Main truck elevated above mark on hull level with eye ....2° 8′ 51″
Pagoda elevated above same mark..0° 40′ 38″

Main truck to water line 95 feet. Bearings magnetic. Var. 9° W. Current against the gunboat throughout 1.2 knots an hour. Error of Pat. Log. one cable additive to each mile it indicates.

Protract dead reckoning track of gunboat on scale of one inch = mile of 2,000 yards.

On separate paper and same scale plot positions of sextant stations, anchorages, and pagoda (plane projection); ruling true and magnetic meridians.

Transfer the dead reckoning track to fit in between positions found by observation, so as to eliminate as far as practicable any error that may have crept in to the dead reckoning which cannot be accounted for.

Track about mid-channel. River a mile wide, with swampy banks. Pagoda on hill having trees around.

Sketch in details. Mark current. Give height of Pagoda above river's surface.

III.—FRENCH.

(Time allowed, 1½ hours.)

I. Translate into French:

St. Paul at Athens.

At Athens, at once the center and capital of the Greek philosophy and heathen superstition, takes place the first public and direct conflict between Christianity and Paganism. Up to this time there is no account of any one of the apostles taking his station in the public street or market-place and addressing the general multitude. Their place of teaching had invariably been the synagogue of their nation, or, as at Philippi, the neighborhood of their customary place of worship. Here, however, Paul does not confine himself to the synagogue, or to the society of his countrymen and their proselytes. He takes his stand in the public market-place (probably not the Ceramicus, but the Eretriac Forum), which, in the reign of Augustus, had begun to be more frequented, and at the top of which was the famous portico from which the Stoics assumed their name. In Athens, the appearance of a new public teacher, instead of offending the popular feelings, was too familiar to excite astonishment, and was rather welcomed as promising some fresh intellectual excitement. In Athens, hospitable to all religions and all opinions, the foreign and Asiatic appearance, and possibly the less polished tone and dialect of Paul, would only awaken the stronger curiosity.

II. Idiom:

1. Translate or explain: "Je ne puis m'en passer"; "cela me passe"; "passez votre chemin"; "il ne m'a pas payé de retour"; "mettez vous en mesure"; "vous n'êtes pas en mesure"; "vous chantez faux"; "nous avons levé l'ancre."

2. Render into French: "Mind what you are about"; "my son is not yet of age"; "mind your business"; "rich as you are you will not succeed"; "you shall have it, at a bargain."

3. State the difference between *plutôt* and *plus tôt*, *dessin* and *dessein*, *veille*, *vieille*, *vielle*.

III. Grammar:

1. Give the present subjunctive, second person singular and plural, of each of the following verbs, *enlever, noyer, jeter, appeler, peler, voir, savoir, mouvoir*.

2. State the rule of formation of the plural of compound nouns, taking as instances, *chou-fleur, grand'mère, sauf-conduit, chef-d'œuvre, tête-à-tête, hôtel-Dieu, contre-danse, arrière-garde, sans souci, sans-culotte, porte-voix, porte-feuille*.

3. State the rule of agreement of the past participle in the reflective verbs. Give instances.

IV.—NAUTICAL ASTRONOMY, &C.

(Time allowed, 2 hours.)

1. On May 20, 1878, in latitude 30° 53′ N., long. 77° 2′ W., the sun had equal altitudes at the following times by chronometer:

A. M.	P. M.
3ʰ 15ᵐ 7.5ˢ.	10ʰ 28ᵐ 22.5ˢ.

Required the error of the chronometer on G.M.T. at noon on May 20th.

2. Explain the causes of twilight. Investigate an expression for finding the sun's declination at a given place, on the day when the twilight is shortest.

3. The sun's declination being 10° 12′ N., find the latitude of the place where it reaches the prime vertical, when two-thirds of the time between sunrise and noon have elapsed.

4. What is meant by the Thermal Equator?

Show how the prevailing winds on the west coast of North and South America are affected by the variations in its geographical position.

5. What are the characteristics of the wind of Switzerland known as the Föhn?

It is supposed by some, that this wind draws its supply of air from the Caribbean Sea.

Discuss this theory.

V.—PHYSICS.

(Time allowed, 1 hour.)

1. Describe some experiments which illustrate the phenomena of diffusion, endosmose, and absorption of gases.

2. Describe accurately the method of taking an observation of the magnetic dip.

The true dip at a certain place is 60°. At what angle to the horizon will the needle set itself if it be placed (1) at angle of 60° to the magnetic meridian, (2) at right angles to the magnetic meridian?

3. Explain how you would determine the numerical value of the latent heat of steam at ordinary atmospheric pressure.

4. A small convex lens, focal length $\frac{1}{4}$ inch, placed close to the eye, is used to form a magnified image of an object $\frac{1}{4}$ in. from the lens. Draw a diagram showing the formation of the image, and calculate the magnifying power.

VI.—STEAM ENGINE.

(Time allowed, 2 hours.)

1. Explain what is meant by *efficiency of the boiler.*

State the theoretical value of a pound of coal, and what percentage of it is usefully employed in propelling a modern ship.

Mention the various causes of the loss of efficiency, throughout the whole propelling apparatus, and the average value of each loss, so that the total theoretical value of the coal may be accounted for.

2. Give a sketch of a large connecting rod, showing one of the ends in detail. .

3. Describe the action of the steam in a Compound Engine, and sketch indicator diagrams illustrating your answer.

4. Explain the action of the link motion when used to produce a variable cut-off.

MISCELLANEOUS EXAMINATION PAPERS GIVEN AT THE SESSION OF 1877-'78.

PHYSICS.

Examiner.—Prof. A. W. RÜCKER, M. A.

Class C.—Probationary lieutenants of Royal Marine Artillery.

1. Describe Nicholson's Hydrometer, and the method of using it to determine the specific gravity of a solid body.

If the sinking weight is 998 grams, and the weights, when a certain body is in the upper and lower pans, are 912.9 and 924.4 grams, respectively, what is the specific gravity of the body?

2. Enunciate the laws of liquid pressure. The fall of the water in a lock is 12 feet, and it is 10 feet deep when the lower gates are open. If the lock is 15 feet wide, and the gates when closed are supposed to be perpendicular to its length, find (1) the liquid pressure on the bottom when the lock is full, (2) the decrease in the pressure on the lower gates when it is emptied.

N. B.—1 cub. foot of water weighs 1,000 ozs.

3. Explain the principle of the barometer, and method of constructing one.

What would the height of a glycerine barometer be if the mercurial barometer stood at 30 inches?

Specific gravity of mercury 13.59
Specific gravity of glycerine 1.26

4. What are the distinctive characteristics of a magnet?
Give some account of a theory of its internal constitution.

5. Describe a method of making a magnet by means of (1) another magnet, (2) an electric current.

6. Distinguish between, and describe experiments illustrating, the different methods by which heat may be transferred from one body to another.

7. Define the specific heat of a body. How many pounds of water at 0° C. would be required to cool 1 cwt. of iron from 94° C. to 19° C. ?
Specific heat of iron = .114.

8. Describe a method of determining the coefficient of linear expansion of a solid. How does change of temperature affect the rate of a clock, and by what devices are the errors thus introduced compensated?

9. Enunciate the laws of refraction.
Why does a body immersed in water appear nearer to the surface than it really is?

10. Draw a diagram showing how to determine the position of the image of an object placed upon the axis of a spherical mirror.
Determine the position of the image of an object placed at a distance of 15 inches from a convex spherical mirror, the radius of which is 10 inches.

HEAT, ACOUSTICS, LIGHT.

Examiner.—Prof. A. W. RÜCKER, M. A.

CLASS A₁.—Students in naval architecture and marine engineering of second and third years. A₂.—Same of first year. B₁.—Lieutenants qualifying for gunnery and torpedo officers. B₂.—Voluntary captains, commanders, and lieutenants.

1. A cylindrical barometer tube extends nine inches above the upper surface of the mercury, and the volume of the "Torricellian vacuum" thus formed is four cubic inches. Into this space a quantity of air is introduced which, at a pressure of 32 inches of mercury, occupies $\frac{1}{4}$ cubic inch. By how much will the mercury be depressed if the temperature remains constant during the experiment?

2. Describe fully a method of determining the specific heat of a body.
Two grams of water at 50° C. are mixed with 15 grams of an oil at 5° C., the specific heat of which is .48. Find the temperature of the mixture.

3. Explain the principles and construction of Regnault's Hygrometer.
An air-tight box, the volume of which is one cubic meter, is filled with moist air at atmospheric pressure and at 26° C. If the dew point in the box is 17° C. and the box and its contents are reduced to 10.5° C., how much moisture will be deposited on the sides as dew?
Weight of one cubic meter of water-vapor at 0° C. and 760 m. m. = .81 kilogram.
Maximum tension at 17° = 14.5 m. m.
Maximum tension at 10.5° = 9.45.
Coefficient of expansion for heat = $\frac{1}{273}$.

4. How is the boiling point of a liquid determined and by what circumstances is it affected?

5. Describe experiments to prove—
(1.) That the power of conducting heat is different in different solid bodies, and—
(2.) That fluids conduct heat badly.

6. Explain and describe experiments to illustrate the statement that heat is a form of energy.
How many units of heat would be required to raise a shot weighing 1 cwt. to the top of a tower 96.5 feet high? If it were then projected in a horizontal direction with a velocity of 1,544 feet per second, how much heat would be generated if it were brought to rest on striking the ground?

Mech. equivalent of heat = 772 foot pounds.

7. Describe a method of producing Lissajous' figures, and explain the modifications they undergo when the tuning forks employed are not quite in unison.

8. How is the velocity of sound in air determined? To what temperature must carbonic-acid gas be heated that the velocity of sound in it may be one-tenth greater than in atmospheric air at 0° C.?

Density of air = 1.
Density of carbonic acid = 1.52.
Coefficients of expansion of both gases = .00366.
Ratio of specific heats of air 1.41.
Ratio of specific heats of carbonic acid 1.26.

9. What evidence have we for the fact that the light emitted by the heavenly bodies travels with the same velocity as that proceeding from terrestrial sources?

10. Define a lens, its axis, and its principal focus.

The image of the sun formed by a certain convex lens on its axis is 10 inches distant from the lens; the refractive index of the glass of which the lens is composed is 1.5, and its two faces are equally curved.

What is—
(1.) the focal length of the lens,
(2.) the radius of the spheres of which the faces are parts,
(3.) the shortest possible distance between an object placed on the axis of the lens and its image?

11. Explain the construction of the astronomical telescope, and show how to determine its magnifying power.

12. Describe the arrangements for obtaining a *pure* solar spectrum, and discuss the special characteristics and properties of its various parts.

MAGNETISM AND ELECTRICITY.

Examiner.—Prof. A. W. RÜCKER, M. A.

A_1.—Students in naval architecture and marine engineering, of second and third years. A_2.—Same of first year. B_1.—Lieutenants qualifying for gunnery and torpedo officers. B_2.—Voluntary students.

1. What is meant by a line of force?
Describe and explain the formation of the curves exhibited by iron filings in the neighborhood of a magnet.

2. How may the dip of the needle be accurately determined?
If the angle of dip is 45°, and a declination needle makes 25 oscillations per minute, how many would it make at a place where the dip is 60°, and the total intensity of the earth's magnetic force is one-fifth greater?

3. Describe experiments to prove that liquids and gases are affected by magnetic influences.

4. Explain the causes of the semicircular and quandrantal deviations of the magnet in iron ships.
If the maximum quadrantal deviation be 4°, what will it be when the ship's head is (1) N. by E., (2) NNE., and (3) NE.
sin 2 pts. = .383.
sin 4 pts. = .707.

5. Describe the construction, and explain the action, of a frictional electrical machine.

6. Explain the protective effect of lightning conductors. Mention any points to which special attention should be paid in their construction.

7. What is meant by specific inductive capacity, and how are the inductive powers of bodies compared?

8. Describe and explain the method of using a tangent galvanometer.

An electric current sent through a tangent galvanometer and a voltameter deflects the needle through $30°$ and electrolyzes .72 milligrams of water per minute. How much water will be electrolyzed in the same time if the strength of the current be increased so that the deflection becomes $60°$?

9. Give a short account of the principal phenomena of thermo-electricity. Describe the thermopile.

10. Under what circumstances are electric currents induced by other currents and by magnets ?

11. A galvanic current was sent through a tangent galvanometer, the resistance of which together with that of the wires which connected it with the battery was five ohms. When an additional resistance of 52 ohms was included in the circuit the tangent of the angle of deflection was reduced to one-third of its former value. Find the internal resistance of the battery.

12. A silver wire is joined at one extremity to an iron wire of the same length, and the other extremities being connected with the poles of a battery, a current of constant strength is sent through them for five minutes. The total quantity of heat generated in the wires is 45 units. How much is produced in the silver and how much in the iron wire?

The specific resistance of iron is six times that of silver, and the diameter of the iron wire is twice that of the silver one.

CHEMISTRY.

Examiner.—Dr. ATKINSON.

All students of first, second, or third years in naval architecture and marine engineering. Lieutenants qualifying for gunnery and torpedo officers. Voluntary students.

1. A specimen of niter is contaminated with common salt ; in what way would you purify it ? Give the tests for nitrates and for chlorides.

2. Describe the successive changes which take place when sulphur is subjected to the action of heat.

3. What simple methods are available for determining the amount of oxygen present in atmospheric air ? What natural processes are at work which tend to maintain uniformity in the composition of the atmosphere ?

4. Supposing that the sole products of the explosion of gunpowder are $3CO_2$, N, and K_2S, of what substances, and in what proportions, must gunpowder consist to yield precisely these products? $C=12$, $O=16$, $K=39$, $S=32$.

5. When chlorine and steam are passed through a red-hot tube the following decomposition takes place :

$$Cl_2+H_2O = 2HCl+O.$$

What volume of oxygen measured at $15°$ C. and under a pressure of 760 m.m. would result from the decomposition, under these circumstances, of 18 grams of steam ?

6. What are the conditions for the production of a voltaic current ? Mention some instances of accidental voltaic combinations.

7. By what tests would you distinguish between nitrogen and carbonic oxide ; between marsh gas and olefiant gas ; and between sulphurous acid and sulphuretted hydrogen ?

8. Describe the preparation and collection of ammoniacal gas. What change takes place when it comes in contact with hydrochloric-acid gas ?

9. How may ozone best be prepared in quantity ? How may it be distinguished from oxygen, and from peroxide of hydrogen ?

10. Describe the principal allotropic varieties of carbon, and their chief uses ; and explain by reference to experiment the antiseptic properties of charcoal.

MARINE SURVEYING.

Examiner.—Lieutenant DAWSON, R. N.

B₁.—Lieutenants qualifying for gunnery and torpedo officers.   B₂.—Voluntary students.

1. Explain the different methods used for measuring a base line in marine surveying.

2. May 8, chronometer fast on mean time, Mauritius, 3ʰ 45ᵐ 46ˢ, daily rate gaining 2ˢ.85.

The same chronometer fast on mean time, Aden, 4ʰ 36ᵐ 21ˢ.75, at noon of May 19, and at noon of May 26 fast on mean time, Aden, 4ʰ 36ᵐ 46ˢ.25.

Ascertain the Aden rate: also the meridian distance between Mauritius and Aden, using the mean of the rates.

3. From summit A, 239 feet above the sea surface, B was elevated 5° 45′ and C depressed 2° 12′.

Horizontal distance between A and B 1,892 feet.

Horizontal distance between A and C 1,530 feet.

Required the heights of B and C above the sea surface.

4. From gunboat in 3½ fathoms, frigate in 6 fathoms bore SSW. (true), distance 2 miles, and sloop in 5 fathoms bore W. ½ S. (true), distance 1.5 miles.

At launch in 2½ fathoms.

Gunboat, 24° 30′.   Sloop, 35° 18′ frigate.

Soundings in fathoms at equal distances between sloop and launch 4½, 4, 3. Reduction to low water 9 feet.

Protract on scale of 2 ins. = a mile, and place position of launch and the *reduced* soundings on the plan.

5. In survey of a harbor △ⁱ were selected at *Windmill* on table land on east side of entrance, at *Church* on hill about a mile inland from the harbor's head, and at *Tree* on a projecting headland on southwest side of entrance.

Working from a measured base line the distance between *Church* △ and *Windmill* △ was calculated to be 18,159 feet.

At *Windmill* △.   Lat. 20° N.

	°	′	
Zero *Church* △	360	0	Magnetic bearing N. 45° W.
Sun's center................................	125	37	} Corrected December, 18° 5′
Sun's center, altitude of			} N. for a. m. true bearing.
Minaret Φ end of sand in East Bay	7	0	
Tree Point	291	0	
Tree △	298	50	
Spur in tree range.........................	310	0	
Cone in tree range.........................	316	10	
South end of reef extending from Black Point..	329	0	
Stony peak Φ Black Point	335	30	

At *Church* △.

	°	′
Zero *Windmill* △	360	0
Windmill Point	15	0
South end of reef.......................	44	0
Black Point............................	48	0
Tree Point.............................	51	0
Tree △	57	30
Bight of West Bay	69	0
Spur in tree range......................	75	10
Cone in tree range	89	30

	°	′
Stony Peak	103	30
Minaret	347	0
End of sand in East Bay	357	50

At *Tree* △.

	°	′
Zero *Church* △	360	0
Black Point	7	0
South end of reef	14	30
Minaret	20	30
End of sand in East Bay	46	0
Windmill △	61	20
Tree Point	120	0
Spur Φ Cone in tree range	290	0
Stony Peak Φ bight of sand in West Bay	329	0

West Bay (shore of sand) curves between Tree Point and Black Point; East Bay (shore of sand) curves between Black Point and end of sand. Black Point is a rocky shelf dividing East from West Bays. Reef of rocks awash extends southward from Black Point. Coast between end of sand and Windmill Point straight and cliffy. Minaret surrounded by scattered houses. Cultivation around the head of the bay. Tree range of hills extends between *Church* and *Tree* △s, decreasing in height towards the latter.

Project and give a sketch of this work on a scale of 1.5 inch = a mile of 6,053 feet, ruling the true and magnetic meridians through *Windmill* △.

N. B.—*Windmill* △ *should be about 3 inches from the right-hand edge of the paper.*

NAVIGATION AND NAUTICAL ASTRONOMY.

Examiner.—H. B. GOODWIN, Esq., M. A.

B₁.—Lieutenants qualifying for torpedo and gunnery officers. B₂.—Voluntary students.

1. Explain the terms: Departure, Longitude, Course, Rhumb Line, Great Circle.

A ship sails a certain number of miles on a given course from a place on the earth's surface, given in position. By what methods may the longitude in be determined, and under what circumstances should one or other of these methods be preferred?

A ship sails from lat. 39° 21′ S., long. 55° 36′ E., first N. by W. 718 miles and then NE. by E. 256 miles. Find the latitude and longitude in.

2. Explain, with the aid of diagrams, the terms: Rational Horizon, Celestial Meridian, Ecliptic and Prime Vertical, Sidereal Time.

At two places on the earth's surface the ecliptic coincides with the prime vertical, at the first when the sidereal time is 6 hours, at the second when it is 18 hours. Required the latitudes of the two places.

3. A ship that can sail 5 points off the wind 6¼ knots wishes to reach a point 137 miles NW. by W. of her in two boards. She starts on the port tack, the wind being NW. by N. After what interval should she go about?

4. Prove the rule for finding latitude by meridian altitude of a heavenly body.

At the summer solstice of the northern hemisphere the meridian zenith distance of the sun was three times as great as the latitude at two different places. Find their respective latitudes.

5. Explain what is meant by the Polar Angle in a Double Altitude observation, and give the method of finding the Polar Angle in a Double Altitude of two stars taken at different times.

On December 22, 1878, in latitude by account 45° N., the following double altitude of stars was observed:

True altitudes.	Time by chronometer.	Bearing.
Regulus 49° 29′ 30″	6ʰ 15ᵐ 27ˢ	SW. ¼ W.
Arcturus 61 11 0	7 29 59	SE. by S.

Required the true latitude.

6. November 19, 1878, in lat. 25° 39' S., and long. 45° 7' E., at 8ʰ 40ᵐ p. m., the altitude of Jupiter's center (W. of meridian), observed by artificial horizon, was 63° 22' 20'', when a chronometer showed 6ʰ 12ᵐ 11ˢ, the index correction being — 2' 5''. Required the error of the chronometer on G.M.T., and its rate, having given the error on G.M.T. at G.M. noon, October 30, 35ᵐ 17ˢ fast.

7. Explain the method of finding variation by amplitude. Why cannot this observation be depended upon in high latitudes?

Find the compass bearing of the sun at rising, at a place in latitude 27° 40' S., when the declination is 7° 40' N. (Variation 18° W., deviation 7° E.)

8. Explain the method of finding error of chronometer by equal altitudes, and investigate an expression for the equation of equal altitudes.

9. Explain the meaning of the terms: Latitude and longitude of a heavenly body. Find the true distance between the sun and moon from the following data:

    Latitude moon 4° 49' N.   Longitude sun 323° 42'
    Longitude moon 77 0

10. In what latitude north is the day 1ʰ 48ᵐ longer when the sun's declination is 20° N. than when it is 10° N.?

FORTIFICATION.

Examiner.—Capt. B. G. HALL, R.M.A.

Gunnery Lieutenants, Marine Officers, Probationary Lieutenants R. M. A.

1. Given a plan of a house without any scale attached. On inspection you find that the front of the building is figured 90 feet, and measures 4.8 inches. Construct a scale of feet for the plan.

2. When suitable cover does not exist naturally, what is the readiest means of obtaining it for infantry in line?

Draw a section—adding dimensions—of the cover you would construct for this purpose, provided half an hour only were available—scale 10 feet to 1 inch. Also state the number of men and tools per 100 feet linear of cover required for its construction.

Indicate, generally, its position with reference to the crest line of the hill, when it is thrown up for the defense of an elevated position with steep and well-rounded slopes, similar to those met with in the chalk hills of the south of England. State your reasons.

3. State the general principles which should be observed in preparing brick walls, not exposed to artillery fire, for defense.

Illustrate them by a sketch showing a wall from 10 to 12 feet high thus prepared. What modification would you propose in the above if liable to be cannonaded?

4. Show, by a hand sketch, the plan and section, through the embrasure, of a Battery for two guns, suitable to the field artillery now in use in our service, and applicable to a site which has little command over the surrounding country.

Give approximately the time and the number of men required for its construction.

5. A field-work parapet has to be constructed on level ground with a "command" of 8 feet, and of sufficient thickness to resist modern field artillery.

Show, by hand sketch, the profile you would adopt, and mark the various dimensions and slopes, which are to be those ordinarily used in average soil.

Show also the calculations by which the width of the ditch has been determined.

If the above parapet had to be constructed in haste, and labor were abundant, how far would you modify the profile, retaining the same amount of cover?

6. Enumerate the principal obstacles used in conjunction with field works.

Where are they usually placed, and what conditions should they fulfill to render them efficient?

If the obstacle could not be protected from artillery fire, which would you prefer to use?

7. What is a revetment in field fortification?

Where and why is it required?

Describe, and illustrate by hand sketch, a Gabion revetment of the interior slope of a parapet provided with a banquette.

8. Having regard to the very destructive effect of modern artillery shell fire, how would you protect the garrison of a field redoubt from suffering inordinate losses?

Show, by hand sketches, sections of the constructions you would employ for the purpose.

9. What is the object of the Glacis in permanent fortification, and how are its height and slopes regulated?

10. What is a "ravelin"?

State the advantages of large "ravelins" over small ones.

11. Give the advantages claimed for the polygonal over the bastioned system.

12. Describe the construction of a bridge of boats across a river, and illustrate, by hand sketch, how the boats should be fitted.

How would you calculate the available buoyancy of a boat? and what modifications in construction would you make in a tidal river?

PERMANENT FORTIFICATION.

Examiner.—Capt. B. G. HALL, R. M. A.

Marine Officers, Probationary Lieutenants R. M. A.

1. Define the terms Terreplein of the Rampart, Covered Way, Full Revetment, Demi-Revetment, Outworks, and Detached Works.

2. What rules should be attended to in determining the height of the escarp of a fortress? A marked feature of works of modern construction is a deep but narrow ditch. To what do you attribute it?

3. Give a hand sketch of a "detached escarp." What are its advantages and defects?

4. What are the principal defects of the bastioned system?

5. Upon what does the efficiency of the flank defense of the enceinte of a fortress depend? Describe how this has been obtained in the polygonal system.

6. State the defects of the ordinary embrasure of earthen parapets. Why do you find them still retained in many of our coast batteries?

7. What are the considerations which have led to the adoption of a chain of detached forts for the defense of our dockyards? State the principal advantages of this system of defending important points; and what were the reasons why the forts around Paris in 1870–71 failed to secure to the full these advantages?

8. In what positions are casemated batteries used? Describe briefly the modern type now used for sea defenses.

9. Explain, generally, the different methods by which the capture of a fortress may be attempted. Judging from the experience of late wars, which method do you think would, under ordinary circumstances, be most likely to be attended with success against a well-found modern fortress? State your reasons.

10. Describe and illustrate by sketch a screen gun battery.

HEAT, STEAM, AND COMBUSTION.

Examiner.—Prof. ALEX. B. W. KENNEDY.

CLASSES A₁ and A₂.

1. The efficiency of a boiler is said to be the product of the efficiencies of its furnace and its heating surface. Explain this, and state what quantities would have to be measured in practice in order to determine each.

2. The following is the analysis of a piece of Northumberland coal: what is (ap-

proximately) its theoretic calorific value if one-fifth of the carbon be burnt only to CO, the remainder to CO_2?

C	82. 4
II	4. 8
O	8. 1
N	2. 4
S	0. 3
Ash	2. 0
	100. 0

3. If 20 lbs. of air pass through the furnace per pound of fuel, and if the air enter the furnace at 60° Fahr., and the gases of combustion leave the boiler at 460°, what will be the available calorific value of the fuel?

4. How is the economic value of a bituminous coal, as well as the amount of smoke produced by its combustion, affected by the times and places at which air is admittted to the furnace?

5. How much work would be done in a perfect engine working between 1,250° Fahr. and 285° Fahr., for the expenditure of 40,000 thermal units?

6. State the conditions under which air and dry saturated steam, respectively, can undergo the following changes, or such of them as are possible for a given weight of both fluids:

(a) Change of volume at constant temperature.

(b) Change of pressure at constant temperature.

(c) Change of temperature at constant volume.

7. What is the total amount of heat required to raise 2.5 lbs. of water from 65° Fahr. to 302° Fahr., and to evaporate it at that pressure?

8. Explain the meaning and use of "pressures equivalent to" quantities of heat or work.

9. What is meant by saturated steam?

10. The mean effective pressure during a stroke is frequently taken as

$$= p_1 \left(\frac{1 + \text{hyp. log } r}{r} \right) - p_3,$$

where p_1 and p_3 stand for the initial pressure and back pressure, respectively, r being the ratio of expansion. Explain how this expression is obtained, and state to what extent it represents the mean pressure which would probably be obtained in practice.

11. Two engines of the same size work with the same initial and back pressures, one without expansion and one cutting off at .25 of the stroke. Compare the work done and the amount of steam used in the two machines.

12. Taking the diameter of the cylinders as 12 inches, and the piston speed as 320 feet per minute, and assuming $p_1 = 70$ pounds and $p_3 = 3$ pounds per square inch, calculate the horse power for the two cases mentioned in the last question.

ENGINE DESIGN AND CALCULATION.

Examiner.—JAMES WRIGHT, Esq.

CLASSES A₁ and A₂.

1. What amount of grate surface, of tube surface, and of total heating surface would you allow for the boilers of a compound engine to develop, at most 1,000 I. II. P., steam of 75 cwts. absolute, to be expanded 8 times?

2. Give an outline design of the boilers in three parts on the low cylindrical plan; size of shells, furnaces, and tubes to be given in figures.

3. Give sketches, with dimensions, showing how the combustion chamber and the ends of one of these boilers should be stayed.

4. Show the boilers placed in the vessel, and indicate the positions of the various fittings and mountings.

5. Give a sketch of a pair of spring-loaded safety-valves for one of these boilers, and state what the diameter of the valves should be.

6. Give a sketch of the stop-valve and its box, state the size of the valve and steam-pipe, and show how the steam-pipes for the set of boilers could be arranged.

7. What would be the proper dimensions of the high pressure and the low pressure cylinders of the engines to develop 1,000 I. H. P., assuming the piston speed to be 460 feet per minute?

8. Give a section of the high-pressure cylinder, showing the steam and exhaust ports, and state the dimensions of the ports.

9. Draw the valve diagram for the high-pressure valve, and give the travel, lap, and lead.

10. Give a formula for finding the strength of the crank-shaft, and give a sketch of the crank-shaft suitable for this engine, with figured dimensions of the principal parts.

11. Give a sketch of the crank end of the connecting rod, and state the diameter of the rod and of the bolts.

12. What amount of cooling surface would you give for the surface condenser of the above-mentioned engine? Give an outline sketch of the kind of surface condenser you would propose.

13. How would you determine the size of the air-pump and the area of the passages through its valves?

14. Give a sketch showing the valves at one end of the air-pump, and show how they should be arranged to insure efficient working of the pump.

15. Suppose the engines above mentioned would give the vessel a speed of 12 knots when developing 1,000 I. H. P., what diameter and pitch would you give to the propeller? Explain how you arrive at your conclusions.

STEAM AND STEAM ENGINE.

Examiner.—JAMES WRIGHT, Esq.

CLASSES A₁ and A₂.

1. What are the principal heat-producing constituents of ordinary coal? Give a short description of the process of combustion in the furnace of a marine boiler.

2. How much air is required for the complete combustion of 1 lb. of average coal? In what manner is the necessary quantity of air admitted for the combustion of the gaseous portions and of the solid carbon?

3. What is the evaporative power of 1 lb. of the best kinds of Welsh and North Country coal? How much water can be evaporated by 1 lb. of each respectively in a marine boiler under the most favorable conditions?

4. Explain what is meant by the efficiency of the furnace, the heating surface, and the boiler; and distinguish between efficiency and evaporative power of boiler.

5. For what purpose is the steam blast fitted in the chimney? When should it be used? What injurious effects are likely to arise from using it too freely?

6. Why is the steam blast generally preferred to blowing fans for producing a draught in the boilers? How many lbs. of coal can be burnt per foot of grate per hour in the furnace of a marine boiler with the ordinary chimney draught, and how much with the steam blast in use?

7. Which are the most valuable surfaces in the boiler for producing evaporation? Give reasons fully.

8. What is meant by the efficiency of the steam? What is its approximate value in the best types of marine engines?

9. What is the difference in the total quantity of heat contained in 1 lb. of steam at 70° lbs. pressure and in 1 lb. at 30° lbs. pressure?

10. Explain clearly why it is advantageous to use steam of high pressure in preference to steam of low pressure?

11. When is the greatest efficiency of the steam obtained at high, moderate, or low

powers? Distinguish between most efficient rate of working of engines and most economical rate of speed of ship.

12. Give a short description of the manner in which heat passes from the steam to the cylinder and back again to the steam during each stroke in an unjacketed cylinder.

13. How are the losses, due to the initial liquefaction at high rates of expansion, partially prevented by the adoption of the steam jacket? Explain fully.

14. How are the difficulties attending the use of a high rate of expansion with high-pressure steam overcome by the adoption of the compound engine?

15. State fully what are the merits and demerits of compound and simple expansive engines, respectively, for ships of war.

16. Enumerate the causes which produce loss of efficiency in the machinery and propellers, and give approximately the amount of loss due to each of these causes.

17. What does the total efficiency of boilers, steam-engines, and propellers amount to in the best types of marine engines?

18. What difficulties are experienced in the use of expansion valves? In cases where they are dispensed with how is the steam used for low powers?

19. In what manner is a gain realized by the substitution of surface condensation for jet condensation? Show what is the gain in any particular case, and state why it is necessary to use surface condensation with high-pressure steam.

20. What influence has the introduction of surface condensation had on the durability of boilers? In what way was the corroding action at first supposed to be set up, and what should be done to lessen the deterioration of boilers, both when in use and when not in use on board ship?

PRACTICAL ENGINEERING.

Examiner.—JAMES WRIGHT, Esq.

CLASSES A₁ and A₂.

1. Sketch and describe the apparatus usually fitted to the furnace bridges of a boiler for the admission of air, also a good arrangement of furnace door adapted for smoke burning.

2. Give the dimensions of an ordinary fire bar, show how it is placed in the furnace, and state what is the usual width of the space between the bars.

3. Make a vertical section of a rectangular boiler through the furnace and combustion chamber. Explain how the evaporating power of the heating surface is increased by the usual method of constructing the combustion chamber.

4. What inconvenience is frequently experienced when steaming with new boilers or with clean heating surface and forcing the fires? For what reason does the inconvenience cease to be felt after a time?

5. What advantages are gained by the use of the double-ended cylindrical boiler? What disadvantages compared with the single-ended boiler are introduced by the adoption of the double-ended one?

6. When a boiler at work is priming, what is usually done to stop the priming, and to prevent the damage arising from it?

7. What defects of construction, as distinguished from defects of treatment, may lead to priming in a boiler?

8. Make a vertical section of a low cylindrical boiler through one of the furnaces. Why is a hanging bridge fitted in the combustion chamber?

9. What pressure of steam is supplied to the jackets of the high and low pressure cylinders of a compound engine? How are the jackets usually formed?

10. How are the pistons of large horizontal engines supported, and what means are usually applied to facilitate the adjustment of the pistons?

11. To what causes may the heating of a piston-rod be due when the engine is at work, and what would you do to cool it without stopping the engines?

12. Why are mineral oils now used, instead of tallow or vegetable oil, for lubricating the internal working parts of the engines?

13. Why does not the vacuum line of an indicator diagram correspond with the vacuum as shown by the gauge on the condenser? On what, to a great extent, does the vacuum in the cylinder depend?

14. If a bad vacuum were found to exist in a surface condenser, what would you do to ascertain the cause of it, and how would you improve it, if it were not caused by defects of construction?

15. Give a sketch of the usual kind of plunger feed pump worked by the main engine, and show the suction and delivery valves.

16. What arrangement is usually fitted to a feed pump, taking water from a surface condenser, to prevent waste of fresh water?

17. Give a sketch of a thrust bearing, and explain how it can be adjusted.

18. Give a sketch and description of the bearings of a screw propeller fitted to admit of being hoisted.

STEAM AND STEAM ENGINE.

Examiner.—JAMES WRIGHT, Esq.

B₁.—Lieutenants qualifying for gunnery and torpedo officers. B₂.—Voluntary students.

1. What are the respective qualities of Welsh and North Country coal for steaming purposes in marine boilers; under what circumstances is it desirable to use a mixture of the two kinds, and what proportion of each constitutes the mixed coal used in Her Majesty's service?

2. Describe the process of combustion of coal, formation of smoke and flame. To what extent does loss arise in ordinary practice through incomplete combustion?

3. What are the methods in use on board ship for obtaining an air supply for the furnaces, what is the effect of the admission of air by other means than through the furnaces, and how would you estimate the intensity of draft produced by rarefaction in the chimney?

4. Explain how heat is transmitted from the hot furnace gases to the water in the boiler, and state how under similar conditions the rate of transmission varies in relation to difference of temperatures.

5. What conditions in ordinary practice affect the rate of transmission of heat to the water, and how is the efficiency of the heating surfaces of the boiler affected by their arrangement and position?

6. Give the average efficiency and evaporative power, respectively, of—
   (a) the ordinary rectangular boiler,
   (b) the high-pressure cylindrical boiler.

7. Give outline sketches and a description of the above-mentioned two kinds of boiler, as usually made for ships of war.

8. Name the various fittings and mounting applied to a marine boiler, and explain the purpose for which each is used.

9. Describe the course taken by the steam, in passing through a compound engine, from the time it leaves the boiler until it returns again to the boiler in the condition of feed water.

10. Explain why it is advantageous to use steam of high pressure instead of steam of low pressure; to what extent are we at present limited with regard to the maximum pressure which can be employed for marine engines and boilers? Give reasons.

11. What are the reasons for the alleged superior economy of the compound engine, and what are the advantages and disadvantages of this kind of engine for ships of war?

12. Sketch the indicator diagrams you would expect to obtain from the high and low pressure cylinders of a compound engine, and explain how the mean pressure of each is calculated.

13. The high-pressure cylinder of a compound engine is 40 inches, and the low-pressure cylinder is 70 inches in diameter; the mean pressure in the former is 30 pounds,

iu the latter 10 pounds per square inch. The length of the stroke being 3 feet, aud the number of revolutious per minute 60, what is the indicated II. P. ?

14. In what manner is a gain realized by the substitution of surface for jet condeu-. sation ? Why is it necessary to use surface condeusation with high-pressure steam?

15. What iufluence has the introduction of surface coudeusation had on the durability of marine boilers; iu what way was the corroding actiou at first supposed to be set up, and what should be doue to lesseu the deterioration of the boilers, both wheu in use aud when not in use ?

16. Enumerate the causes which produce loss of efficiency in the machiuery aud propeller of a ship, aud give approximately the amount of loss due to each of these causes.

STEAM AND THE STEAM ENGINE.

Examiner.—J. WRIGHT, Esq.

CLASS A₂.

1. Describe the actiou of the steam in au ordinary engine, aud also iu a compound eugine, and give sketches of indicator diagrams iu illustration of your answers.

2. Whether is it more necessary (*sic*) for the sake of ecouomy of fuel to steam-jacket the cylinders of an ordinary engine or those of a compound engine, the iuitial pressure of steam being the same in both cases ? Give the reasons for your answer.

3. Give a sketch of an iudicator diagram from an ordinary engine, explaiuiug by means of it the action of the steam in the cyliuder, and give sketches also of the diagrams you would expect to obtain if the engiues were out of order from various causes.

4. Give sketches of the indicator diagrams from a compoñud eugine with oue highpressure and one low-pressure cylinder and the cranks at right angles, and show how you would calculate the horse-power of the engine from the diagrams.

5. The indicator diagram of an ordinary engine shows that the initial pressure of steam was 30 lbs. above and the pressure at release 5 lbs. below the atmospheric liue, back pressure 3 lbs. What would be the mean pressure, and how much would the efficiency of the steam be increased by expansion ?

6. Communication valves and pipes are now sometimes fitted to compound eugines to allow the steam to be admitted directly to both cyliuders. Give a sketch of any arrangement of this kiud you are acquainted with, and state under what circumstances it would be used.

7. What are the advantages derived .from surface condensation as compared with jet condensation, and what disadvantages may arise from the use of the former ?

8. State what ecouomy you would expect to realize by heating the feed water from 120° to 200° by meaus of the waste heat going up the chimney.

9. Explaiu the actiou of the link motion when used to produce a variable cut-off. Give an explauatory diagram, aud state the disadvautages arisiug from the use of the link for this purpose.

10. What is meant by superheated steam, and why is ecouomy obtaiued by superheatiug ? To what extent may superheating be safely carried, aud why are superheaters now seldom fitted to high-pressure boilers ?

11. What is the usual amouut of heating surface iu a boiler, aud the amount of cooliug surface in a surface coudenser per I. H. P. ?

12. What are the several sources of loss aud waste in boilers and engiues, and what is doue in each case iu practice to reduce the loss aud waste as much as possible ?

PRACTICAL ENGINEERING.

Examiner.—J. WRIGHT, Esq.

CLASS A₂.

1. Give a sketch of a double-ported slide valve and the relief riug ou its back. Show also the ports iu the cylinder and the valve in the proper position when the piston is at the eud of its stroke.

2. Explain what is meant by outside lap, inside lap, lead, and angular advance.

3. Give a sketch and description of the link motion, and state how you would proceed to set properly a slide valve worked by this motion.

4. What description of expansion valve do you consider the best for the engines of a ship of war? State the reasons for your opinion.

5. Give a sketch and description of any plan of auxilary valves for facilitating the starting of engines.

6. Give a sketch of a connecting rod, showing the crank-pin end, brasses, and bolts in detail, with approximate figured dimensions for a crank-pin 12 inches diameter.

7. What would you do to ascertain whether a line of propeller shafting is true, and also whether the propeller shafting is true with the crank shaft of the engines?

8. In the same line of propeller shafting one part is of iron solid, and another of steel hollow, and both parts are of the same external diameter. What should be the diameter of the hole through the steel shaft to make it of the same strength as the iron shaft, the relative strengths of the materials being as two to one?

9. Give a sketch and description of the kind of lubricator you consider the best for the crank-pin of a horizontal engine.

10. If a bad vacuum were found to exist in a surface condenser when the engines are at work, what would you do to endeavor to find the cause of it?

11. How would you clean a surface condenser in which the steam is condensed inside the tubes when it becomes dirty?

12. Why is mineral oil now used for the lubrication of slide valves and cylinders, and what drawbacks attend the use of this oil?

13. Give a rough sketch of a circular high-pressure boiler with two furnaces and the tubes over the furnaces, and show on it the positions of the various mountings, including two water gauges and two steam gauges.

14. Give a sketch, with figured dimensions, showing the stays you would fit for the back and top of the combustion chamber of such a boiler as the above.

15. Give a sketch of the usual kind of spring-loaded safety valve now adopted for high-pressure boilers. Why is a spring better than a lead weight?

16. When boilers supply steam for engines with surface condensers, what causes prevent the return to the boilers of all the water evaporated, and how is any deficiency of feed water made up?

17. What measures should be adopted for the preservation of marine boilers when they are not in use?

18. How can the rate of combustion of coal in marine boilers be regulated and how can it be forced? What weight of coal per square foot of grate surface can be burned per hour at the highest rate?

THE THEORY OF THE STEAM ENGINE.

Examiner.—Prof. ALEX. B. W. KENNEDY.

CLASS A₁.

1. How much heat must be expended in raising the temperature of one pound of air from 700° to 1200° Fahr. at constant pressure, and how much of this heat goes to increase the internal energy of the gas?

2. Given the values of the specific heat of air at constant volume and pressure, deduce from them the value of J. (One cubic foot of air at atmospheric pressure and 32° Fahr. weighs 0.0807 pounds.)

3. Show how to represent graphically the whole amount of heat (spent in internal as well as in external work) expended in any operation performed on a given quantity of air. Show how your construction applies to the cases in which either external or internal work becomes zero.

4. The efficiency of an air engine is not affected by the ratio of expansion used in it, a quantity which is, however, of vital importance in connection with the efficiency of a steam engine. Explain this.

5. Supposing steam to be a perfect gas, give an expression for the constant pv; compare values thus obtained with those given by the formulæ representing the results of actual experiment.

6. The saturation curve of a given weight of steam may be drawn as au hyperbola; give the formula which shows this, and say how the asymptotes may be determined in a given case.

7. If steam be admitted to a cylinder and expanded r times, and the equation to the expansion curve be $pv^{\frac{m}{n}} = $ constant, show that the total work which can be done during admission and expansion may be represented by—

$$ p_1 \left(m - n \sqrt[n]{1} \right) - rp_3, $$

where p_1 and p_3 are the initial and back pressures respectively.

8. Give some explanation of the immense rapidity with which steam is condensed on the walls of a steam-engine cylinder.

9. It is required to carry out a thorough test of a common steam engine working with a jet condenser. What are the principal quantities which will have to be measured for the purposes of the test?

10. If the temperature of the injection in such a case be 45°, and that of the discharge 88° Fahr., the total discharge 200 pounds per minute, and the absolute pressure (obtained from the vacuum gauge) 2.9 pounds per square inch, find approximately the quantity of steam condensed per minute.

11. State what you know as to the nature and effect of the free expansion which usually occurs during some part of the stroke of a compound engine.

12. Describe accurately what takes place when steam is "wire-drawn," and investigate this phenomenon in connection with its drying effect upon wet steam.

STRENGTH OF MATERIALS AND STRUCTURES.

Examiner.—Prof. W. C. Unwin.

CLASS A₁.

1. Let AB, BC be two straight or curved rafters of a roof, assumed to be hinged at the springings A and C and at the ridge B. A force acts on AB, its direction passing between A and C. Find, graphically, the reactions at the springings, and, for the case when the rafters are straight, the bending moment at any point of the rafters.

2. Find by a diagram the center of gravity of a 5-sided frame of bars of uniform section.

3. Let AB, BC be the rafters of a braced roof truss; ADEC the polygonal tie rod; DB, EB, the bracing bars, and DF, DG, struts at the centers of the rafters. Suppose a horizontal force P, applied at F the center of the rafter AB. Draw a stress diagram, distinguishing tension bars from compression bars.

4. Let AB be two points on the same level between which a cord is suspended. Weights W_1, W_2, W_3, W_4 are hung on the cord at given distances from A, measured horizontally. Draw the position of the cord when the tension at A is equal to the sum of the loads.

5. Show that the best form of cross-section for a deep sewer is an ellipse. If ϕ is the angle of repose of the soil, show that the horizontal and vertical axes of the ellipse should have the ratio

$$ \sqrt{\left(\frac{1 - \sin \phi}{1 + \sin \phi} \right)} $$

6. State what forms of cross-section are adopted for the struts of wrought iron and cast iron, and give reasons for the selection of these forms.

7. Find the greatest twisting moment which can be applied to a shaft 100 feet long and 3 inches diameter, the angle of torsion not exceeding 1°. Modulus of transverse elasticity 10,300,000.

8. Show that for equal values of the limiting stress the resistance of a cylinder to twisting is double its resistance to breaking across.

9. Two bars of equal section and of the same material are employed to support a load. One bar is attached to an unyielding support 10 feet above the load, the other to a similar support 25 feet above the load. Determine the ratio of the stresses in the bars.

10. Give a short account of Wöhler's experiments on the ultimate resistance of materials, and state what inferences have been drawn from them as to the safe working stress of structures.

11. How can columns be divided into classes according to the proportion of length to diameter? State how in each case the column gives way when loaded, and give the formulæ suitable for calculating their strength.

12. State the conclusions at which Styffe has arrived as to the influence of the percentage of carbon in steel on the strength, ductility, and capability of welding. State also the influence of phosphorus, sulphur, and silicon on the properties of steel.

13. What is the resistance of a bar of wrought iron 10 feet long and 2 inches diameter, the proof load being 15 tons per square inch, and the modulus of elasticity 29,000,000 lbs. per square inch?

14. Would the deflection be greater for a round or square beam, of equal sectional areas, and loaded to the same intensity of stress?

STRENGTH OF MATERIALS AND STRUCTURES.

Examiner.—T. A. HEARSON, Esq., R.N., F.R.S.N.A.

Lieutenants qualifying for gunnery and torpedo officers. Voluntary students.

1. Describe the five simple straining actions or ways in which a piece of material may be broken.

If a bar be fastened into the ground in a sloping direction and a weight suspended from the upper end, what are the straining actions at any transverse section of the bar?

2. A simple triangular frame with tie bar horizontal, carries a load at the vertex. Describe how to draw a diagram representing the forces on the different parts of the frame, and show that the supporting forces are also represented in the same diagram. Show that the tension of the tie rod

$$= \tfrac{1}{4} \text{ load} \times \frac{\text{length of tie rod.}}{\text{height of frame.}}$$

3. Describe and make sketches showing three different ways in which a loaded beam supported at the ends may be strengthened by supporting it at more than one intermediate point.

4. Describe the common suspension bridge.

The platform of a bridge 200 feet long and 24 feet wide is loaded with 47 pounds per square foot. It is carried by a pair of suspension chains, which, at the center, dip 18 feet below the tops of the piers. Find the tension of the chain at the lowest point.

5. A beam is supported at the ends and loaded uniformly; show that the greatest bending moment on it is half what it is when the same load is concentrated at the center.

Draw a diagram representing the bending moment at any point of the beam in each case.

6. A steady load of 3 tons is carried by a bar $1\frac{1}{4}$ inches diameter. What is the intensity of the stress produced?

Suppose an additional load of 2 tons suddenly applied, what will be the total stress produced?

7. Draw and describe a diagram representing the extension of a wrought-iron bar under a gradually increasing load, stating the proof load and the breaking load.

Point out on the diagram how much work is required to stretch a bar to its elastic limit, and how much is required to produce rupture.

APPLIED MECHANICS: FIRST PAPER.

Examiner.—Prof. Alex. B. W. Kennedy.

Students in naval architecture and marine engineering of first year.

1. Compare a suspension bridge, a bowstring girder, and an arch as to their stress conditions under a stationary load, uniformly distributed.

2. There is in general both shear and bending moment at every section of a beam. What is the relation between them? What is the condition that there may be (a) only shear, (b) only bending moment at any particular section?

3. How must a pair of pulleys on shafts which are neither parallel nor intersecting be placed so that one may drive the other direct by a strap?

4. Given the position of two intersecting axes, show how to find the diameters of a pair of bevel wheels to transmit a given velocity ratio from one to the other.

5. The tup of a steam hammer weighs 500 pounds. It strikes a piece of iron 2 inches thick at a velocity of 12 feet per second and compresses it to 1 inch. Find the mean pressure on the iron while it was being compressed.

6. Supposing a crank to revolve uniformly, how is the velocity of a pump plunger, worked by it, affected by the length of the connecting rod by which it is driven? Show how to obtain the relative velocities of the plunger and the crank pin at any instant.

7. What is meant by the "efficiency" of a sliding piece? Show how to obtain the efficiency of a piston-rod guide block, fully proving the construction which you use.

8. Distinguish between the effects of (a) a stationary load, (b) a load applied suddenly but without impact, and (c) a blow, in straining a beam.

9. A bar of wrought iron 2 inches square is 120 inches long when unstrained. The modulus of elasticity of the material is 22,000,000 pounds per square inch. What will be the length of the bar when a load of 25 tons is suspended from it?

10. A floor joist 12 feet long carries a uniformly distributed load of 9 cwt. The joist is 12 inches deep by 3 inches thick. Find the maximum intensity of the stress in it.

11. The area of the stream driven backwards by a pair of paddle wheels is 24 square feet, and its velocity, relative to the ship, 7 feet per second. What is the resistance of the vessel?

12. Water issues from an orifice at a velocity of 12 feet per second, under a head of 5 feet. Find the coefficient of resistance.

APPLIED MECHANICS: SECOND PAPER.

Examiner.—Prof. Alex. B. W. Kennedy.

Students in naval architecture and marine engineering of first year.

1. Show how to construct a polygon whose ordinates represent the bending moments along a beam under a given system of concentrated loads. What is the relation of this polygon to the form which any flexible chain, hanging from the supports of the beam, would take under the same loads?

2. Show that the shear in a beam changes sign where the bending moment is a maximum.

3. Prove the form of uniform strength for a beam supported at one end and loaded uniformly (a), if its depth be constant, (b) if its breadth be constant.

4. How would you compare, as to relative strength and weight, two beams of the same length and the same extreme breadth and depth, but one solid and the other I-shaped?

5. A shaft 2.5 inches diameter and 50 inches long works under a twisting moment of 13,200 inch-pounds. What will be the linear displacement, under this moment, of a point at a radius of 10 inches, upon a lever fixed to the further end of the shaft?

6. What are the essential conditions which should be fulfilled by the surfaces of wheel-teeth?

7. A Cornish (non-rotary) engine has a cylinder 40 inches diameter, and in its downward stroke lifts pitwork which weighs 15,080 pounds. The mean steam pressure on the piston during the period of acceleration is 25 pounds per square inch, and during the retardation 12 pounds per square inch. The maximum velocity is attained when the piston has traveled 2 feet. Find (a) the length of the stroke made, (b) the maximum velocity of the piston. (Neglect all weights except that of the load mentioned.)

8. Distinguish between the action of a governor and a fly-wheel as regulators of the speed of an engine. What are the chief characteristics of a good governor?

9. On the assumption that the crank-pin of an engine rotates uniformly and that its connecting-rod is infinitely long, show that the resistance due to the acceleration of the reciprocating parts varies as the cosine of the angle between the crank and the axis of the engine. How is this modified if a short connecting-rod be employed?

10. What are the relative positions of crank, connecting-rod, and cross-head in an ordinary engine when (a) the velocities of the crank-pin and cross-head are equal, (b) the velocity of the cross-head is a maximum?

11. Show how to determine the efficiency of a square-threaded screw. Prove that it becomes infinite when the pitch angle is either 0 or Φ (the angle whose tangent is equal to the coefficient of friction).

12. Calculate the quantity of water delivered per second by a pipe 20 inches in diameter and 2,000 feet long, under a head of 25 feet. State the conditions to which the coefficient of friction which you adopt applies.

APPLIED MECHANICS. I.

Examiner.—Prof. W. C. UNWIN.

Students in naval architecture and marine engineering of second and third years.

1. An oscillating engine has a stroke of 10 feet, the trunnions being at the center of the cylinder, and distant 15 feet from the center of the crank-shaft. Velocity of crank-pin 420 feet per minute. Find the angular velocity of the cylinder when the crank is at one dead point, at 45° from the dead point, and at right angles to the piston rod.

2. Give a graphic construction for the curve of kinetic energy of a heavy piston driven by a uniformly rotating crank.

3. Suppose that the velocity curve of a heavy machine part is drawn. Show that the subnormal of the velocity curve is proportional to the acceleration of the machine part at the corresponding instant of the motion.

4. Suppose the work done on the piston of an engine to be represented by an indicator diagram. Show how you would infer from it the weight necessary in the fly-wheel rim to control the fluctuation of speed within a given limit.

5. What are the advantages and disadvantages of belting as a means of communicating motion? Sketch the different forms of pulley used with different kinds of belt.

6. Show that a crossed belt adapted for one pair of pulleys of stepped cones, is of the right length for any other pair, if the sum of the radii of each pair is the same.

7. A belt connects two shafts, and when running the tensions are P and 2 P on the tight and slack sides of the belt. Assuming the radii of the shafts to be ρ_1 and ρ_2, and those of the pulleys to be R_1 and R_2, show that the fraction of the work transmitted which is wasted in journal friction is—

$$3 \mu \left(\frac{\rho_1}{R_1} + \frac{\rho_2}{R_2} \right)$$

8. The gearing of a crane is arranged thus:

Winch handle 17 ins. radius on first shaft;

Pinion 4½ ins. diam. on first shaft gearing into wheel, 36 ins. diam. on second shaft;

Pinion 6 ins. diam. on second shaft gearing into wheel, 36 ins. diam. on barrel shaft;

Chain barrel 16 ins. diam.

Find the weight which would be raised by two men exerting a mean force of 30 lbs. each at the winch handle, and the pressure on the belt of each pair of wheels. Friction may be neglected.

9. A lock gate is supported on a flat pivot under the heelpost, and held in place by a cylindrical bearing at the top of the heelpost. The pivot is 6 ins. diameter, and the cylindrical top bearing 12 ins. diameter. Vertical pressure on pivot 2 tons. Horizontal pressure on pivot and upper bearing 5 cwt. each. Coefficient of friction 0.2. Find the force which must be applied at 12 feet radius to move the gate.

10. Show that when the pitch circles of two wheels with epicycloidal teeth roll in contact through a small arc ds, the tooth of one slides a distance

$$ p \left(\frac{1}{R_1} + \frac{1}{R_2} \right) ds. $$

on the tooth of the other, p being the distance of the point of contact from the pitch point.

11. Assuming the accumulated work in the reciprocating parts of a pair of engines, coupled at right angles, to be approximately constant, find an expression for the weight placed in the fly-wheel rim to which the reciprocating parts are equivalent, as regards their influence in moderating fluctuation of speed.

12. A pendulum governor works at a normal height of h feet, and ranges through a height nh above and below its mean position. Show that the ratio of greatest and least speed is

$$ \frac{N_1}{N_2} = \sqrt{\frac{1 + n}{1 - n}} $$

13. Draw a tooth for a cycloidal-toothed wheel of 12 ins. diameter, with 12 teeth; rolling circle 6 ins. diameter. Addendum and Root circles 14 ins. and 9¼ ins. diameter.

14. Taking the weight of belting at 1¼ lbs. per sq. foot, find the tension in a belt running at 30 feet per second, due to centrifugal force.

15. One spur wheel drives another through an intermediate or idle wheel. Show that the pressure of the intermediate wheel on its supports is proportional to the work transmitted. Hence show that the arrangement can be used as a dynamometer.

APPLIED MECHANICS. II.

Examiner.—Prof. W. C. UNWIN.

Students of naval architecture and marine engineering of second and third years.

1. Distinguish between Young's modulus, and the moduli of direct and lateral elasticity.
Show that in an isotropic solid

$$ E = A - \frac{2B^2}{A+B} $$

2. A bar is subjected to a bending and twisting moment acting at the same time. Deduce a formula for a bending moment equivalent to the two moments.
A shaft, of circular section, weighing G lbs. per cubic foot, transmits a twisting moment T. Express in a formula the bending moment equivalent to the combined bending and twisting moments to which it is subjected.

3. Find the stress at any point of a thick hollow cylinder under internal fluid pressure.
A thick cylinder 1 foot in internal and 2 feet in external diameter is subjected to an internal fluid pressure of 20 tons per sq. in. Find the change of external and internal diameter. Modulus of elasticity 30,000,000 lbs. per sq. in.

4. The velocity of water issuing from an orifice is found to be 30 feet per second, the head above the center of the orifice being 20 feet. Find the coefficients of resistance and velocity.

5. State how the friction of water flowing over solid surfaces varies with variation of area and quality of surface and with the velocity.

6. The frictional resistance of a pipe is 1 lb. per square foot of surface at a velocity of 10 feet per second, and varies as the square of the velocity. Thence determine the flow through a circular pipe 1 foot in diameter connecting two reservoirs, 1 mile apart, and having 5 feet difference of surface level.

7. Show that on a probable assumption as to the relative friction of the fluid filaments, the curve of velocities for a vertical in an indefinitely wide stream, is a parabola.

8. A jet of water 4 ins. diameter impinges on a fixed cone, the axis coinciding with that of the jet, and the apex angle being 30°, at a velocity of 10 feet per second. Find the pressure tending to move the cone.

9. Given the speed, the form of vanes, and dimensions of a turbine wheel, and the direction of motion and velocity of the entering water; show how to determine the absolute path of the water through the wheel.

10. An overshot wheel is to be designed for a supply of 30 cub. ft. per second on a fall of 24 feet. State how you would determine the diameter and the size of the buckets of the wheel, and what power you would expect to obtain.

11. Show that in any turbine

$$\eta\, g\, H = w_1\, V_1 - w_0\, V_0$$

where η is the efficiency, H the fall, w_1 and V_1 the whirling velocity of the water and the velocity of the wheel at the point where the water enters, and w_0, V_0, the same quantities at the point where the water is discharged. Hence show that in well-designed turbines

$$\eta\, g\, H = w_1\, V_1 \text{ very nearly.}$$

12. Show that the difference of pressure head at two points in a forced vortex, in which all the particles have equal angular velocities, is—

$$\frac{v_2^2 - v_1^2}{2g}$$

v_1, v_2 being the velocities of the water at the two points in the vortex.

SHIP DESIGN AND CALCULATION.

Examiner.—N. BARNABY, Esq.

CLASSES A₁ and A₂.

(The numbers in parentheses indicate the relative values of the questions.)

1. Describe the process by which the position of the center of buoyancy is obtained. (20)

2. What are the relative positions of the center of buoyancy and the center of pressure of the vertical fluid forces in a vessel floating at rest? (10)

3. How would you calculate the position of the center of pressure of the vertical fluid forces in a ship? (30)

4. Show how the relative positions of the metacenter, center of buoyancy, and center of gravity, under different conditions of lading, may be conveniently represented on a diagram. (15)

5. Investigate and explain fully the formula for dynamical stability now in common use, first determined by Canon Moseley. (35)

6. Describe in general terms the nature of the calculation for ascertaining the amount of statical stability of a ship, in foot-tons, at a given angle of inclination. (40)

7. A certain ship reaches her position of greatest stability at 35°, and her stability vanishes at 85°. If she be inclined under her canvas to 25°, about how far would she require to roll in smooth water in order to capsize under the pressure of the canvas, if the reduction in the pressure of the wind due to greater inclination be disregarded? (20)

8. A curve of stability being given, how would you ascertain the reserve of energy beyond a given angle of heel, under sail, to resist upsetting? (40)

9. Distinguish between the statical and dynamical stabilities of a vessel, and express one in terms of the other. Find the dynamical stability of a prismatic vessel at a given angle of heel, the section being a semicircle, and the density of the prism ⅔ of the fluid in which it floats. (50)

10. How is the stability of a vessel affected by the presence of free water in the bilge? (15)

11. What is the nature of the fluid resistances encountered by a ship passing through smooth water? What is your estimate of their relative importance? (40)

12. Describe the peculiarities of various forms of screw-propeller with which you may be acquainted, and give the reason for any advantage they may appear to possess. (40)

13. Having the following data, determine the measured mile speed of a ship. (20)

Displacement .... 8,000 tons.
Area of immersed midship section .... 1,200 square feet
Indicated horse-power 7,500
Coefficients of performance { for displacement 170
                             { for midship section 515

14. Calculate the horse-power that must be developed by the engines of this ship when she is steaming 12 knots per hour, supposing the coefficients of performance to be 180 and 550, respectively. (20)

15. Describe the operations of "tacking" and "wearing" in a vessel with ship rig. (30)

16. Describe the broad features of the design of some one modern ironclad ship, giving her name; and also of some one unarmored ship of war. (40)

LAYING OFF.

Examiner.—N. BARNABY, Esq.

CLASS A₂.

(The numbers in parentheses indicate the relative values of the questions.)

1. Describe the method of getting in the deck lines at the ship. (30)

2. Describe the method of laying off a harpin placed just above the knuckle, and having the same round-down and sheer as the knuckle. (35)

3. How would you obtain on the sheer drawing the projections of the transverse sections and water lines at the outside of the planking, from those ordinarily given at the inside of the planking or outside of the timbers? (20)

4. How do you obtain a point in the middle of the rabbet of a tapering stem in a wood ship? (15)

5. When level lines and sheer lines are ended in the half-breadth plan at the back of a circle swept with a radius equal to the thickness of the plank, what is the nature of the inaccuracy in the formation of the rabbet of stem so obtained, and with what form and direction of the fore edge of rabbet of stem would this plan be strictly accurate? (15)

6. How would you define the "joint" of a frame, and how would you obtain the bevelings of the timbers on each side of a joint in the square body of a wood ship? (10)

7. What sections are shown in their true form in the half-breadth plan, and what in projection only? (15)

8. Show by a sketch (in half-breadth plan) how the frames at the fore and after parts of an iron ship are disposed. (30)

9. State what means are adopted for obtaining an account of the plates of the bottom of an iron ship for the purpose of demanding them of the manufacturer. (30)

10. How are armor plates shaped and bent to form? Give the complete account of the operation both for a plate with no a'l curvature, and for a plate with considerable

curvature and twist, commencing with the preparation of the specification for the manufacturer. (45)

PRACTICAL SHIPBUILDING.

Examiner.—N. BARNABY, Esq.

Lieutenants qualifying for torpedo and gunnery officers. Voluntary students.

(The numbers in parentheses indicate the relative values of the questions.)

1. Show, by sketches, the nature of the receptacles in which bilge water will accumulate, if allowed to do so, in a wood ship, a composite ship, and an iron ship. (50)

2. What are the injurious effects of such accumulations in the three cases? (40)

3. Show, by sketches, how the ribs and skins of composite ships are fastened together, and state the nature of materials used, and their dimensions. (50)

4. What is the nature and composition of the cement and paint used in the bilges of iron and of composite ships? (50)

5. Name and sketch the different parts of anchor gear required for letting-go, riding, and weighing. (50)

6. Show, by sketches, the most approved plans of lowering and releasing boats. (50)

7. Compare, for a given time, the quantities of water which a 9-inch Downton's pump worked by hand will throw, and the quantity which would flow in by a round hole one foot in diameter, 10 feet under water. (50)

8. State briefly the ventilating arrangements in an iron steamship of war, including those for the magazines. (35)

9. Describe the arrangements for steering by steam, and the mode of passing from steam to hand power. (35)

10. How are lightning conductors fitted? (40)

11. Show the positions, by sketch, of the transverse water-tight bulkheads in a composite ship. State the thickness of plates used, and how the bulkheads are put together and stiffened. (50)

12. How are the stringers and gutter waterways placed in a composite ship; what are their uses; and how are they secured in place? (50)

13. What are the structural differences between the "Opal" and "Comus" classes of corvettes in Her Majesty's navy, and what do you consider the relative advantages to be? (50)

PRACTICAL SHIPBUILDING.

Examiner.—N. BARNABY, Esq.

Students in naval architecture and marine engineering.

(The numbers in parentheses indicate the relative values of the questions.)

Engineers are only expected to answer the questions marked with a star. Naval architects may answer all if time permit.

1.* Sketch (full size) a section along and across the line of rivets of an ordinary lap of double-riveted bottom plating, and of a high-pressure and low-pressure boiler. (40)

2. Show by sketches how a stealer is worked in an inside and in an outside strake of plating. (40)

3.* In any case of doubt as to whether the surfaces at the riveted work are well closed by the rivets, how would you test the work? (25)

4.* Supposing two plates to be united by double butt straps, treble riveted, the intermediate rivets in the front and back rows being omitted, show by sketches the several ways in which the plates may be torn asunder, and point out in each case the amount of iron broken and sheared. (35)

5.* What are the practical limitations to the proportion between the diameter of rivets and the thickness of plates they connect? (25)

6.* Describe by sketches a drift punch, a rimer, and a rose drill, and describe their uses in plating. (30)

7.* What is the process of annealing a steel plate; under what circumstances is this operation performed, and what are the effects produced? (25)

8. In what cases would you treble rivet the butts of bottom plating in an iron ship? (15)

9.* What are the uses of an inside middle-line keel in an iron and in a composite ship? (30)

10. Name and sketch the different parts of anchor gear required for letting-go, riding, and weighing. (45)

11. Sketch the different kinds of side-lights and ventilating scuttles in use in iron ships. (40)

12. Describe the several fittings which the shipbuilder has to make and fit in connection with a heavy broadside gun at the port and upon the deck. Give dimensions and nature of materials for an 18-ton gun. (40)

13.* Show by a sketch a midship section, and a portion of the longitudinal elevation of any composite vessel with which you are acquainted, and state her general dimensions. (40)

14. Describe the fastenings which would be employed in such a vessel, and show by a sketch how they would be arranged. (25)

15.* In applying forge tests to a piece of cold sheared iron or steel plate, may any difference in result be expected according to which side is uppermost? If so, to what cause do you attribute this difference? (20)

16.* In attaching a forging to a plate by means of rivets through a flat palm, supposing there were two rows of rivets, one behind the other, say two in front and three behind, by what considerations would you decide how much the palm might be thinned down across the hinder row of rivets? (40).

17. State of what materials and in what manner a scupper from the main deck of an iron ship is formed and fitted. (30)

18. Give a sketch of a boat's davit for a 28-feet cutter with figured dimensions of the parts. (30)

19. How are lightning conductors fitted? (40)

20.* State briefly the ventilating arrangements in an iron steamship of war, including those for the magazines. (35)

21.* Describe the arrangements for steering by steam and the mode of passing from steam to hand power. (35)

STABILITY AND OSCILLATIONS OF SHIPS.

Examiner.—J. R. PERRETT, Esq., F.R.S.N.A.

CLASS A₁.

1. State the conditions that must be fulfilled in order that a body may float freely in equilibrium in still water, point out the different kinds of equilibrium it may possess, and show how to determine all the possible positions of equilibrium of a floating body, which turns about a longitudinal axis fixed in direction.

2. Define "statical stability." Obtain an expression for the initial transverse statical stability of a ship, and estimate what change will be made in it by the addition of weights to the ship.

The weights being small compared with the weight of ship, where should they be placed in order that the initial stability shall remain unchanged?

3. Having given the curve of stability of a ship at any one draft, explain how the curve of stability at any other draft can be obtained.

4. Describe Mr. Barnes's method of graphically constructing metacentric diagrams for a ship at different immersions, and prove that at any draft of water for which the metacentric curve has a horizontal tangent the center of curvature of the curve of flotation is coincident with the metacenter.

5. Obtain the equation $r = \dfrac{di}{dV}$, where r is the radius of curvature of the surface of

flotation, V the volume of displacement, and i the moment of inertia of the plane of flotation.

6. Show how to determine the position of the center of gravity of a ship by experiment when she is afloat in still water.

What advantage is gained by having an automatic time-record of the angles of heel?

7. Explain how you would determine the angle to which a ship would heel if her broadside guns were fired simultaneously.

8. Investigate a formula which will determine the period of unresisted rolling of a ship in still water, having given her curve of statical stability.

Using this formula, prove that when the statical stability is directly as the angle of inclination the period of ship is isochronous.

9. A ship is moving with a given speed V at right angles to the lines of troughs and crests of a series of uniform waves whose period relatively to the ship is P_1, obtain an equation which will determine P the absolute period of the waves.

Distinguish between the possible cases (1) ship meeting waves, (2) ship overtaking waves, and (3) waves overtaking ship.

10. What is meant by the effective wave slope? Obtain a formula which determines its position in a wave, assuming the presence of the floating body causes no distortion in the wave strata.

11. Explain the terms "Curve of Extinction" and "Rate of Extinction," and show that the rate of extinction is usually expressible in the form

$$- \frac{d\Theta}{dn} = a\,\Theta + b\,\Theta^2.$$

12. Obtain the differential equation to the rolling of an isochronous ship in still water

$$- \frac{d^2\theta}{dt^2} = \frac{\pi^2}{T^2}\ \theta + K_1 \frac{d\theta}{dt} \pm K_2 \left(\frac{d\theta}{dt} \right)^2 \bigg\}$$

and prove that $K_1 = \frac{2Ta}{\pi^2}$ and $K_2 = \frac{4}{3} \frac{T^2}{\pi^2} b.$

a and b being the coefficients in the previous question.

13. Obtain Froude's Equation to the unresisted rolling of a ship among waves. State clearly his assumptions, and show that the angle of roll at any instant is that which the wave slopes would have imposed on the ship independently of her initial inclination and velocity, added to that which her initial conditions would have imposed on her in still water.

A ship being initially stationary and upright, prove that the successive ranges of rolling are given by the formula

$$\Theta = \frac{\pi}{2}\cdot \frac{H}{L}\ \frac{1}{1-\frac{T}{T_1}}\ \sin \frac{2n\,\pi}{1+\frac{T_1}{T}}$$

where n has the values 1, 2, 3, &c.

When $T = T_1$ prove that the increment of roll for each oscillation is equal $\frac{\pi}{2}$ Θ^1, Θ^1 being the maximum slope of waves.

14. Obtain an expression for the angle of "Permanent" rolling of an isochronous ship amongst uniform waves of given maximum slope and period.

15. When a ship is rolling in still water, prove that a short period pendulum suspended at her center of gravity, will, instant by instant, indicate with very considerable exactness her angle of inclination, and further prove that when the pendulum is suspended at some distance (l) directly above or below the center of gravity, the angles (θ_l) that it will indicate will be approximately related to the true angles of inclination (θ_0) by the expression

$$\frac{\theta_l}{\theta_0} = \frac{L \pm l}{L}$$

where L is the length of the simple pendulum keeping time with the ship.

THEORY OF WAVES AND MARINE PROPULSION.

Examiner.—J. R. PERNETT, Esq., F.R.S.N.A.

CLASS A₁.

1. Describe the geometry of trochoidal wave motion, and prove that the normal of the trochoid passes through the vertex of the rolling circle, and that there is a point of inflexion when the normal is a tangent to the tracing circle.

2. Prove that the surface of a trochoidal wave is in dynamical equilibrium.

3. Regarding the trochoidal surface as an infinitesimal layer, prove that, in order that the pressure on the lower side may be uniform, the thickness at each point of the layer must be inversely proportional to the normal drawn from that point to the vertex of the rolling circle.

4. If a series of particles be in a vertical plane when the fluid is undisturbed, prove that in the wave motion this series will be again vertical at the transits of crest and trough, and at every other period they will lie on a curve, such that if a tangent to it be drawn at the point where it intersects any surface of equal pressure, the tangent will pass through the lowest point of the corresponding rolling circle.

5. Prove that the horizontal momentum of the particles moving forward in a trochoidal wave is equal to the horizontal momentum of the particles moving backward, each being approximately

$$= mRHV$$

where R is the radius of the rolling circle, H is the height of wave from trough to crest, and V is the speed of wave.

6. The mechanical energy of a trochoidal wave is half actual and half potential. Prove this, and obtain the total energy of a wave which has a period of 5 seconds, and a height from crest to trough of 1 foot.

7. State the "Law of Comparison" which governs the relation between the resistances of ships and their models.

For what element of resistance is this law exact, and what corrections are needed in comparing the other elements of resistance?

At a speed of 300 feet per minute in fresh water, a model 10 feet in length, with a wet skin of 24 feet, has a total resistance of 2.39 lbs., of which 2 lbs. is due to surface friction resistance, and .39 lbs. to wave-making resistance, the eddy-making resistance being nil. What will be the total resistance, at a corresponding speed in salt water, of a ship 25 times the size of the model, having given that the surface-friction resistance per square foot of the skin of a ship at that speed is equal 1.3 lbs.?

8. Explain why it is that a ship traveling against a stream, with a given speed through the water, experiences a greater resistance than if she were traveling at that speed in still water.

9. Explain the term "Plane Water-Line in Two Dimensions," and obtain the general equation of the water-line curves generated by a circle.

Describe briefly their graphic construction.

10. Prove that a perfect fluid, flowing through a smooth pipe of any form whatever, will not tend to push the pipe endways, provided the two ends of the pipe are in the same straight line and have the same sectional area.

11. Prove that the steady motion of a perfect liquid past a solid free from discontinuity of curvature and perfectly smooth must be irrotational.

12. What is meant by "Effective Horse Power," and what is approximately the ratio between it and the Indicated Horse Power when the engines are working at full power? Why is this ratio not constant at all speeds?

13. If a vessel be propelled by screw or paddle, prove that the forward momentum generated by the passage of the ship is exactly balanced by the backward momentum generated by the action of the propeller.

Explain what is meant by the "Frictional Wake" of a ship, and show the relationship which exists between it and the ship's surface friction resistance.

14. Explain the nature of the action of a jet propeller, and prove that if the friction of engine and water resistance of passages be neglected, the efficiency of the propeller is

$$= \frac{2V}{v + V}$$

where v is the velocity of discharge of water from orifices, and V the speed of ship.

15. State the chief assumptions in the theory of the screw propeller, and indicate the method adopted to obtain an expression for the thrust of a screw of uniform pitch working in undisturbed water.

Prove that the efficiency of the screw is equal to—

$$1 - \frac{\text{slip of propeller}}{\text{speed of propeller}}.$$

MARITIME INTERNATIONAL LAW.

Examiner.—The Right Hon. MOUNTAGUE BERNARD, D. C. L.

Lieutenants qualifying for gunnery and torpedo officers. Voluntary students.

1. A cruiser seizes at sea, for a cause which the commanding officer judges to be sufficient, a vessel of neutral or uncertain nationality. What are the captor's duties?

2. If the vessel seized be an enemy's, what is then the duty of the captor? In what sense is it true that adjudication by a prize court is necessary in this case?

3. What is meant by a "convenient port" for sending in a prize?

4. Explain the meaning of the rule that, when a capture has been made without "probable cause," a prize court will decree payment of costs and damages.

5. In a war between England and Russia, two British merchantmen, A and B, are captured by Russian cruisers, and carried, the one into San Francisco, the other into New Orleans, the United States being neutral. A is condemned by a Russian vice-admiralty court, sitting in San Francisco; B (whilst still lying at New Orleans) by a prize court at St. Petersburg. Both are sold to citizens of the United States, and afterwards come, under the American flag, into the port of London. Can both or either be claimed by the original owner on the ground that the condemnation was invalid? Give reasons.

6. On what *evidence* are questions of prize or no prize commonly determined, and what reasons may be given for the established practice in this respect?

7. On what is the right of a captor to have the benefit of his prize founded, and how is it limited according to the law and practice of the British navy?

8. What is a tender to a man-of-war, and what is legally necessary to constitute a tender?

9. What are privateers, and how are they distinguished from public ships of war on the one hand, and from non-commissioned vessels on the other?

10. Explain the phrase "Droits of admiralty," and give examples of seizures which fall within that designation.

11. A and B, two Queen's ships, are cruising in company. A makes out a strange sail, and gives chase, B joining in the chase. A, after nightfall, overhauls the chase and captures her. B claims to share as joint captor. What must B prove in order to substantiate her claim?

GERMAN.

Examiner.—Dr. ROST.

Lieutenants qualifying as gunnery and torpedo officers. Voluntary students.

I. Write out the present indicative of the reflexive verb, sich laben.

II. Give the German names of the days of the week and of the four seasons, adding to each of the latter an appropriate adjective.

III. Conjugate the singular of the present indicative, and give the past participle, of können, mögen, essen, nehmen, fallen, greifen.

IV. Translate into German : I have taken a walk. We found nothing to eat. The moon shines brightly. The ship is in the harbor. The night was very stormy. There was a loud cry in the street. I saw them lying on the sand. They found fresh water on the island. We have had a long voyage. The weather is much finer to-day.

V. Translate: Wie lange sind Sie schon hier? Welche Sprachen sprechen Sie? Wann sind Sie zuletzt krank gewesen? Wie viele Meilen können Sie an einem Tage marschiren? Welche fremde Länder haben Sie besucht? Wo hat es Ihnen am besten gefallen? Wo haben Sie Deutsch zu lernen angefangen? Wie lange fährt ein Dampfschiff von hier nach Copenhagen?

VI. Answer, in German, any five of the above questions.

VII. Express in German words : 139 degrees, 45 minutes; a quarter to 7 ; half past 7; the 12th of March; sixfold; $\frac{2}{7}$; the first time; 15 times.

VIII. Decline the relative pronoun der, die, das.

Translate: In what, into what, for what, with what.

IX. Mention ten German prepositions, with the case or cases each governs.

X. Give the English for der stille Ocean, die Ostsee, die Nordsee, das Mittelländische Meer, das südliche Eismeer, der Wendekreis des Steinbocks, Grönland, Norwegen, Griechenland, Hinterindien.

XI. Dictation.

XII. Translate into German :

1.

MALTA, *May* 28.

II. M. S. Black Prince proceeds to Suda Bay on Thursday, to join the squadron of Admiral Lord John Hay.

PORTSMOUTH, *Tuesday.*

The Sultan, 12, armor-plated ship, Capt. E. K. Howard, bearing the flag of Rear-Admiral W. McDowell, C. B., and lately commanded by H. R. H. the Duke of Edinburgh, arrived at Spithead to-day. The Warrior, 32, steamed out of harbor and anchored at Spithead to take in her powder and shell.

2.

Of the three rivers Mekong, Shweli, and Salwen, the Salwen is, in the parallel at which we crossed, beyond question the largest. The "Topography of Yünnan" does not give its breadth, but draws special attention to its evil reputation for malaria. The Lu River, anciently called the Nu, is met with twenty miles south of Yung-chang. The mountains on both banks are exceedingly steep, and its exhalations are so poisonous that it is impassible during summer and autumn.

XIII. Translate:

Die Papua-Inseln bilden noch ein unbegrenztes Arbeitsfeld für den Naturforscher. Diese Inseln allein sind die Heimath der Paradiesvögel, dieser Wundererscheinungen der Vogelwelt, die an Pracht des Gefieders nichts übertrifft. Wallace ist es, welcher die ersten Paradiesvögel lebend nach Europa brachte, und nicht einmal in ihrer Heimath hatte er sie kaufen können, sondern zufällig in Singapore getroffen, aber mit Vergnügen die geforderten hundert Pfund Sterling dafür bezahlt. Vor ihm hat schon der französische Arzt und Naturforscher Lesson, welcher auf seiner Erdumsegelung 1824, kurze Zeit auf den Papua-Inseln verweilte, etwa ein Dutzend Paradiesvogelarten in unversehrten frischen Bälgen zusammengebracht, die ersten zuverlässigen Mittheilungen über das Freileben dieser Vögel gegeben, und damit die alten Fabeln und Märchen, welche über dieselben verbreitet waren, zerstreut.

Vor Kurzem wurde in Kiel die Taufe des jüngsten Gliedes unserer Flotte, der Panzercorvette Bavaria vollzogen. Dieses Schiff ist mit Bezug auf seinen Panzer und seine Artillerieausrüstung das stärkste der Marine. Die Länge der Bavaria beträgt 91 m., die Breite 18 m.; die Panzerhaut ist 16 Zoll eng. stark; die Artillerieausrüstung wird aus sechs Geschützen von schwersten Kaliber bestehen. Dies schiff gehört zur Klasse der Thurmschiffe und ist ausserdem noch mit einem mächtigen Rammsporn versehen. Beim Schlusse der kurzen Taufrede schleuderte der Vice-Präsident die an schwarz-weiss-rothen Bändern pendelnde Flasche Champagner gegen den Bug der Corvette; das Commando "Kapp ab" ertönte, einige Secunden vernahm man das Durchhauen des Stoppers, und langsam majestätisch unter den Klängen der National-hymne, dem Jubelrufe der Tausende von Zuschauern glitt die Corvette in die blauen Wellen der Ostsee hinab.

XIV. Reading.

XV. Conversation in German.

All students except acting sub-lieutenants.

I. Translate into English:

Le Chien.

Au commencement Dieu créa l'homme, et, le voyant si faible, il lui donna le chien. Il chargea le chien de voir, d'entendre, de sentir, et de courir pour l'homme. Et le chien, qui est le plus docile et par conséquent le plus intelligent des animaux, n'eut garde de désobéir à la volonté de Dieu. Il se fit le serviteur dévoué, l'agent de police de l'homme.

Le chien est, dans toute société fondée sur la propriété individuelle, comme la nôtre, le gardien vigilant, et le défenseur héroïque de ce qui s'appelle l'ordre public et la propriété. Les bêtes sont ce que les hommes les font. Les chiens de la tribu Arabe, or-ganisés pour la défense de la commune, considèrent comme dégradant le service d'un homme seul, et ils ont bien raison. Cependant, le chien n'entre pas dans la discussion de la question de droit ; son devoir est d'obéir et de se taire : il obéit sans murmurer. Le chien est la plus belle conquête que l'homme ait jamais faite, car cette conquête a donné à l'homme, dit M. de Buffon, des sens qui lui manquaient. Le chien est le pre-mier élément du progrès de l'humanité.

"Sans le chien point de sociétés humaines" écrit le livre sacré des anciens Parsis. Sans le chien, en effet, pas de troupeau. Sans le troupeau, pas de subsistance assurée, pas de viande à volonté, pas de laine, pas de temps à perdre, pas d'observations astro-nomiques, pas de science, pas d'industrie. C'est le chien qui a fait à l'homme tous ses loisirs. L'Orient est le berceau de la civilisation, parce que l'Orient est le patrie du chien.

Les indigènes de l'Asie, qui avaient le chien, ont été dispensés de se livrer aux péni-bles travaux qui absorbaient tout le temps et toutes les facultés des Peaux-Rouges In diens d'Amérique. Ils ont eu du temps de reste, et ils ont pu l'employer à créer l'in-dustrie.—ALPHONSE TOUSSENEL.

II.—Grammatical questions based on the foregoing passage.

(N. B.—Choose between the two sets of questions A and B, and answer only the ques tions contained in the set you select.)

A.

1. Write the feminine form of the nouns *le chien, le serviteur, le gardien*, and of the adjectives *public, premier, docile ;* the two masculine forms of the adjective *belle ;* the plural of *le troupeau* and *le berceau ;* the singular of *les animaux*, and the two plural forms of *le travail*, explaining the difference in the use of both.

2. Write the first person singular of the conditional present and the subjunctive present of the verbs *voir, entendre, sentir, faire, courir, employer*. When is the *y* replaced by an *i* in all verbs ending in *oyer* ?

3. Conjugate in full the affirmative imperative of the reflective verb *se taire*, and the negative imperative of *se livrer*.

4. *La nôtre.* When is *notre* written thus, without a circumflex accent over the *o* ? Give the plural forms of *notre, le nôtre*, and translate as instances : "our house and yours; our houses and yours."

5. *Intelligent, conséquent, vigilant, public, seul, premier, humain.*
Give the adverbs in *ment* formed from these adjectives, and state the rules.

6. *Parce que.* Explain the difference between *parce que* and *par ce que*, and translate into English: *par ce que je vois ; parce que je vois.*

Progrès. When do you put a grave accent over the *e* in the final syllable *es ?*

B.

1. *La plus belle conquête que l'homme ait jamais· faite.* Why is the subjunctive mood required in this sentence? Why is the participle past *faite* in the feminine? Why is *jamais* used without *ne ?* Explain the difference between *jamais* and *ne jamais.*

2. *Les peaux-rouges Indiens.* Why is *peaux-rouges* masculine here? What would be the meaning of *des peaux rouges* in the feminine?

De reste. State the difference between *de reste* and *du reste.*

3. *Le défenseur.* Explain the difference in the meaning of this noun and of *le défendeur*, and give the feminine form of the latter.

4. State the difference between *un livre* and *une livre*, *la tribu* and *le tribut*, and between *fonder* and *fondre ;* conjugate in full the indicative present of both verbs.

5. *Un homme seul.* Explain the difference between this expression and *un seul homme.*

Temps. When do you translate the English word *time* by *temps*, when by *fois*, when by *mesure*, and when by *pas ?*

6. *Ils ont raison.* Translate into English: *Vous avez raison. Vous avez une raison. Vous avez de la raison ;* and into French: "I have reason to believe. It stands to reason. State your reasons."

III. Translate into French :

(N. B.—It will be sufficient to translate two of the three following passages, A, B, C.)

A.—*The siege of Acre.*

There was no hope of carrying the place by a "coup de main." The French, remote as they were from France and Egypt, could not afford fresh losses ; they had already twelve hundred wounded, and the plague was in the hospitals. Accordingly, on the twentieth of May, the siege was raised. The resistance was no doubt due to the bravery of the gallant English admiral, Sir Sidney Smith, of whose courage and character Bonaparte spoke in the highest terms. The French general attributed the failure of the attack to the circumstance that Sir Sidney took the French battering-train, which was on board of some small vessels in the harbor. Until this period, Bonaparte had never experienced any reverses ; he had continually marched from triumph to triumph, and therefore he confidently expected the taking of Saint Jean d'Acre. In the letters which he addressed to the generals in Egypt, he had even fixed the twenty-fifth of April for the accomplishment of that event. He said afterwards: "The slightest cir- cumstances produce the greatest results ; had Saint Jean d'Acre fallen, I would have changed the face of the world ; the fate of the East depended on this small town."— CUNNINGHAM.

B.—*Letter addressed by the Duke of Wellington to the Duke of Berry, nephew of King Louis XVIII, before the battle of Waterloo.*

We have had a very sanguinary battle near the farm of Quatre-Bras, and the Prus- sians near Ligny. I had very few men and no cavalry ; yet I repulsed the enemy and had some success. The Prussians have suffered greatly and retired during the night ; and I was obliged to do the same in the course of yesterday. The Prussians have been joined by their fourth corps, containing about thirty thousand men ; and I have almost all my people with me. I hope, and I have reason to believe, that all will go on well ; but we must be prepared for everything. It is on this account that I beg your Royal Highness to do what is written in this letter. Let the King start for Antwerp (An- vers), not on a false alarm, but on receiving certain intelligence that the enemy has entered Brussels (Bruxelles) in spite of me.

C.—*Colloquial sentences.*

I am the son of a rear-admiral, and my youngest brother is already a midshipman.— I have cruised during a whole year on board an iron-plated frigate.—The mouth of a cannon was visible in every port-hole.—How many cartridges have you in your cartridge box?—The sailor stumbled over a cable and fell down on the deck, breaking his leg.—What are you doing?—Our boat does not contain more than eight men.—When are you going on leave?

Frederick the Great formed his cavalry two deep, and made it charge at a gallop with the greatest order; out of twenty-two pitched battles, this cavalry decided the victory fifteen times.

IV. Oral examination.

V. Writing from dictation.

SPANISH.

Examiner.—Señor CARRIAS.

Lieutenants qualifying as gunnery and torpedo lieutenants. Voluntary students.

1. In what cases do the adjectives *bueno*, *malo*, and *postrero*, drop the final *o*?
2. Explain and exemplify the use of *Don* and *Señor.*
3. Write in the feminine *un enemigo leal*; *los mejores actores.*
4. Translate, "I fear she will see you." "I do not fear he will see you."
5. When is *than* generally translated by *que* and when by *de?*

Translate into Spanish:

6. It is curious to observe a literary man's boyhood.

7. My condition soon changed, and life became a period of enchantment.

8. Every word was distinctly heard, I believe, by every person in the room.

9. In consequence, however, of some slight illness it was advised, and secretly determined, that he should be sent to school at the sea-side; and one morning Mr. Hayes mounted his horse, and James his gray pony. Mile after mile was left behind, and they came at nightfall to an inn where the boy's luggage had previously been forwarded. The next day brought the equestrians to Dublin, and their destination then turned out to be Dr. Miller's famous school at Blackrock.

(First Division only.)

10. If anything could have consoled Ellen for the sacrifice she had made, it would have been the unwonted look of contentment and peace upon her father's face as he placed his best loved child in the carriage which was to bear her from the home of her youth, beside one who was well fitted to be her guide and comforter through life's weary pilgrimage. By a strong exertion of determination she had gone through the day's proceedings, as only her sex can on occasions when such an effort is necessary, and even controlled her emotions when receiving her sisters' tender congratulations and loving farewell kisses, but fairly broke down on parting with her father and mother.

Translate into English:

11. Tal fué el lenguage usado ante el rey por Don Francisco, despidiéndose bien á su pesar de la corte, donde habia hecho en todo aquel reinado la primera figura. Desde 1,603 hallábase viudo de Dona Catalina, con quien tuvo dos varones y tres hembras. Hácia 1,610 estuvo muy próximo á pasar á segundas bodas con la Condesa de Valencia, señora de edad proporcionada á la suya.

12. Excelente efecto causó la desapasionada lectura de los seis tomos de las *Memorias.* Su autor se atrajo desde luego no escasas simpatias con el nobilísimo porte de guardar silencio profundo, que, á pesar de ir ya muy á viejo, y de serle necesario vindicar su honra zaherida por calumnias infames, nunca hubiese roto si le sobreviviera el primogénito de sus coronados valedores.

S. Ex. 51——19

ARITHMETIC.

Examiner.—Prof. C. Niven.

Lieutenants qualifying for gunnery and torpedo officers. Voluntary students. Probationary lieutenants of Royal Marine Artillery.

1. Find, by Practice, the price of 14 cwt. 2 qrs. at 3s. 2d. per lb.

2. Reduce the fraction $\frac{1400}{3333}$ to its lowest terms (explaining the process employed), and the resulting fraction to a decimal.

To $\frac{1}{15}$ of a crown add $\frac{7}{18}$ of a pound and $\frac{5}{23}$ of $\frac{1}{9}$ of $\frac{3}{4}$ of a guinea.

3. Divide .00355 by .0568, and the quotient by .000025.

What decimal can be subtracted 123 times from .0401 so as to leave a remainder of .000125 ?

4. Two trains 400 and 150 feet long are traveling at the rate of 30 and 50 miles an hour, respectively, in opposite directions; how long do they take in passing each other?

5. A watch is set right at noon on Monday; at 9 o'clock on Tuesday morning it is found to be 3 minutes slow. What is the true time when, on Wednesday afternoon, it points to half past four?

6. Reduce $\frac{1}{7}$ and $\frac{3}{7}$ to circulating decimals.

Show why the division need be carried out only to 3 places in the first case, and to 8 places in the second.

Extract the square root of 161.1716, and show that the number of square inches in a square mile is $3^4 . 2^{12} . 10^2 . 11^2$.

8. Grape sugar consists of carbon, hydrogen, oxygen in the proportions of 12, 14, 14 times their combining weights, which are respectively 6, 1, 8 grains. How much is there of each constituent in 1 lb. of sugar?

9. A bill of 470l. 10s. is payable three months after date; what is the discount on it at $3\frac{1}{2}$ per cent. ?

10. The price of three per cent. consols is $91\frac{1}{4}$; what sum must be invested in order to purchase enough stock to yield an income of 56l. a year?

What is the rate of interest on the money invested?

11. If workmen's wages constitute half the cost of production, and the manufacturer charge a profit of 20 per cent. on his outlay, what will his profit amount to per cent. if, in order to meet a rise of 5 per cent. in the rate of wages, he increases the selling price 5 per cent. ?

12. ABC is a semicircular window, of which AB, the diameter, is 7 ft. 7 in. If C be a point in the circumference 2 ft. 11 in. from the end B, find how much glass will be required to complete the window after the triangular space CAB has been filled in.

(The area of a circle $= \frac{22}{7}$ (radius)^2.)

ALGEBRA.

Examiner.—T. S. Aldis, Esq., M. A.

Lieutenants qualifying for gunnery and torpedo officers. Voluntary students. Probationary lieutenants of Royal Marine Artillery.

1. Simplify $\dfrac{x^2 + x - 6}{x^2 + 4x + 3} + \dfrac{x^2 - 5x + 4}{x^2 - 1} - \dfrac{x^3 + x^2 + x}{x^3 - 1}$.

2. State and prove the rule for finding the G. C. M. of two algebraical expressions.

Ex. $x^3 - 17x + 16$ and $x^3 - 3x^2 + x + 1$.

3. Solve the equations—

(1.) $\dfrac{x-a}{b} + \dfrac{x}{c} = \dfrac{2x-a}{c} + \dfrac{x-a}{d}$.

(2.) $\dfrac{a}{x-b} + \dfrac{b}{x+a} = \dfrac{a+b}{x-c}$.

(3.) $\sqrt{x^2 + 3x - 1} + x - 5\frac{1}{2} = 0$.

4. A number of three digits equals 11 times the sum of its digits. The second digit equals the sum of the other two, and by adding 693 to the number we reverse the order of its digits. Find the number.

5. Solve the equation—

$$(1.) \quad \frac{x-a}{x-b} + \frac{x-b}{x-a} = \frac{a^2+b^2}{ab}.$$

$$(2.) \begin{cases} x^2 + y^2 = 8\frac{1}{4}. \\ x^2 - y^2 = 4. \end{cases}$$

6. A rectangular field of two acres has a wall a quarter of a mile long around it. Find the length and breadth of the field.

7. Simplify $\left\{ \dfrac{1 + \dfrac{1}{\sqrt{2}}}{1 - \dfrac{1}{\sqrt{3}}} - \dfrac{1 - \dfrac{1}{\sqrt{3}}}{1 + \dfrac{1}{\sqrt{2}}} \right\}^2$.

8. Show how to sum a given geometric series to n terms.
What do you mean by the sum of such a series to infinity? Illustrate your answer by the series $1 - \dfrac{1}{\sqrt{3}} + \dfrac{1}{3} - \dfrac{1}{3\sqrt{3}} + \text{etc.}$

9. Prove the binomial theorem for a positive integer.

Prove that $\left\{ 1 + n + \dfrac{n(n-1)}{1.2} + \text{etc. to } (n+1) \text{ terms} \right\}^2 =$

$$1 + 2n + \frac{2n(2n-1)}{1.2} + \text{etc. to } 2n + 1 \text{ terms.}$$

10. Define a logarithm, and show that $\log \left(\dfrac{a}{b} \right) = \log a - \log b$.

If $\log_{10} 2 = .30103$, find $\log_{10} 64$, $\log_{10} 5$, and $\log_{10} 4 \sqrt{5}$.

11. Resolve into partial fractions—

$$\frac{x^2 - 7}{x^3 + 8} ; \ \frac{x^2 - x + 1}{(x-2)^3} ; \ \frac{2}{35}.$$

12. Find the coefficient of x^{12} in the expansion of—

$$\frac{1}{x^3 - bx^2 - a^2x + a^2b}.$$

GEOMETRY.

Examiner.—Prof. C. NIVEN.

Lieutenants qualifying for torpedo and gunnery officers. Voluntary students. Probationary lieutenants of Royal Marine Artillery.

1. Bisect a given rectilineal angle.
Into how many parts is it possible, by means of Euclid's constructions, to divide a right angle?

2. If from the ends of the side of a triangle there be drawn two straight lines to a point within the triangle, these straight lines shall be less than the other two sides of the triangle.
A triangle and a quadrilateral, having no re-entrant angles, stand on the same base and on the same side of it, the triangle being entirely outside the quadrilateral; prove that the triangle has the greater perimeter.

3. If a straight line fall upon two parallel straight lines, it makes the alternate angles equal to each other. Prove this; and state clearly the axiom upon which your proof is based, and mention any other which has been proposed instead of it.

4. Triangles on the same base and between the same parallels are equal to each other.
D, E, F are the middle points of the sides BC, CA, AB of a triangle, and AD, BE, CF meet in O; prove that the six triangles into which the figure is divided are equal to each other.

5. Describe a square which shall be equal to a given rectilineal figure.

6. Prove that the opposite angles of a quadrilateral inscribed in a circle are together equal to two right angles.

A circle is cut out of a sheet of paper and a triangle inscribed in it. The segments which lie outside the sides of the triangle are doubled over those sides respectively; prove that the arcs will now meet in a point, which point is the intersection of the perpendiculars from the angles of the triangle on the opposite sides.

7. Describe a circle which shall touch three given straight lines.

Describe a circle which shall pass through two given points and touch a given straight line.

8. If two circles cut one another, the straight line joining their centers bisects the common chord at right angles.

To what theorems of tangency does this result give rise?

9. A, B, C, D are four given points; it is required to find a point P such that

    (i) $AP^2 + BP^2$ may be as small as possible;

    (ii) $AP^2 + BP^2 + CP^2 + DP^2$ may be as small as possible.

10. If the vertical angle of a triangle be bisected by a straight line which also cuts the base, the segments of the base shall have the same ratio which the sides of the triangle have to one another.

If the line AD which bisects the angle A cut the base in D, prove that 2 AD is less than AB + AC.

11. Find a mean proportional to two given straight lines.

AB is the diameter of a semicircle and C is any point on the curve. BC, AC meet the tangents at A, B in D, E; prove that AB is a mean proportional between AD, BE.

12. Similar triangles are to one another in the duplicate ratio of their homologous sides.

TRIGONOMETRY.

Examiner.—Prof. C. NIVEN.

Lieutenants qualifying for gunnery and torpedo officers. Voluntary students. Probationary lieutenants of Royal Marine Artillery.

1. Express the cosine of an angle in terms of the tangent.

If $\tan A = \dfrac{3}{4}$, $\tan B = \dfrac{5}{12}$, find the values of $\cos B$, $\cos \overline{A-B}$, $\cos 4A$.

2. Find the sines of 60°, 45°, 15°.

Prove that $(\sec 15^\circ + \operatorname{cosec} 15^\circ)^2 = 24$.

3. Prove the formulæ—

$$\operatorname{Sin}(A+B)\sin(A-B) = \sin^2 A - \sin^2 B.$$
$$\operatorname{Sin} 3A = 3\sin A - 4\sin^3 A.$$
$$\operatorname{Sec}^2 A + 4\sec^2 2A = 16\operatorname{cosec}^2 4A - \operatorname{cosec}^2 A.$$

4. Solve the equations—

$$\operatorname{Tan}\left(\frac{\pi}{4}+\theta\right) = 3\tan\left(\frac{\pi}{4}-\theta\right)$$
$$\operatorname{Sin}\theta + \sin 2\theta + \sin 3\theta = 0.$$
$$\operatorname{Tan} 3\theta - \tan\theta = \tan 4\theta = \tan 2\theta.$$

5. If $\tan\dfrac{A-B}{2} = \dfrac{\sin\dfrac{a-b}{2}}{\sin\dfrac{a+b}{2}}$, and $\tan\dfrac{A+B}{2} = \dfrac{\cos\dfrac{a-b}{2}}{\cos\dfrac{a+b}{2}}$,

prove that $\cos A \tan a = \cos B \tan b$, and $\sin A \sin b = \sin B \sin a$.

6. Express the cosine of half any angle of a triangle in terms of the sides.

If in a triangle $a = 5$, $b = 6$, $c = 7$, prove that

$$\sin\frac{A}{2} = \frac{1}{\sqrt 7}, \ \sin\frac{B}{2} = \frac{3\sqrt 3}{\sqrt{.5}}.$$

7. If, in a triangle, a^2, b^2, c^2, be in arithmetical progression, prove that cos 2A, cos 2B, cos 2C, are also in arithmetical progression.

8. Investigate formulæ for solving a triangle in which are given two sides and the included angle.

Example $a = 1\frac{1}{4}$, $b = 3$, $C = 60°$.

9. Find an expression for the radius of a circle inscribed in a triangle.

If a, β, γ, δ be the radii of the four circles which touch the sides of a triangle; prove that the area of the triangle $= \sqrt{a\,\beta\,\gamma\,\delta}$.

10. An observer, the height of whose eye is h, standing by the bank of a river, observes that the altitude of a tree on the opposite bank is a, and that the depression of the top of its image in the water is β; find the height of the tree and the breadth of the river.

11. State and prove Demoivre's Theorem when the index is a whole number. State it completely when the index is a vulgar fraction $\frac{m}{n}$.

Sum the series—

$$\cos \theta \cos \Phi + \frac{\cos 2\theta \cos 2\Phi}{1.2} + \frac{\cos 3\theta \cos 3\Phi}{1.2.3} + \ldots \text{ to } \infty.$$

12. Expand $\sin \theta$ in ascending powers of θ.

If θ represent the number of minutes in the angle, what does the series become ? Employ the equality—

$$\sin\left(\frac{\pi}{6} + \theta\right) + \sin\left(\frac{\pi}{6} - \theta\right) = \cos \theta \text{ to prove that}$$

$$\pi = 3 + \frac{6}{1.2.3}\left(\frac{\pi}{6}\right)^3 - \frac{6}{1.2.3.4.5}\left(\frac{\pi}{6}\right)^5 + \text{etc.}$$

CO-ORDINATE GEOMETRY.

Examiner.—T. S. ALDIS, Esq., M. A.

Lieutenants qualifying for gunnery and torpedo officers. Voluntary students.

1. Find the co-ordinates of a point which divides the line joining two given points in a given ratio.

Hence find the locus of the middle point of the line drawn from the origin to any point in the straight line $3x + y = 7$.

2. Find the equation to a straight line in terms of the intercepts it makes on the axes.

Points are taken on the axis of x at distances a and $- a$ from the origin and on the axis of y at distances a and $2a$ from the origin. Find the co-ordinates of the points of intersection of the lines which join these points, two and two.

3. Show how to find the length of the perpendicular from a given point upon a given straight line.

$$3x + 4y = 7\,;\ 5x + 12y = 3\,;\ 8x + 15y = 13$$

are the equations to the sides of a triangle. Find the radius, and the co-ordinates of the center of the circle inscribed in it. Explain the ambiguities of sign which arise.

4. Show how to transform co-ordinates from one pair of rectangular axes to another with the same origin and inclined to them at a given angle. What does the equation $x^2 - y^2 = 1$ become when the axes are turned through an angle of 45° ?

5. Find the equation to the tangent at any point of the curve $r = \dfrac{l}{1 - e \cos \theta}$ and the locus of the foot of the perpendicular upon it from the origin.

6. Prove that the subnormal in the parabola is constant.

7. Find the locus of the middle points of parallel chords in an ellipse, and prove that the tangents at the points where the locus cuts the ellipse are parallel to the chords.

8. Show that the equation to the normal at any point of an ellipse can be expressed in terms of the tangent of the angle it makes with the axis of x.

9. Find the locus of a point, tangents from which to a given ellipse are at right angles.

10. Find the locus of a point, the product of whose distances from two given straight lines is constant.

11. Find the locus of the middle points of parallel chords in the curve represented by the general equation of the second degree.

Find the condition that all such loci may be right lines, parallel to a fixed one, whatever the direction of the chords may be. What will the equation represent in this case?

12. Find the conditions that the general equation of the second degree may represent a circle.

STATICS.

Examiner.—T. S. ALDIS, Esq., M. A.

Lieutenants qualifying for torpedo and gunnery lieutenants. Voluntary students. Probationary lieutenants of Royal Marine Artillery.

1. What do you mean by the resultant of two forces? Forces of 5 lbs. and 4 lbs. act at an angle whose sine is ⅔. Find the magnitude and direction of the resultant.

2. What is the principle of the triangle of forces? Prove its truth.

Forces act along the perpendiculars drawn from the angular points of a triangle on the opposite sides, and inversely proportional to them. Show that they are in equilibrium.

3. Show how to calculate the ratio of the power to the weight in the inclined plane. What force applied horizontally will keep a weight of 5 lbs. from slipping down a smooth plane inclined at an angle of 30°?

4. What is friction, and the coefficient of friction? How would you determine the latter in any given case?

A train will just rest on an incline of 1 in 300. The engine which just weighs the tenth part of the train will rest, with its breaks on, on an incline of 1 in 8. Find approximately the steepest incline on which the engine and train together can rest.

5. Show how to find the resultant of two unlike parallel forces acting in one plane. Explain your result when the forces are equal.

6. A girder 100 feet long weighs 100 tons, and weights of 30, 50, 70, and 90 tons are placed along it at intervals of 20 feet. Find the pressures on the abutments.

7. Show how to find the resultant of any number of forces in one plane.

Forces proportional to 1, 2, 3, 4, 5, 6, act along the sides of a regular hexagon taken in order. Determine the resultant.

8. Which system of pulleys is most commonly used, and why? Calculate the ratio of the power to the weight in it.

A man with a handle 2 ft. long winds a rope on an axle a foot in diameter. The rope passes through one movable pulley attached to the object to be moved. Compare the power with the strain on the object.

9. Show how to calculate the limits of the power to the weight for equilibrium in a rough screw.

10. Two heavy stone slabs, 10 feet square, are placed on a rough plane with their upper edges resting against each other. If the coefficient of friction be ⅓, find the greatest distance that their lower edges can be apart without slipping.

11. Find the center of gravity of a semicircular lamina.

12. State Guldinus's properties.

Find the volume of a ring of given radius and section.

Lieutenants qualifying for gunnery and torpedo officers. Voluntary students. Probationary lieutenants of Royal Marine Artillery.

1. Define the specific gravity of a substance, and show how to find the specific gravity of a mechanical mixture of a number of known substances.

Gold and copper, whose specific gravities are 20, 9, are mixed in the proportions of 10 : 1, and form an alloy whose specific gravity is 19. Show that the volume of the alloy is less than the sum of the volumes of its components by $\frac{1}{19}$ of that sum.

2. Explain the statement that water always seeks its own level. Prove the statement.

3. Two liquids, which do not mix, are in equilibrium in a bent tube; prove that their heights above the common surface of separation are inversely as their densities.

A uniform tube in the form of an equilateral triangle is fitted with equal volumes of three liquids which do not mix, and whose densities are as 3 : 4 : 5. It is then held with one side vertical, so that the lightest liquid occupies the highest part of the tube, and the heaviest the lowest part; prove that each side is divided equally between the two fluids which it contains.

4. Find the pressure at any point of a heavy liquid, and the *whole* pressure on any surface immersed in it.

A heavy cone, whose vertical angle is 60° and height 3 inches, is lowered into water by means of a string attached to its vertex. Find the depth of its vertex when the whole pressure on the curved surface is six times the weight of water displaced.

5. Explain what is meant by the center of pressure.

A jar, standing on a table, is filled with equal volumes of three liquids whose densities are as $\frac{1}{4}$, $\frac{1}{2}$, 1, the lightest being uppermost, and the heaviest being at the bottom. The contents are now thoroughly mixed; find the change of pressure on a vertical plane bisecting the jar, and the change of the center of pressure of this plane.

6. Determine the conditions of equilibrium of a floating body.

A block of ice, a yard cube, floats with .08 of its volume above water, a piece of granite being embedded in the ice. Find its size, the specific gravities of ice and granite being .918 and 2.65.

7. A uniform triangular plate ABC, movable about a fixed hinge at the angle A, which is a right angle, rests in a liquid whose surface passes through B and bisects AC. Show that, if ρ be the ratio of the density of the liquid to that of the solid,

$$\tan {}^2B = 4 \cdot \frac{2-\rho}{4-\rho}.$$

8. Explain how Nicholson's hydrometer is used to find the sp. gr. of a powder not soluble in water.

9. Describe the experiment by which it is shown that the elastic force of a gas at a given temperature varies as its density.

Also state fully the law of expansion of a gas due to an increase of temperature, and deduce the relation between the elastic force, density, and temperature of a gas.

10. OACD is a uniform tube, bent up vertically and closed at O. A certain quantity of mercury being poured in, the quantity of air in OA ($= a$) is made such that the mercury has the same level in both branches. An additional quantity of mercury is now added which will be sufficient to fill b of the tube. Find the final density of the internal air compared with that outside, the height of the barometer being h.

11. Describe and explain the air pump, and find an expression for the density of the air in the receiver after n strokes.

How is ice formed by means of the air pump?

KINEMATICS AND KINETICS.

Examiner.—T. S. ALDIS, Esq., M. A.

Lieutenants qualifying for gunnery and torpedo officers. Voluntary students. Probationary lieutenants of Royal Marine Artillery.

1. What do you mean by the resolution of velocities? A man in a ship walks round the capstan in a circle of 12 feet radius once in a minute, in which time the ship moves forward a foot. Calculate the man's absolute velocity when in the line of the ship's motion.

2. Show how to calculate the space described in a given time under a uniform acceleration, the particle starting from rest.

I drop a stone down a well, and after three and a half seconds hear it splash. Find approximately the depth of the water from the surface of the ground.

3. Show how to find the range of a particle projected with a given velocity at a given angle to a horizontal plane. What angle of projection gives the greatest range?

4. Show that when a body moves under the action of a central force the areas swept out are proportional to the time.

5. Calculate the motion of a particle sliding down a cycloidal arc.

Hence deduce the time of oscillation of a simple pendulum.

6. How is the value of g found from observations with a pendulum?

A pendulum that swings in a second at the surface of the sea, loses 12 seconds a day on the top of a hill. Calculate the height of the hill, neglecting its attraction.

7. State the three laws of motion. Explain the relation between force and acceleration.

Equal weights are suspended to a wheel and circle whose diameters are as 1 to 2. They start from rest. Find the velocity acquired in three seconds, neglecting the mass of the wheel and axle.

8. Describe Atwood's machine. How do you directly obtain from it the acceleration produced in a given mass by a given weight? What effect has the pulley on your result?

9. A ball is projected with a velocity of 20 feet per second at an angle of 45° to a horizontal plane. The coefficient of elasticity is $\frac{1}{4}$. Determine the subsequent motion.

10. Show how to find the time of describing any arc of a parabola from the vertex, the focus being the center of force.

11. Show how to find the force to the center when a particle describes a circle with a given velocity.

How could Newton prove that gravity, as shown at the earth's surface, diminished according to the square of the distance retains the moon in her orbit?

12. If the earth be taken as spherical, compare the time of oscillation of a pendulum (1) at the pole (2), at the equator (3), in latitude 60°.

DIFFERENTIAL AND INTEGRAL CALCULUS.

Examiner.—Prof. C. NIVEN.

Lieutenants qualifying for torpedo and gunnery officers.

1. Define a differential coefficient, and illustrate it geometrically.

Investigate, directly from the definition, the differential coefficient either of e^x or of $\log_e x$; given one of these differential coefficients, deduce the other.

2. Find the differential coefficients of—

$$\frac{\sqrt{1+2x}}{(2x-1)^2}, \quad \log \frac{x}{\sqrt{1+x^2}}, \quad \tan^{-1}\sqrt{\frac{x}{a}}(\sin x)^x.$$

3. State and prove Leibnitz's Theorem.

Apply it to find the nth differential coefficient of $(x^2-1)\sin x$.

4. State and prove Taylor's Theorem.

Prove that log tan $\left(\dfrac{\pi}{4} + x\right) = 2x + \dfrac{4}{3} x^3 + \dfrac{4}{3} x^5 + \ldots$

5. Explain how to find the value of a function which, for a particular value of the variable, assumes the form $\dfrac{0}{0}$; and illustrate geometrically the result obtained.

Find the value of $\dfrac{e^{ax} + e^{-ax} - 2}{1 - \cos x}$ when $x = 0$.

6. Explain how the maxima and minima values of $\phi(x)$ are found.

Of all cones which circumscribe a given sphere find that whose curved surface is the least.

7. Given $x = r \cos \theta$, $y = r \sin \theta$, where r and θ are functions of t, prove that—

$$x \frac{d^2x}{dt^2} + y \frac{d^2y}{dt^2} = r \frac{d^2r}{dt^2} - r^2 \left(\frac{d\theta}{dt}\right)^2.$$

8. Prove that the sine of the angle between the tangent to a curve and the radius vector to the point of contact is $\dfrac{x \dfrac{dy}{dx} - y}{\sqrt{1 + \left(\dfrac{dy}{dx}\right)^2}}$; and transform this expression to polar co-ordinates.

Prove that tangents to the curve $r = a (1 + \cos \theta)$, at the extremities of a focal chord, are at right angles.

9. Define an asymptote, and explain a general method for finding all the asymptotes of a curve.

Find the asymptotes of the curve $xy (y - 2x) = a (y^2 - x^2)$, and find the nature of the curve near the origin.

10. Trace the curve $y = \tan x$.

11. Investigate the rule for integration by parts.

Integrate the following functions:

$$\tan x, \frac{1}{a^2 + x^2}, (x \log x)^2, \frac{1}{x^3 + x}, \frac{1 + \sin x}{1 + \cos x},$$

and find the value of $\displaystyle\int_0^{\frac{\pi}{2}} \sin^6 x \, dx$.

12. Find by integration the volume generated by the revolution of a sector which is the sixth part of a circle round one of its bounding radii.

PURE MATHEMATICS: FIRST PAPER.

Examiner.—T. S. ALDIS, Esq., M. A.

Students in naval architecture and marine engineering of first year.

1. Show how to inscribe three equal circles in a given circle so that they may touch it and each other.

2. Parallelograms of the same altitude are to one another as their bases.

Why is it not sufficient, in the demonstration of this proposition, to prove that if the multiple of the first be equal to that of the second the multiple of the third will be equal to that of the fourth?

3. Solve the equations—

(1.) $\dfrac{3}{x-1} - \dfrac{4}{x+1} = \dfrac{1}{12}$

(2.) $\begin{cases} \dfrac{x^2}{a^2} + \dfrac{y^2}{b^2} - \dfrac{z^2}{c^2} = 1. \\ \dfrac{x}{a} + \dfrac{y}{b} - \dfrac{z}{c} = 1. \\ x + y + z = a + b + c. \end{cases}$

4. Find the nth term of, and sum the series—

(1.) $12 - 8\sqrt{3} + 16$ etc.

(2.) $1 + 4 + 12 + 32 + 80$ etc.

5. If of 80 persons born together one dies every year till all are dead, what is the chance of a person aged 36 reaching the age of 63 years?

How could you calculate the chance of a boy who is 15 years old surviving his father who is aged 40?

6. Show how to extract the square root of an expression of the form $a + \sqrt{b}$.

Find the fourth root of $\cdot\frac{17}{4} + 3\sqrt{2}$.

7. Give the rth term of the expansion of $(1 - 2x)^{-}$ in its simplest form; also sum

the series $\frac{1}{2}\left\{ 1.2 - 2.3x + 3.4x^2 - 4.5x^3 + \text{etc.} \right\}$ to infinity.

8. Prove by means of a figure that

$$\sin(A + B) = \sin A \cos B + \cos A \sin B,$$

when A and B are each greater than 45° and less than 90°.

9. The apparent diameter of the sun being 31', after looking at it for an instant, I glance at a house a mile away and see that the impression of the sun on the retina just extends from the top to the bottom of the house. Find approximately its height.

10. The sides of a triangle are 300 and 400 feet, and the included angle is 65°. Find the remaining angles. Given L cot 32° 30′ = 10.19581, log 7 = .84510, L tan 12° 38′ = 9.35051, and L tan 12° 39′ = 9.35111.

11. Find the equation to the straight line which bisects the angle between two given straight lines.

How can the ambiguous sign be determined?

12. Find the equations to the tangents to the circle $x^2 + y^2 = c^2$, which pass through the point $4c, 3c$, and the equation to the chord of contact.

13. Find the polar equation to a conic section, the focus being the pole, and the equation to the tangent to it at any point.

Find the locus of the foot of the perpendicular from the focus on the tangent.

14. Prove that the locus of a point the sum of whose distances from two given points is constant is such that the distance of any point in it from a fixed straight line bears a constant ratio to its distance from one of the given points.

PURE MATHEMATICS : SECOND PAPER.

Examiner.—Prof. C. NIVEN.

Students in naval architecture and marine engineering of first year.

1. Find the equation of a tangent to a parabola, and prove that it may be put into

the form $y = mx + \frac{a}{m}$.

Find the locus of the intersection of tangents which inclose a constant angle a.

2. Determine the equations of the lines which join the extremities of the major axis of an ellipse to the extremities of one of the latera recta, and find the poles of these lines.

3. Prove that the portion of the tangent to a hyperbola between the asymptotes is bisected at the point of contact.

Find the tangents to the ordinary and conjugate hyperbolas which pass respectively through the foci of the conjugate and ordinary hyperbolas; and show that, if these tangents be at right angles, the hyperbolas are rectangular.

4. Find the center, asymptotes, and eccentricity of the conic

$$(x - 2y + 1)(x + y - 2) = 3x - 2y.$$

5. Investigate, directly from the definition, the differential co-efficient of x^n.

Differentiate $\sqrt{\dfrac{x+a}{x-a}}$, $e^{\tan^{-1}x}$, $\log\left(\dfrac{\sqrt{1+x^2}-x}{x}\right)$.

Prove that, if $(a+\cos\theta\,\sqrt{a^2-1})\,(a+\cos\phi\,\sqrt{a^2-1})=1$, then

$$\frac{d\,\theta}{d\,\phi}=\frac{1}{a+\cos\phi\,\sqrt{a^2-1}}.$$

6. If the fraction $\dfrac{f(x)}{\phi(x)}$ take the form $\dfrac{\infty}{\infty}$, when $x=a$, show how to find its true value.

Find the value of $\dfrac{\log_e\left(1-\dfrac{x^2}{a^2}\right)}{\cot\dfrac{\pi x}{a}}$ when $x=a$,

and of $\dfrac{1}{2x^2}-\dfrac{1}{2x\tan x}$ when $x=0$.

7. Find the maxima and minima values of $x^4\sqrt{a^2-x^2}$.

If lines be drawn through a point between two given straight lines, find that which cuts off the least area.

8. If x and y are connected by the relation $\phi(x,y)=0$, show how to find $\dfrac{dy}{dx}$, $\dfrac{d^2y}{dx^2}$.

9. Determine the perpendicular from the origin on the tangent to a given curve at any point.

The perpendicular on the tangent to the curve $(x^2+y^2)^2=2a^2xy$ is $\dfrac{(x^2+y^2)^{\frac{1}{2}}}{a^2}$.

10. Find the radius of curvature of the ellipse $4x^2+11y^2=48$ at the point (1,2).

11. Explain what is meant by the envelope of a system of curves, and show how to find it.

Circles are drawn having their centers on the arc of a parabola and touching the tangent at the vertex; prove that the envelope of these circles is the given tangent and a certain circle.

12. Integrate the expressions $\dfrac{x}{\sqrt{1-x-x^2}}$, $\dfrac{1}{\sqrt{a^2+x^2}}$, $e^x\cos^2 x$,

and evaluate $\displaystyle\int_0^a \sqrt{\dfrac{a+x}{a-x}}\,dx.$

13. Find an expression for the volume generated by the revolution of a plane curve round a line in its own plane.

Find the area between the curve $\sqrt{x}+\sqrt{y}=\sqrt{a}$ and the co-ordinate axes, and the volume generated by the revolution of this area about either. Find also the length of the curve between the axes.

PURE MATHEMATICS.

Examiner.—Prof. C. NIVEN.

Students in naval architecture and marine engineering of second and third years.

1. Draw the common tangents to two given circles.

If a quadrilateral circumscribe a circle, the sum of the angles which either pair of opposite sides subtend at the center is the same.

2. Find an expression for the radius of a circle inscribed in a triangle.

If r be the radius of the inscribed circle, and r_1, r_2, r_3 the radii of the escribed circles, prove that

$$rr_1+r_2r_3=bc.$$

3. Find the equation of the tangent to an ellipse at any point in the form $y = mx + \sqrt{m^2a^2 + b^2}$.

Show that the locus of the intersection of tangents, to the ellipse $\dfrac{x^2}{a^2} + \dfrac{y^2}{b^2} = 1$, which inclose a constant angle a, is

$$(x^2 + y^2 - a^2 - b^2)^2 = 4 \cot^2 a \, (b^2x^2 + a^2y^2 - a^2b^2).$$

4. For all rectangular transformations of co-ordinates in the equation $Ax^2 + 2Fxy + By^2 + 2Ex + 2Dy + C = 0$, the following functions of the coefficients remain unchanged, viz: $A + B$, $AB - F^2$.

Prove this, and investigate what is known about the curve when one of these expressions varies.

Find the locus of P such that, A and B being fixed points, $PAB \sim PBA = a$.

5. Show how to reduce the integral:

$$\int x^m \, (a^2 + x^2)^{\frac{n}{2}} \, dx.$$

Find the following integrals:

$$\int \frac{x}{(x+1)^2 \, (x^2 + 1)} \, dx, \quad \int_0^a \sqrt{\frac{2a}{x} - 1} \, dx.$$

6. Investigate an expression for the volume of a surface whose equation is given in polar co-ordinates.

By this, or any other method, find the volume cut off from the solid bounded by the surface of revolution $r = a \, (1 + \cos \theta)$, by the sphere $r = \dfrac{a}{2}$.

7. Show how to differentiate an integral with regard to a constant contained in the subject of integration.

By this means find the integral $\displaystyle\int_0^{\frac{\pi}{2}} \frac{\cos 2\theta \, d\theta}{(a^2 \cos^2 \theta + b^2 \sin^2 \theta)^2}$.

8. Solve the following differential equations:

$$x \frac{dy}{dx} + y = y^2 \log x.$$
$$(x + 1 - 2y)dy = (2x + y + 1)dx.$$

9. The equation $(x^2 + 2ax - 2y\sqrt{x^2 + 2ax})dy + y(x + a)dx = 0$ has an integrating factor of the form $\phi(x)$; find it, and solve the equation.

Solve the equation $\dfrac{d^2y}{dx^2} + a^2y = \sin ax$, by the method of the "Variation of Parameters."

Solve also the equation $\dfrac{d^2y}{dx^2} + a^2y = x + \sin ax$.

11. Find the relation between a and b, that the ellipse $ax^2 + by^2 = 1$ may envelop the parabola $y^2 = 4cx$.

12. Investigate the theory of the number of constants in the solution of a system of ordinary simultaneous equations.

Solve the equations: $\dfrac{d^2x}{dt^2} - 2a \dfrac{dy}{dt} + c^2x = \cos nt$

$$\dfrac{d^2y}{dt^2} + 2a \dfrac{dx}{dt} + c^2y = 0.$$

APPLIED MATHEMATICS: FIRST PAPER.

Examiner.—Prof. C. NIVEN.

Students in naval architecture and marine engineering of first year.

1. State the parallelogram of forces, and deduce from it that, if three forces acting on a particle be in equilibrium, each is proportional to the sine of the angle between the other two.

A small ring of given weight rests on the arc of a smooth circular hoop which is fixed in a vertical plane, being attached to the highest point by a string whose length is equal to the radius of the hoop; find the tension of the string and the pressure on the hoop.

2. Investigate the conditions of equilibrium of a particle acted on by any number of forces in one plane.

Three equal spheres rest in one vertical plane against each other, being suspended from a point by strings each equal to the radius of one of the spheres and attached to points in their surfaces. Find the tensions of the strings and the pressures between the spheres.

3. Define the center of gravity of a body, and prove that the center of gravity of a uniform triangular plate coincides with that of three equal particles placed at its angles.

Two uniform rods of the same substance and thickness, and of lengths 5 and 3 inches, are rigidly connected at one end A so as to be at right angles, and are suspended by a string so that the other ends are in the same horizontal plane. Show that the distance from A of the point on the longer rod at which the string is attached is 1.0225 inch.

4. Determine the resultant of a number of forces in one plane.

Prove that the equation of the line of action of the resultant is

$$x'\Sigma(Y) - y'\Sigma(X) = \Sigma(xY - yX).$$

5. State the principal laws of friction which have been deduced from experiment. How can the coefficient of friction be found?

6. Determine the center of gravity—
    (i) of a sector of a circle.
    (ii) of a segment of a circle.

7. Describe the different systems of pulleys, and find whether a weight of 15 lbs. will be able to support 2 cwt. in a system in which there are four movable pulleys, the strings around which are attached to a fixed beam, and each of which weighs one pound.

8. Explain the principle of the screw, and determine the mechanical advantage gained.

9. Explain the difference, in dynamics, between a ton and the weight of a ton. What are their numerical values on the foot-pound-second system? If the weight of a ton be the unit of force, and a minute and yard those of time and length, what will be the unit of mass?

10. Establish the equation $v^2 = V^2 + 2gx$ for the motion of a falling body, and express it in the language of the science of energy.

A ball of 10 lbs. is dropped from a height of 289.8 feet, but, after falling half-way, it explodes into two equal parts, one of which is reduced by the explosion to rest. Find the subsequent motion of each part, and determine the kinetic energy developed by the explosion.

11. Find the range of a projectile on a horizontal plane passing through the point of projection.

A particle projected from a point, A, in the floor of a room returns to A after striking one of the opposite walls and the floor successively. Prove that if it strike the wall at right angles and the floor once its elasticity $= \frac{1}{2}$, and that it strikes the floor half-way between the foot of the wall with half its original velocity and exactly opposite to its original direction.

12. Two balls, A, B, whose masses are as $2 : 1$, and which are moving in opposite directions, collide. If the first ball be brought to rest, and the coefficient of elasticity be $\frac{3}{4}$, prove that their original velocities are as $7 : 5$.

APPLIED MATHEMATICS: SECOND PAPER.

Examiner.—T. S. Aldis, Esq., M. A.

Students in naval architecture and marine engineering of first year.

1. Define fluid, vapor, gas.
Show that the pressure at any point in a fluid at rest is the same in every direction.
2. What do you mean by specific gravity? How would you compare the specific gravities of (*a*) two coins, (*b*) two samples of milk?
3. What is the center of pressure and the total resultant pressure? Find them in the case of an equilateral triangle one of whose sides (12 feet long) is on the surface and the opposite angular point 10 feet beneath it.
4. A hollow cylinder, a foot long, closed at the upper end, weighs as much as half the water it will hold. It is sunk, with the closed end uppermost, in a vertical position in water. How high will the water rise within it when the top is a foot below the surface? At what depth will it rest in equilibrium?
5. Explain how a ship can tack against the wind. A Chinese junk will run before the wind faster than an English ship. In what case, and why, will the ship outsail the junk?
6. In passing from the freezing to the boiling point, air expands .366 of its volume. A cubic foot of air at 60° F. (the barometer standing at 29 in.) weighs 527 grains. What will be the weight of a cubic foot of air when the thermometer is at 90° and the barometer at 28¼ in.?
7. Determine the conditions of equilibrium of a floating body. Why is an ironclad with a low free-board specially unfitted to carry sail?
8. In Atwood's machine equal weights of 10 ozs. are suspended to the string which passes over the pulley and a bar of 1 oz. weight is placed across one. This, after falling through the space of a foot, passes through a ring which removes the 1 oz. weight. How far will the 10 oz. weight descend in the next minute?
9. Show how to calculate the space described in a given time under the action of a uniformly accelerating force, the motion being in a straight line.
A stone thrown down a rough board inclined at an angle of 30° neither gains nor loses velocity in its descent. What velocity will it gain by falling down the board (which is 20 feet long) when it is inclined at an angle of 60°?
10. A particle revolves in an ellipse about a center of force in the focus. Calculate the law of attraction.
11. Explain the action of the governor of a steam-engine. Show how to calculate the position it will assume for a given number of revolutions per minute, neglecting all weights but those of the balls.
12. A smooth bead slides down the arc of a cycloid; determine the motion.

APPLIED MATHEMATICS.—FIRST PAPER.

Examiner.—T. S. Aldis, Esq., M. A.

Students in naval architecture and marine engineering of second and third years.

1. State and prove, for direction, the principle of the parallelogram of forces, explaining clearly the assumptions you make.

ABCD is a quadrilateral figure. Forces act along BA, BC, DA, and DC proportional to them, show that their resultant is a single force represented by four times the straight line which joins the middle point of BD to the middle point of AC.

2. Show how to find the resultant of any number of forces acting on a rigid body.

A cube has forces proportional to 1, 2, 3, 4 acting along the edges of one face taken in order. Forces proportional to 4, 1, 2, 3 act along the corresponding edges of the opposite face in the opposite direction. Find the resultant.

3. Show how to find the C. of G. of a solid of revolution, and find it in the case of a hemisphere.

4. State the laws of friction.

Show how to calculate the total friction in the case of a rope stretched round a rough cylinder.

5. Explain the principle of the arch, and show how you would calculate the curve required for a given arrangement of the load upon it.

6. What are the laws of motion? Give an experimental illustration of each.

Investigate formulæ for the motion of a particle on an inclined plane under the action of gravity.

7. Calculate the motion of a body projected obliquely and acted on by gravity.

A building 20 feet high, 20 feet wide, and 30 feet long is surmounted by a gable roof rising 20 feet higher. A smooth stone is projected horizontally with a velocity of 2 feet per second just along one side of the ridge from one end of it. Find where it will strike the ground.

8. Calculate the motion of a particle acted on by a central force varying as the distance.

A weight hangs from a peg by an elastic string which it stretches to double its unstretched length. If the weight be slightly displaced, find the time of a small vertical oscillation.

9. A block of wood thrown on ice with a velocity of 10 feet per second is brought to rest after passing over 30 yards. A bullet of equal weight with the block is then shot into it with a velocity of 100 feet per second. Determine the subsequent motion.

10. A ball is dropped from a height of 10 feet on a plane inclined at an angle of 30°; the coefficient of elasticity is $\frac{1}{2}$; find the points where the ball will again strike the plane.

11. Two perfectly elastic particles are revolving in the same direction and in the same plane round a center of force varying inversely as the square of the distance. One is moving in a circular orbit, the other in a parabola whose latus rectum equals the diameter of the circle. They collide as the second particle is approaching the center of force. Determine the subsequent motions.

APPLIED MATHEMATICS: SECOND PAPER.

Examiner.—Prof. C. Niven.

Students in naval architecture and marine engineering of second and and third years

1. Determine the general equations of equilibrium of a fluid; and show that, when the external forces are such as arise from a potential, the surfaces of equal potential, of equal density, and of equal pressure coincide.

A heavy liquid is contained in a vessel and is also under the action of two centers of force which are in the same vertical line, and which exert equal forces at equal distances, but one of which is repulsive and the other attractive. The law of force being directly as the distance, prove that the free surface is a horizontal plane, and find the pressure at any point.

2. Define the whole pressure and resultant pressure on a surface immersed in a fluid; and show how to calculate them.

Prove that the total normal pressure on a spherical surface immersed to any depth in water is the same as that on the circumscribed cylinder immersed to the same depth.

3. Find the center of pressure of a circle immersed in water to any depth.

4. Find the form of the free surface of a fluid which rotates uniformly, in relative equilibrium, round a vertical axis.

A cylindrical jar whose weight is $\frac{1}{m}$th of the weight of water which it would contain, is filled $(1-\frac{1}{n})^{\text{th}}$ full and is then placed, mouth downwards, on a horizontal table which is made to rotate uniformly round a vertical axis coinciding with the axis of the jar. Prove that the angular velocity necessary to cause the fluid to escape is the same as if the jar weighed $\frac{1}{n}^{\text{th}}$ of the water it would hold and were $(1-\frac{1}{m})^{\text{th}}$ full; and find this angular velocity.

5. Investigate the conditions of stability, for small displacements, of a body floating in water.

A pyramid on a square base, whose other faces are equilateral triangles, floats in water with its vertex immersed and base horizontal, find the condition of stability. How will the stability be affected by tilting it round different axes?

6. Investigate the law of density of a vertical column of still air of uniform temperature.

Find the law of density on the hypothesis that the temperature diminishes in harmonical progression as the height increases in arithmetical progression, the variation of gravity in ascending being disregarded.

7. State the hypotheses upon which the equation of fluid motion $\frac{p}{\rho} = C + gz - \frac{r^2}{2}$ is founded; and prove the equation.

8. Define the component velocities at any point of a fluid in motion; and, in the case of motion in one plane, find an expression for the quantity of fluid which flows, in given time, in through the boundary of a circle of radius a whose center is at the origin.

9. Given a plane figure of any form; find the line round which it has the least moment of inertia.

The diagonals of a square plate being drawn, the two opposite triangles are cut out; find the principal axes and moments of inertia of the remaining figure, and the moments of inertia about each of the edges of the figure.

10. State D'Alembert's principle, and investigate any conclusions which can be drawn from it for the motion of a rigid body under no forces.

11. State and prove the principle of the convertibility of the centers of suspension and oscillation of a pendulum.

A pendulum is formed of two uniform rods of equal lengths, but of different materials and thicknesses, connected at one end so as to be in the same straight line. Their masses are m, m', and the axis of suspension passes through the middle point of m; find the time of oscillation of the pendulum.

12. State and prove the equation of Vis Viva.

A rod AB is capable of turning round A in a vertical plane, the other end being attached to an elastic string BC which is fastened to a fixed peg vertically above A, and such that $AC = AB$. The elasticity of the string is such that a weight equal to that of the rod would stretch it to three times its natural length AB. If the rod be started from its position of stable equilibrium with an angular velocity $= 2\sqrt{\dfrac{3g}{AB}}$, find the subsequent motion until the string becomes slack.

NOTE H.

ADMIRALTY CIRCULAR IN REGARD TO PRIVATE STUDENTS IN NAVAL ARCHITECTURE AND MARINE ENGINEERING.

A limited number of students unconnected with the naval service will be permitted to receive instruction at the Royal Naval College, in the course laid down for acting second-class engineers and dockyard apprentices.

The full course will be for three sessions, of nine months each.

The fee (payable in advance before entry) is £30 for each session, or £75 for the full course. Students who have already paid one fee of £30 will be allowed to compound for the next two sessions by a payment of £50 at the commencement of the second session.

Proportionate fees will be paid by students attending special classes only.

Students not connected with the naval service will reside outside the precincts of the college.

Facilities for visiting the royal dockyards will be offered to all private students, being British subjects.

Applications for admission should be addressed to the secretary of the Admiralty, Whitehall.

My lords reserve entire discretion in the selection of the candidates to be admitted.

ENTRANCE EXAMINATIONS.

Private students will be examined before entrance, in accordance with the programme laid down in the general regulations established for the admission of students to the Royal Naval College, as follows, viz:

1. The ordinary rules of arithmetic.

2. Algebra up to quadratic equations, the three progressions, the binomial theorem, and the theory of logarithms.

3. The subjects of the first four books of Euclid's Elements; proportion and similar figures, or the definitions of the fifth book and the proportions of the sixth book of Euclid's Elements.

4. The definitions and fundamental formulæ of plane trigonometry, including the solution of plane triangles. De Moivre's formula and its principal applications.

5. Elements of statics, dynamics, and hydrostatics.

6. Co-ordinate geometry, up to the equations of the conic sections.

7. Geometrical drawing.

ANNUAL EXAMINATIONS.

All private students will be examined at the end of each session. Certificates of proficiency in the various subjects they may have studied will then be awarded.

NOTE I.

(Page —.)

EXAMINATION PAPERS, GUNNERY SHIP EXCELLENT.

SUB-LIEUTENANTS.

(July, 1876.)

1.—Explain *fully* how to divide an arc into degrees.

What is the angle between the axis of the gun and the keel line, when the pointer on the slide coincides with the zero mark?

S. Ex. 51——20

2.—Show how to calculate the correction which has to be applied to the bearing at any particular gun in order that its fire may be directed on the same point as the center gun.

Give the form in which these corrections are tabulated.

Is this method absolutely correct for all bearings?

If not correct, express the error in terms of (d) the distance from, and (θ) the angle of training of, the center gun.

3.—What adjustments are necessary when placing the director in position? Explain fully how to test their accuracy.

4.—What are the essential points of the Fraser system of gun construction? Why is steel preferable to wrought, iron for the inner barrel?

5.—Draw a diagram of the 12″ 35-ton M. L. R. gun, distinguishing between the coiled, forged, and steel portions.

State generally how a double coil is formed.

6.—The deflection scale is marked to 30′, calculate the maximum speed for which it can be used with the 7″ M. L. R. gun when firing at an object distant 1,000 yards; the time of flight being 2.2 seconds.

7.—Define the following terms, and give diagram explaining:

> Line of fire.
> Line of sight.
> Trajectory.

What forces act upon a projectile *during* flight?

8.—What are the advantages of elongated over spherical projectiles?

9.—Why is it necessary to give rotation to elongated projectiles?

Upon what does the velocity of rotation required by a projectile depend?

10.—Describe the manner in which a charge of powder is consumed in the bore of a gun. How does the size of the grain affect the action?

11.—How is the perforating power of a projectile measured?

12.—Find the thickness of armor plate which can be penetrated by a projectile whose weight is 250 lbs. and diameter 8.92 inches, when moving with a velocity of 1,200 feet per second.

$$\text{Given }\ R = 3.138\ T^{\frac{3}{2}}$$

Where R = energy in foot tons per inch of shot's circumference.

Where T = thickness of plate in inches.

13.—Explain, with diagram, the general distribution of the armor in the *Hercules* and *Shannon*.

SUB-LIEUTENANTS.

(January, 1877.)

1.—In what plane should racers be laid? Give full reasons for your answer.

Explain how to calculate the correction which has to be applied to the bearing at each gun, except the center one, when concentrating a broadside.

Give the form of the table.

2.—The director is exactly over the center gun and the broadside is converged on the beam for a distance of 400 yards, but when the guns are fired the object is distant 200 yards. Will the shot from the center gun pass above or below the point aimed at, and at what approximate vertical distance?

Given axis of telescope above axes of guns = 10 feet.

> Elevation for 200 yards = 13′.
> Elevation for 400 yards = 30′.

3.—With reference to what plane is a gun given elevation when laid by scale, as in broadside firing—gun directing?

Explain, with a figure, why the heel scale is required in addition to the elevating scale when laying by director.

How would you test the accuracy of the elevating scales?

4.—Explain the meaning of the term "steel."

In what does it essentially differ from wrought iron?

What are its advantages and disadvantages as a gun material?

5.—Explain, with a diagram, the principal parts of a 9″ M. L. R. gun with two double coils.

Describe the vent bush with which the Woolwich guns are vented.

6.—A gun is laid, with the sight close down, for an object distant 120 feet. How far below the point aimed at will the shot strike? .

Initial velocity 1,200 feet per second.

Tip of center fire-sight to axis of gun, 25 inches.

7.—Does the axis of a rifled elongated projectile remain parallel to itself during flight? If it does not, give some idea of the motion, together with its cause.

8.—Compare the lengths of the dangerous spaces, with battering and full charges, when firing a 9-inch M. L. R. gun at an object of which the height is 20 feet and the distance 1,500 yards.

Given—

$$\text{Tan } \theta = \frac{\text{R tan } \overline{\varphi' - \phi}}{\text{R}' - \text{R}}$$

Where θ = Angle of descent.

R = Range = 1,500 yards.

R′ = R + 100.

ϕ = Elevation due to R = 2° 17′ for bat. charge.

3 16 for full charge.

ϕ' = Elevation due to R′ = 2 28 for bat. charge.

3 30 for full charge.

Explain, with a figure, the meaning of the term "dangerous space."

9.—When a charge of powder is exploded, are the products of combustion liquid, solid, or gaseous? And in what proportion by weight?

By which products is the shot propelled from a gun?

10.—How are combustion and ignition affected by the size of the grain and by the density and hardness of the powder?

11.—Compare the perforating powers of the following guns at 1,000 yards:

Gun.	Total energy at 1,000 yards.
9-in. M. L. R	2,648 feet tons.
8-in. M. L. R	1,837 feet tons.

12.—Compare the resisting power of a 6-inch plate, when fired at direct, with that of a 9-inch plate inclined at an angle of 60° with the line of fire.

13.—Describe generally the armored side of the Warrior.

SUB-LIEUTENANTS.

(April, 1878.)

1.—Explain fully how to divide any arc into degrees.

No. 1 of the 1st gun (5th gun being the center gun) in firing an electric broadside applies his correction the wrong way; where will the shot from his gun fall?*

2.—Explain how to test the accuracy of the racers and the director without using a spirit-level.

3.— What are the principal points of the English gun manufacture? Give a diagram of the 9-inch R.

4.—The deflection scale is marked to 30′; for what speed can it be used, distance of object 1,000 yards, time of flight 3″?

* Data for first question: distance of object 300 yards on the beam; distance between each pivot 20 feet.

5.—Define range, trajectory, line of sight, line of fire, angle of departure and descent.

6.—Compare the dangerous spaces when firing a 9-in. M. L. R. gun at an object 15 feet high and the distance 2,000 yards.

Charge.	Elevation.
Battering..	3° 17'
Full..	4° 30'

The angle of descent is about ¼ greater than the angle of elevation.

7.—The initial velocity of the projectile of the 10-in 18-ton gun is 1,364 f. s.; what will be its velocity at 1,500 yards?

Weight of shot, 400 lbs.; diameter of bore, 10 inches.

8.—Explain and describe the action of the crusher-gauge.

Did it register the best results with pebble powder or R. L. G. and why?

9.—Compare the perforating powers of the following projectiles:

Diameter.	Weight.	Velocity.
9.92 inches..................	400 lbs	1,118 f. s.
8.92 inches..................	250 lbs	1,236 f. s.

10.—Explain with a diagram the general distribution of the armor of the Triumph.

SUB-LIEUTENANTS.

(June, 1878.)

1.—In what plane is the director placed with reference to the racers and to the ship?

2.—Calculate the vertical and horizontal corrections for the starboard director.

Distance above center gun	20 feet.
Distance abaft center gun	80 feet.
Speed of ship	10 knots.
Distance of target.:.......................................	800 yards.
Time of flight...	1.8''
Correction of rifling.......................................	5'

3.—The charge of the Bacchante's guns has been altered from 14 lbs. to 22 lbs.; why does that alter the correction table?

4.—What metals are used in the construction of the Woolwich guns? Why are the Woolwich guns preferred to Krupp's?

5.—Give a short account of the manufacture of a 10-inch gun, giving a diagram.

6.—Define "angle of elevation," "line of sight," trajectory, and deviation.

The ship is heeling 10° to port and you are in charge of the bow gun, firing at an object right ahead, distant 1,000 yards; what error is introduced by the heel, and how would you correct it?

7.—Describe the armor and offensive powers of the Glatton. Is she different from the Hecate, and in what respects?

8.—Explain the meanings of the terms "tenacity," "elastic limit," "tensile strength," and how are they respectively measured as applied to metals?

9.—Give a description of the process by which cast iron is obtained from the ore. How may the cast iron so obtained be converted into steel?

10.—Explain by a diagram what is meant by a theoretically perfect gun.

SUB-LIEUTENANTS.

(July, 1878.)

1.—Explain what marks are placed on the racer and how it is done.

2.—How do you practically test whether the director is in adjustment with the guns?

3.—Work out the correction tables for a director 30 feet above and 60 feet to left of midship gun.

Correction for rifling...........................	5'
Speed of ship....................................	15 knots.
Distance of object...............................	500 yards.
Time of flight....................................	2''

4.—Define "line of fire," "line of sight," "angle of departure," "jump," and "trajectory." What forces act on the projectile after it leaves the gun?

5.—Describe how a double coil is made and give a diagram of the 10-inch gun.

6.—What is a vent like?

7.—Describe the armor and armament of the Belleisle.

8.—Define the terms "elastic limit," "malleability," "durability," and "tenacity," as applied to metals.

9.—What are the properties of "wrought iron," "steel," and "cast iron"? Give a definition of the term "steel."

Why is this metal used for the inner tubes of our heavy guns?

10.—How are the muzzle velocities of the projectiles of our service guns ascertained? The muzzle velocity of the 9-inch 12-ton gun is 1,420 f. s.; find the energy per inch of shot circumference at 1,000 yards:

Diameter of shot ....8.92 inches.
Weight of shot.... .. ..250 lbs.
Using Bashforth's tables.

GUNNERY LIEUTENANTS.

(July, 1876.)

FIRST PAPER.

1.—What is the difference between a statical and a dynamical strain? Explain the terms "malleability" and "weldability." Give a full explanation of the terms "elasticity" and "ductility"; and by means of a diagram show that of two metals, having equal "limits of fracture," that which possesses the greater ductility will absorb the larger amount of work before rupture.

2.—Distinguish between cast iron, wrought iron, and steel. Why is the classification which has reference to the proportion of carbon no longer admissible? State and explain the general effect of quick cooling on an iron casting.

3.—What are the general properties possessed by wrought iron? How is its structure affected by rolling into bars? Why is it unsuited for the inner barrel?

4.—For what material is the term "steel" now usually reserved? Give the leading points in the Bessemer process. What is the object of cooling in oil? State the arguments for and against steel for the exterior of guns.

5.—What are the principal strains to which a gun is subjected? Give an investigation of the ratio in which the tangential strain is transmitted. Explain that system of increasing the circumferential strength which is known under the name of the system of varying elasticity.

6.—Draw diagrams of the latest pattern 7″ and 10″ M. L. R. guns, distinguishing between the welded, coiled, and steel portions. Also draw a diagram of and explain the Woolwich groove. What is meant by the term "clearance"?

7.—Explain, with diagram, the form of chamber:
          1. In the earlier Woolwich guns.
          2. In the Fraser guns.

What is the reason for the difference? What alteration in this respect is now being tried in the 81-ton gun? Fully explain the object aimed at in the experiment, also any possible disadvantages which may arise.

8.—Describe the different kinds of vent bushes. Of what material are they made?

How are vents examined?

And when would a gun be condemned for reventing?

9.—A 10" M. L. R. gun was laid, with the sight close down, for the center of a target, of which the dimensions were 7 ft. 6 in. wide by 3 ft. 3 in. deep, the distance being 120 feet:

How far below the target did the shot pass?

Also how should the gun be laid in order that the shot may strike the center of the plate?

> Initial velocity 1,200 feet per second.
>
> Tip of center fore-sight to axis of gun 25 inches.

10.—What considerations determine the proof to which a gun should be subjected?

How are heavy guns proved?

What are the objections against—

> 1. A very heavy charge and service projectile?
>
> 2. A very heavy projectile and service charge?

11.—Give a brief outline of the phenomenon of explosion of gunpowder.

What is, according to Noble, the pressure when the powder entirely fills a closed vessel?

When fired from heavy guns, what is the limit laid down by the gunpowder committee?

12.—To which products of combustion is the work done on the projectile attributed?

13.—Give an outline of the methods adopted by the gunpowder committee to determine the action of gunpowder when fired from heavy guns.

Why were two descriptions of instruments necessary?

Questions set one month before the examination, and on which papers should now be handed in.

Give your opinion (supporting it, if possible, by the results of actual practice) on the following points:

The distance at which it would be desirable to engage, and the description of fire which it would be desirable to use, in single actions between unarmored ships.

Can you suggest any modifications in, or additions to, the descriptions of fire already laid down?

Or any alterations in the mode of carrying out the quarterly practice?

Could any experiments be usefully carried out to throw light on the question? If so, what?

GUNNERY LIEUTENANTS.

(July, 1876.)

SECOND PAPER.

1.—Define the following terms:

> Line of fire.
>
> Line of sight.
>
> Trajectory.
>
> Range.

2.—To what principal motions is an elongated projectile subjected in its passage through the air?

Explain the effect of the resistance of the air on elongated projectiles of the service form.

3.—Explain, with a figure, the great advantage of a flat trajectory.

What other advantages necessarily follow?

What points have to be considered in order to attain it?

4.—Investigate the following expression for the angle of descent:

$$\text{Tan. } \theta = \frac{R \text{ Tan. } (\phi' - \phi)}{R - R'}$$

And calculate the length of the "dangerous" space or error which can be allowed in estimating the distance, when

R = range = 1,500 yards.

R' = R + 100 yards.

ϕ = angle of elevation due to R = 2° 17'

ϕ' = angle of elevation due to R' = 2° 28'

5.—What is the effect of the following projectiles on the unarmored side of a ship of the Shannon class?

.65-inch Gatling bullet.

64-pdr. common shell.

64-pdr. shrapnell.

9-inch case.

6.—How are the perforating powers of guns compared in this country?

In what ratio do the resisting powers of armor plates vary when fired at direct?

7.—What properties are most desirable in the iron for armor plates?

What is the use of an iron skin?

Explain the peculiarity of the Palliser armor bolt.

8.—Explain, with diagrams, the general distribution of the armor in the Dreadnought, Inflexible, and Nelson.

9.—In what plane is it necessary to lay the deck on which a turret revolves?

Give reasons for your answer.

Explain the method of marking a turret in degrees.

10.—Show how to calculate a correction table for concentrating slide guns.

Investigate, in terms of (d) and (θ), an expression for the error arising from using the table when trained before or abaft the beam (d being the distance from the center gun and θ the angle of training).

Of what use are converging plates?

11.—Investigate an expression for the error in direction due to the sights being inclined, in terms of (θ) the inclination of the sights and (a) the angle of elevation.

12.—What adjustments are necessary when placing the director in position?

Explain fully how to test their accuracy.

13.—Investigate the horizon method of determining the distance of an object.

Are there any objections to its use?

If so, what?

GUNNERY LIEUTENANTS.

(February, 1877.)

FIRST PAPER.

1.—Explain the meaning of the terms "stress," "ultimate tensile strength," and "limit of elasticity."

2.—Distinguish between cast iron and wrought iron.

Give a brief description of a method of obtaining wrought iron.

3.—Define the term "steel."

Distinguish as clearly and fully as time permits between wrought iron and steel.

What have been hitherto the chief obstacles to an extended use of the latter?

4.—No possible thickness can enable a cylinder to bear a pressure from within greater on each square inch than the ultimate tensile strength of the material; prove this.

5.—Explain, with diagram, the principal parts of a 9" M. L. R. gun with two double coils.

Describe the bushes with which guns are vented.

How are vents examined?

6.—With the uniform twist compare the pressure required to produce rotation with that necessary to give translation.

Distinguish between uniform and increasing twist.

Which absorbs the most work?

7.—Describe the elevating gear proposed for the guns of the Inflexible.

What are its advantages?

Point out the disadvantages of the capstan-head elevating gear.

8.—Explain, with a figure, the forces which act on the discharge of a gun mounted on a carriage and slide; also point out the advantage of a low carriage.

9.—When a charge of powder is exploded, what are the proportions by weight of gases and solid products?

Give an outline of the method by which the volume of gas was determined.

10.—Establish a relation between the pressure and density when a charge of powder is exploded in a closed vessel.

11.—Calculate the velocity of a projectile whose weight is 180 lbs. when fired from an 8″ M. L. R. gun :

Given length of bore .... 14.8 calibers.

Charge .... 35 lbs.

Gravimetric density of powder 1

Factor of effect 78.4

Number of volumes of expansion.	Total work in foot tons that the powder is capable of realizing per lb. burned.
5.0000	83.53
5.8824	89.35
6.2500	91.46
6.6667	93.64

12.—What is the meaning of the term "factor of effect"?

Explain fully the object of "air-spacing" the charge, and of chambering the gun.

Fourteen days allowed for this question.

Give your opinion about the mounting, working, and use of machine guns—

1. On board the ship.

2. In the boats.

3. In landing operations.

GUNNERY LIEUTENANTS.

(April, 1877.)

FIRST PAPER.

1.—Explain the meaning of the terms *elasticity* and *elastic limit*, as applied to metals. How are they respectively measured?

2.—Summarize the properties of wrought iron.

3.—Explain the meaning of the term *steel*.

4.—In cylinders of metal, the power exerted to resist pressure from within by the different parts varies inversely as the square of the distance of the parts from the axis; prove this.

5.—The angular velocity imparted to a projectile depends upon the length of twist and the muzzle velocity of translation; prove this.

6.—Explain, with diagram, the principal parts of a 64-pdr. Mark III gun.

What considerations determine the length of the bore of a gun?

7.—Calculate the angle of deflection corresponding to a speed of 14 knots, when firing at an object distant 1,000 yards, time of flight 2.5 seconds.

Does the amount of deflection required to counteract the effect of the wind vary with the distance? Explain your answer.

8.—Find the mean resistance offered by the compressor to the recoil of 9-pdr. 8-cwt ⋅ gun—

    1. When the recoil is .... 3 feet.
    2. When the recoil is .... 1 inch.
    Given the weight of projectile.................... 9¼ lbs.
    Initial velocity of ditto......................... 1,380 feet.
    Weight of carriage............................... 275 lbs.

9.—State the conditions which must be satisfied by a good elevating arrangement.

0.—Explain fully how a projectile is propelled from a gun.

11.—Calculate the velocity of a projectile whose weight is 89 lbs. when fired from a 64-pr. M. L. R. gun:

    Given the caliber............................. 6.3 inches.
    Length of bore............................... 15.5 calibers.
    Charge .... 12 lbs.
    Gravimetric density.......................... 1.
    Factor of effect .... 76.

Number of volumes of expansion.	Total work in foot tons that the powder is capable of realizing per lb. burned.
8.3333.....	101.00
9.0909	103.82
10.0000.....	106.87
11.1111	110.18

12.—What means have been successively adopted to reduce the pressure in the bore? And how is it that it has been possible to combine increased velocity with a lower maximum pressure?

Question set on April 6, and papers given in at 9 a. m. on 5th May.

Examine the question of the armament of the unarmored steel corvette Mercury, whose displacement is 3,700 tons. It is supposed that she will be classed as a fifth-rate, that her complement will be 250, and that the weight allotted to armament is 112 tons.

GUNNERY LIEUTENANTS.

(April 1, 1878.)

FIRST PAPER.

1.—Explain the terms, "tenacity," tensile strength," "limit of elasticity" as applied to metals. How are they respectively measured?

2.—Give a description of the manufacture of a steel tube for a heavy gun from the time the ingot leaves the manufacturer's hands until it is ready for the B tube. How is the temperature at which the A tube shall be toughened determined?

3.—Explain the meaning of the term "initial tensions and varying elasticities" as applied to gun construction.

How is the principle carried out—

        (a) at Woolwich.
        (b) by Whitworth.
        (c) by Palliser.
        (d) by Rodman.

4.—Enumerate the general properties of wrought iron, and give a description of the process at Woolwich by which it is obtained from obsolete cast-iron material.

5.—What is meant by the term "jump" of a gun? A 9-pounder was laid accurately horizontal for a wooden target 200 feet distant; the height of the level of the axis of the gun was marked on the target; on firing the gun, the shot struck 10 inches above this level. The mean velocity was found to be 1,370 f. s.: Calculate the angle of departure or the "jump" of the gun.

6.—How is the velocity of rotation of a projectile measured

Find approximately the number of revolutions per second and also the angular velocity of the projectile of the 7-inch 6½-ton gun at the muzzle:

    Muzzle velocity........................... 1,525 f. s.
    Caliber .... 7 inches.
    Twist .... 1 in 35 calibers.

7.—What means have been adopted to ascertain the pressure, volume of gas, temperature of explosion, and products of combustion, when a charge of powder is exploded in a closed vessel?

8.—What is the actual work realized by 110 lbs. of B. powder, in the 12-inch 35-ton gun; length of bore=16.5 calibers; gravimetric density of powder=1; factor of effect 93.1. Hence find the muzzle velocity of projectile; weight 700 lbs.

Number of volumes of expansion.	Total work that gunpowder is capable of realizing per lb. burnt, in foot-tons.
6. 2500	91. 45
6. 6667	93. 64
7. 1429	95. 94
7. 6923	98. 39
8. 3333	100. 00

9.—What is considered the limiting angle of penetration of our service projectiles? Give approximately the percentage of work lost on impact, owing to the conversion of work into heat in the case of projectiles made of—

    (a) Wrought iron.
    (b) Cast iron.
    (c) Hard-tempered steel.

In recent experiments, what material appears to recommend itself most as a metal for projectiles?

10.—State briefly what experiments have been made in England with gun-cotton shells, and with what results?

11.—What method was adopted by the Italian commission in their recent experiments at Spezia, to determine the velocity of shock actually necessary for perforation?

If V = velocity of shock on impact with target (as observed)
v = velocity of exit after perforation
Show that
Velocity of shock actually necessary just to perforate $= \sqrt{V^2 - v^2}$

GUNNERY LIEUTENANTS.

(April, 1878.)

SECOND PAPER.

1.—Define, with diagrams where necessary, the following terms: "Angle of fire," "angle of descent," "line of sight," "terminal velocity," "high angle fire."

2.—Describe the effect of the resistance of the air combined with the rotation of projectiles of the service form.
What theories are advanced to explain the deviation of these projectiles to the right?

3.—Explain, with diagram, the principle of the Boulengs chronograph, and describe how a velocity is obtained with it.
What advantage has it over the Bashforth chronograph?

4.—An experiment is made with a Bashforth chronograph to ascertain a velocity at one of the screens distant (x) from the muzzle of the gun. The time (t) is end off from the diagram on the cylinder, and also the time of passing the intermediate screens, which are a known distance (h) apart.
Show how an equation may be formed, giving the velocity required.

5.—The velocity of the projectile of the experimental 16-in. 80-ton gun was found to be 1,480 f. s. at 400 yards from the muzzle of the gun. Calculate its initial velocity, using Bashforth's tables.

Weight of shot ....1,700 lbs.
Diameter of shot ....15.92 inches.

6.—Assuming that the second differences of the times between the Bashforth screens are constant, and that the resistance of the air varies as the (velocity)³. Prove Helie's formula:

$$v = \frac{V}{1 + c \, V \, x}.$$

When v = velocity at any point.
V = muzzle velocity.
x = distance from muzzle in feet.
c = constant, depending on form, weight, and velocity of shot.

7.—What influence has the form of the base of a shot on the total resistance offered to it by the air?
Calculate the resistance offered by the air to the motion of an "ogival-headed shot."

w = weight of projectile ....180 lbs.
v = velocity ....1,200 f. s.
d = diamemter of shot ....7.92 inches.
k for that velocity ....109.5.

8.—In what plane should the racers be laid?
Explain how you would test the accuracy of the director and racers.
State the advantage of the—
        (1) Correction plates.
        (2) Correction tables.
        (3) Both plates and tables.

9.—Calculate the total horizontal correction to be applied to the director when placed on the starboard side.

d = distance abaft center gun.................. 60 feet.
R = range..................................... 800 yards.
t = time of flight .... 2 seconds.
s = speed .... 10 knots.
θ = permanent angle of deflection.............. 1° 10'.
ϕ = angle of elevation........................ 1° 15'.

Why will this table be inaccurate for the port director?

10.—The charges of the guns of the Bacchante have been altered from full to battering; why is it necessary to alter the correction tables for her director?

11.—State the advantages of the Dreadnought over the Thunderer, as regards her offensive and defensive strength.
Give a diagram of the former's armor plating.

NOTE J.

WRITTEN EXAMINATION.

FRENCH COMPOSITION.

(June 11, 8 to 10 a. m.)

Siége de Gergovie par César.

Vercingétorix n'ayant pu sauver Avaricum, la capitale des Bituriges, ni empêcher César de passer l'Allier, était allé s'établir à Gergovie, la capitale des Arvernes, pour laisser aux conjurés le temps de se réunir et de s'organiser avant de livrer une bataille décisive aux envahisseurs de la Gaule.

Gergovie était une place très-forte, assise sur une hauteur et entourée presque de tous côtés d'une ceinture de montagnes dout elle était séparée par une plaine étroite.

Vercingétorix avait rassemblé sur ce point des forces nombreuses qui, couronnant toutes les hauteurs, dominaient entièrement la plaine.

César enleva un des plateaux qui faisaient face à la ville, et lui livra plusieurs attaques, mais ne pouvant engager l'ennemi à une bataille et impatient d'obtenir un succès pour prévenir la défection de ses alliés, il tenta une surprise et fit donner un assaut; il laisse entendre dans ses Commentaires qu'il éprouva un échec considérable. Suétone avoue sans détour que les Romains furent repoussés avec des pertes énormes.

LATIN.

(June 11, 1 p. m. to 2.30 p. m.)

Cæsar, ut Brundisium venit, contionatus apud milites, quoniam prope ad finem laborum ac periculorum esset perventum, æquo animo mancipia atque impedimenta in Italia relinquerent, ipsi expediti naves conscenderent, quo maior numerus militum posset inponi, omniaque ex victoria et ex sua liberalitate sperarent, conclamantibus omnibus, imperaret, quod vellet, quodcumque imperavisset, se æquo animo esse facturos; pridie nonas ianuarias naves solvit, impositis, ut supra demoustratum est, legionibus septem. Postridie terram attigit Cerauniorum. Saxa inter et alia loca periculosa quietam nactus stationem, et portus omnes timens, quos teneri ab adversariis arbitrabatur, ad eum locum, qui appellatur Palæste, omnibus navibus ad unam incolumibus milites expósuit.—(Cæsar, de bello civili, lib. iii, cap. 6.)

ENGLISH VERSION.

(June 11, 3.15 p. m. to 4.15 p. m.)

On voyait la côte et le sinistre cap de Trafalgar qui a donné son nom à la bataille. Un vent dangereux commençait à se lever, la nuit à devenir sombre, et les vaisseaux anglais, manœuvrant difficilement à cause de leurs avaries, étaient obligés de remorquer ou d'escorter dix-sept vaisseaux prisonniers. Bientôt le vent acquit plus de violence, et aux horreurs d'une sanglante bataille succédèrent les horreurs d'une affreuse tempête; comme si le ciel eut voulu punir les deux nations les plus civilisées du globe, les plus dignes de le dominer par leur union, des fureurs auxquelles elles venaient de se livrer.

ARITHMETIC AND GEOMETRY.

(June 12, 8 a. m. to 11 a. m.)

1.—State and explain the theory of the periodical fractions and apply to $\frac{7}{11}$ and $\frac{43}{71}$.

2.—Prove that periodic fractions, derived from irreducible fractions of the same denominator, have the same number of figures in a period. Take as an example $\frac{7}{11}$ and $\frac{43}{71}$.

3.—Prove that the expression for the volume generated by a circular segment is $\frac{1}{2} D^2 H$; D being the chord and H its projection upon the diameter. As an application inscribe a right cone in a sphere such that its volume shall be one-half of the spherical segment in which it is inscribed.

DESCRIPTIVE GEOMETRY.

(June 12, 1 to 2.30 p. m.)

A point situated in the first dihedral angle is situated 5 cm. from the horizontal plane and $4\frac{1}{2}$ cm. from the vertical plane.

This point is the center of a regular hexagon, whose plane is parallel to the horizontal plane, and one of whose sides, parallel to the vertical plane, is 3 cm. in length·

This hexagon is the common base of two regular pyramids, of which one has its vertex in the horizontal plane, and the vertex of the other is situated at a distance of 7 cm. from this plane. It is required to construct the shadow of this pyramid upon the planes of projection, knowing that there exists in the first dihedral angle a source of light which sends its rays parallel to a right line, making an angle of 19° with the horizontal plane and 33° with the vertical plane.

ALGEBRA AND TRIGONOMETRY.

(June 13, 8 to 11 a. m.)

1. Find the maxima and minima values of the function

$$y = \frac{2x^2 - 8x + 8}{x^2 - 5x + 4}.$$

Examine the variation of this function for all values of x from $-\infty$ to $+\infty$, and trace the corresponding curve.

2. Resolve the triangle ABC, having given

$$a = 3875.475 \text{ m.}$$
$$b = 4637.095 \text{ m.}$$
$$c = 6143.877 \text{ m..}$$

Determine the surface in hectares.

ORAL EXAMINATIONS.—PARIS.

NOTE.—Three sets of questions have been selected from a large number given at the Paris examination. They have been taken at random, and they are fair examples of the questions at all the centers of examination.

EXAMINATION OF CANDIDATE A.

1.—ARITHMETIC.

1.—Let $A = a^p b^q c^r$, a number of which, a, b, and c, are the prime factors. Find the number of divisors of A.

2.—In how many ways can the number A be decomposed into two factors which are prime to each other ?

3.—Explain the rule of simple interest.

2.—GEOMETRY OF SOLID BODIES.

4.—Give the theory of symmetrical figures.

5.—Determine the volume generated by a segment of a circle which revolves about an axis passing through the center of the circle.

6.—The continuous trace of an ellipse. The length of the *radii vectores* passing through the extremity of the shorter axis. Draw a tangent to the ellipse (1) through a point on the curve (2) through an exterior point.

3.—ALGEBRA.

7.—Give the sum of the terms of a geometrical progression.

8.—What does the formula $s = \frac{lq-a}{q-1}$ become when q becomes equal to 1 ?

9.—Give the formula for computing interest.

10.—How long is it necessary that a capital, A, should remain at interest, at a given rate, in order to become C ?

11.—Investigate the sum of the squares of two numbers, x and y, whose sum is a constant.

12.—Define a maximum and a minimum.

4.—TRIGONOMETRY.

13.—Determine x in the equation

$$5 \sin x + 3 \cos x = 7.$$

14.—Is there a rapid solution of the following general equation?

$$a \sin x + b \cos x = C.$$

5.—DESCRIPTIVE GEOMETRY.

15.—Having given two right lines, of which one is perpendicular to the horizontal plane and the other to any line whatever, it is required to find the shortest distance between them.

16.—Are there several methods of finding the shortest distance between two right lines?

6.—FRENCH.

DICTATION.—Ce fut dans cette place qu'il conçut le dessein de détrôner le roi de Pologne par les mains des Polonais mêmes. Là étant un jour à table, tout occupé de cette entreprise, et observant sa sobriété extrême, dans un silence profond, paraissant comme enseveli dans ses grandes idées, un colonel allemand, qui assistait à son dîner, dit assez haut pour être entendu, que les repas que le tsar et les rois de Pologne avaient faits au même endroit, étaient un peu différents de ceux de Sa Majesté.—(Voltaire.)

Continue the reading of the extract with explanations.

7.—GEOGRAPHY.

Draw a map of the basin of the river Po, describing its northern tributaries.

8.—HISTORY.

1.—The Visigoths: Their origin; their manners and customs; their wars.

2.—The government of Saint Louis: Give an account of everything bearing on the subject.

9.—ENGLISH.

1.—ENGLISH VERSION.—

Un astrologue un jour se laissa choir
Au fonds d'un puits; on lui dit: "Pauvre bête,
Tandis qu'à peine à tes pieds tu peux voir,
Penses-tu lire au-dessus de la tête?"

2.—Dictation at the blackboard of a passage from Irving's Columbus, with explanation.

3.—Give the meaning of the verb *get;* name its derivatives.

Give the meaning of *next.*

10.—LATIN.

1.—Translate and explain Book I, chapter 49, of Cæsar's Gallic War, from *Ubi eum. castris se tenere* to *ad eum locum venit.*

2.—Explain *delegit, intellexit.* Give the French noun and adjective derived from the verb *prohibere.* Explain *acie.*

11.—GREEK.

Translate Book I, chap. 10, of Xenophon's Anabasis from Ἐνταῦθα δὴ Κῦρον to στρατόπεδον.

Give the French derivatives of Κεφαλή.

EXAMINATION OF CANDIDATE *B.*

1.—ARITHMETIC.

1.—Show that when a number divides the product of two factors and is prime to one of them, it divides the other.

2.—A number, N, has as many divisors above the square root as below. If the number of divisors is odd, N is a perfect square.

3.—Extract the square root of a whole number.

2.—GEOMETRY OF SOLID BODIES.

1.—Determine the volume generated by a triangle turning about an axis which passes through one of its vertices.

2.—Prove that two triangular pyramids of equivalent bases and of the same altitude are equivalent.

3.—Two tetrahedrons are similar when they have equal dihedral angles comprised between two similar faces.

4.—The property of the tangent to a parabola. Draw a tangent (1) through a point on the parabola; (2) through an exterior point.

5.—The generation of a helix and the property of a tangent to this curve.

3.—ALGEBRA.

1.—Discuss the equation

$$ax^2 + px + c = 0.$$

What value do x' and x'' approach when x approaches 0?

2.—Determine the rectangular parallelopiped of maximum volume having a given surface.

4.—TRIGONOMETRY.

Resolve a triangle having given two sides and the angle opposite one of them.

5.—DESCRIPTIVE GEOMETRY.

1.—Find the angle between two planes whose traces are in a given right line.

2.—Determine the intersection of two planes.

3.—How many lines are there perpendicular to the ground-line?

6.—FRENCH.

DICTATION.—Un fanfaron, amateur de la chasse, venant de perdre un chien de bonne race, qu'il soupçonnait dans le corps d'un lion, vit un berger. "Enseigne-moi, de grâce, de mon voleur," lui dit-il, "la maison, que de ce pas je me fasse raison." Le berger dit, "C'est vers cette montagne. En lui payant de tribut, un mouton par chaque mois, j'erre dans la campagne comme il me plaît, et je suis en repos."
Continue the reading of the extract with explanations.

7.—GEOGRAPHY.

1.—Give a sketch of the French shore of the Channel.

2.—Draw a map of Russia. Give an account of its physical geography and of the cities in the southern part of the country.

8.—HISTORY.

1.—Give an account of the form of government and of the legislation of the Teutonic nations that invaded the empire in the fifth century. State the principal differences between the Teutonic and Roman laws. Give an account of the Salic law.

2.—Name the first six or seven kings of France. Give the dates and duration of their reigns.

3.—Give an account of the enfranchisement of the communes, stating the causes which led to it. Why were there no communes in the royal domain?

9.—ENGLISH.

ENGLISH VERSION.—Les montagnes voisines étaient convertes de pampres verts qui pendaient en festons; le raisin, plus éclatant que la pourpre, ne pouvait se cacher sous les feuilles et la vigne était accablée sous son fruit.

Dictation at the blackboard of a passage from Irving's Columbus, with explanations.

Explain *rather*. Give the plural of *gulf*. Give the rules for the formation of the plural of nouns. Name the nouns in *f* which take *ve* in the plural. Conjugate *standing*.

10 —LATIN.

Translate Book I, verses 740–746, of the Æneid.

Explain *geminosque Triones*. Give the derivation of (the French word) *septentrion*; the gender, number, and case of *cythara*.

Give the construction and force of *docuit*, *canit*, *pecudes*. Decline *imber*.

11.—GREEK.

Anabasis, Book I, chapter 6, from Οὗτος Κύρῳ to διαγγεῖλαι. Decline ἵππος, ἱππέας, βασιλεῖ. Inflect the last word.

EXAMINATION OF CANDIDATE C.

1.—ARITHMETIC.

1.—Two irreducible fractions having the same denominator have the same number of decimal figures in the period when the fractions are reduced to a decimal form. Example: $\frac{7}{17}$ and $\frac{13}{17}$.

2.—Give the rule of simple interest.

Derive the formula $a = \frac{A\,i\,t}{100}$, and resolve the equation, taking A and t successively as the unknown quantities. Find these two unknown quantities by analysis.

2.—GEOMETRY OF SOLID BODIES.

1.—The volume generated by a triangle turning about an axis passing through one of its vertices. Place upon the surface of the triangle the perpendicular let fall upon the axis through the center of the side opposite to the axis. Determine the position of the axis in order that the volume may be a maximum.

2.—The definition of an ellipse.

3.—How is a tangent to the curve to be drawn (1) through a point on the curve (determine the locus of the foot of the perpendicular let fall from the tangent from a focus), (2) through an exterior point.

3.—ALGEBRA.

1.—What is to be understood by a polynomial, rational and entire in x?

2.—How would you ascertain whether this polynomial is divisible by $x - a$?

3.—In what cases is $x^m - a^m$ divisible by $x + a$?

4.—Decompose the function
$$y = ax^2 + bx + c$$
into two factors of the first degree.

5.—Find the sum of the first n odd numbers.

4.—TRIGONOMETRY.

1.—What does the fraction
$$\frac{x \sin x}{1 - \cos x}$$
become when $x = 0$?

2.—Show how a right line may be determined beyond an obstacle. Suppose, for example, that one is placed at a point A in AC, whence AC may be readily determined. Between C and D is an intervening obstacle. How may DB, the prolongation of AC, be determined?

—Descriptive Geometry.

Find the projections of the intersection of two given planes.

6.—French.

Dictation.—Auguste aima mienx alors recevoir les lois dures de son vainqueur que de ses sujets. Il se détermina à demander la paix au roi de Suède, et voulut entamer avec lui un traité secret. Il fallait cacher cette démarche au sénat qu'il regardait comme un ennemi encore plus intraitable.—(Voltaire.)
Continue the reading with explanations.

7.—Geography.

1.—Draw a map of the Spanish peninsula. Give an account of its physical geography, and fix the position of the principal cities.

2.—Give the political history of the county of Venaissin, and describe its physical characteristics.

8.—History.

1.—Give an account of the Ostrogoths and their origin. Narrate the events of the reign of Theodoric, noting the system of government, the legislation, and the extent of Theodoric's kingdom.

2.—Give a full account of the foundation of the Kingdom of the two Sicilies.
The Norman conquest of Sicily.

9.—English.

1.—English translation.—On exposait une peinture
Où l'artisan avait tracé
Un lion d'immense stature
Par un seul homme terrassé.
Les regardants en tiraient gloire.
Un lion en passant rabattit leur caquet:
"Je vois bien, dit-il, qu'en effet,
On vous donne ici la victoire,
Mais l'ouvrier vous a déçus;
Il avait liberté de feindre;
Avec plus de raison nous aurions le dessus,
Si mes confrères savaient peindre."

2.—Dictation at the blackboard of a passage from Columbus.

3.—Construction and force of the words *no, one, however, indulge, unknown, how, show, blow, height.*

10.—Latin.

Translate Cæsar, B. G., I. 47, from *Biduo post Ariovistus* to *aliquem ad se mitteret,* giving explanations of allusions. Give the construction and force of *perfecta, agi, iterum;* mode and tense of *constitueret, vellet;* force of *cœptæ.* Decline *aliquem, is, hic, suum.*

11.—Greek.

Translate Xenophon, Ἀνάβασις, I, 4, § 11, from "Εντεῦθεν ἐξε αυνει to ὀνόματι.
Conjugate and give construction of ἐξελαύνει. Decline ιῦρος, με; ἀλη.

S. Ex. 51——21

NOTE K.

COURSE IN ENGLISH NAUTICAL LANGUAGE ON BOARD THE BORDA.

[Extracts.]

A ship is said to be sailing before the wind when she has the wind right aft. Both sheets are then hauled aft, the sails are set at right angles with the keel; the staysails are hauled down, and the mainsail is drawn up in the brails, that the wind may act upon the foresail; the helm is put amidships, and the only thing the steersman has to do is to put the helm a little to starboard or larboard, in order to keep the vessel from yawing.

COMMANDS IN TACKING.

Ready about, or Ready all—*Pare à virer.*
Haul over the boom—*Bordez le gui.*
Helm alee gently—*Labarre dessous en douceur.* } About ship—*Envoyez.*
Ease off; or, Let go the jib sheets—*Filez les focs.*
Up tacks and sheets—*Lève les lofs.*
(Stand by, or Get to the) after braces—*Aux bras de derrière.*
Haul mainsail, haul—*Derrière changez.*
Belay—*Amarrez.*
(Stand by the) head braces—*Aux bras de devant.*
Haul off, haul; Let go and haul—*Devant changez.*
(Stand by the) head bowlines—*Boulines devant.*
　　　　after bowlines—　　　*derrière.*
　　or head and after bowlines—*Boulines partout.*
Haul—*Holez;* Belay—*Amarrez.*
Haul aft; or, Sheet the courses; or } *Bordez les basses voiles.*
Fore and main sheets.
Haul taut the weather braces—*Appuyez les bras du vent.*
Sheet the jibs; or　　　　} *Bordez les focs.*
Haul aft the jib-sheets.
Clear or coil the ropes—*Parez les manœuvres.*

A compound engine is one that uses the same steam at both high and low pressure, the object being to take as much elastic force as possible out of the steam before condensing it.

Gun-cotton is easily exploded by percussion or ignition, and any weight of this preparation is a good deal more effective than the equivalent weight of gunpowder. It is well-suited to the working and blowing up of mines, but is not used for military purposes, on account of the facility with which it may be ignited by friction, percussion, and heat.

NOTE L.

SYLLABUS OF LECTURES ON MARINE ENGINEERING.—BORDA.

FIRST YEAR.

I.

Elementary study of the action of a marine engine.

1.—Steam.—Definitions; saturated steam, dry and wet steam, superheated steam; formation, superheating, and condensation of steam. Pressure. Latent heat of steam. Instruments for measuring the pressure and vacuum. Bourdon's manometer. Vacuum indicator. Density of dry saturated steam.

2.—Elementary explanation of the action and working of steam in a cylinder.—Essential parts of an engine. Introduction, exhaustion; absolute pressure; back pressure; effective pressure. Condensing and non-condensing engines. Condenser, vacuum. Back pressure in the cylinder compared with the pressure in the boiler. Double-acting engine; single-acting engine; atmospheric engine. Mechanical work. Kilogrammeter; horse-power. Work of a force acting in the direction of the motion, or otherwise. Graphic representation of the work of a variable or constant force. Mean effort of a variable effort. Moving force, resisting force, and effective force of steam in a cylinder. Advantages of the condenser. Effective power of engines in the different conditions of working.

3.—Transmission of movement from the piston to the shaft.—Connecting rod and crank. Dead centers; travel of the piston from one dead center to the other; top and bottom of a cylinder; names of the dead centers and of the strokes of the piston. Simultaneous positions of the crank, piston, and connecting-rod; half-stroke. Angularity of the connecting-rod; its influence. Summary notions of the correlative movements of the piston and the shaft.

4.—Elementary explanation of the regulation of engines.—Expansion: change of motion of the piston. Limited admission; leads; compression; clearance. Graphic representation of the effective work of the steam in the cylinder, taking into account the adjustment of the valve-gear. Advantage of the exhaust-lead.

Curves of expansion. Graphic construction of the general equation $PV^x = P_0 V_0 =$ constant. Particular case where $x = 1$; that is, the case most often met with. Theoretical expression of the work of steam during expansion. Advantages and inconveniences of expansion. Action of the steam-jacket.

II.

Complete description of a screw-engine. Classification of the principal systems.

1.—General description of an elementary screw-engine.

2.—Mode of action of detail parts.—Transmission of movement. Distribution of steam. Jet and surface condensers. Advantages of surface condensers. Boiler-feeding apparatus. Bilge-pump.

3.—Classification of marine engines, principal systems.—Classification. Geometrical description. Beam engine. Oscillating engine. Direct-acting engine. Back-acting engine. Trunk engine.

III.

Complete exposition of the distribution of the steam by the slide-valve or by separate cut-off gear.

1.—The eccentric.—Its working. Description of the transmission of movement by a fixed eccentric; by eccentric with variable angular advance. Example of system of hooking-on gear.

2.—Slide-valves.—Definitions. Valve-faces; valve-seats. Two kinds of slide-valve. Elementary description of the working of a three-ported slide-valve. D-valve; explanation of its working. Comparison of the working of the two kinds of slide-valves.

Adjustment of a slide-valve; stroke; attachment of the valve-stem. Angle of lead. Lap. Total area of ports. Results of the adjustment of a slide-valve. Steam-lead. Greatest opening of the ports. Wire-drawing steam. Exhaust-lead. Compression. *Résumé* of the functions of the slide-valve.

3.—Mechanism of the reversing-gear.—Principle of the reversement of motion. Means of carrying out this principle. Geometrical study of the reversing mechanism. First system: Single eccentric, with variable advance: (1) by means of a slip-eccentric; (2) by means of an adjustable eccentric. Mazeline's system.

Second system : Two eccentrics, with fixed advance : (1) drop-hook motion ; system of Creusot ; (2) by link-work. Stephenson's link.

4.—Variable cut-offs.—Variable cut-offs driven by cams ; by an adjustable eccentric, the cut-off valve being wide open at the end of a stroke ; at half-stroke. Cut-off driven by a fixed eccentric, the stroke of the expansion-valve being variable, and the valve opening wide at the end of a stroke ; at half-stroke.

IV.

Steam-generators.

1.—Division and classification.
2.—Ordinary rectangular boiler, tubular boiler, return tubular boiler, low-pressure boiler. General description of the accessory parts. Fire-tools. Working of a boiler. Principal dimensions of return tubular boilers, two types. Rectangular boiler. Arrangement of boilers.
3.—Details of construction of boilers and their parts.—Mode of connecting the fixed pieces ; bolts, viz, tap-bolts, stud-bolts, dowels, rivets. Boiler-plates ; mode of assemblage. Angle irons.

Dead plates, grates, bearer-bars, lugs, bridge-walls. Diagonal bracing, longitudinal and bridge-bracing. Man-hole plates. Tubes ; their fastening on the tube-sheet, ferules ; beading : movable tubes. Langloir's system. Gautelme's system. Toscer's system. Fixed and movable pipes. Natural draught produced by the smoke-pipe. Forced draught.

Check-valves. Glass water-gauge ; putting the glass tube in place. Gauge-cocks Blow-off cock. Safety-valve. Dimensions and load of a safety-valve. Safety-valve for small boilers at high pressure. Atmospheric valve. Stop-valve. Communicating-valve. Dry pipes and superheaters.

Pipes and valves. Joints with fixed and movable flanges. Soldered and riveted flanges. Slip-joint. Single-way, two-way, and four-way cocks. Hollow-plug cock. Sea-cocks. Kingston valve.
4.—Cylindrical boilers.—Return tubular and high-pressure boiler. General description of boiler with appendages. Arrangement of brace-tubes. Method of joining the boiler-plates. Principal dimensions.
5.—Short description of the principal types of boilers in use.—Tubular boiler. Side-flue boiler. Cylindrical boiler for launch. Belleville boiler for launch ; working of this boiler. Details of feed regulator.

V.

Management of engines and boilers.

1.—Working the engine.—Preparations for starting. Blowing through and turning over. Direct blowing through of the condenser. Starting. Increasing or reducing speed. Final stop. General care to be given to the engine while in motion.
2.—Fuel and combustion.—Calorific power. Evaporative power. Air of combustion. Space occupied by a given weight of coal. Manner of burning of the fuel. Conditions of good combustion. Imperfect combustion. Smoke, soot, clinker ; slag.

General character of coal ; coal, properly so called ; rich and poor coals ; hard or compact coals. Anthracite. Lignite. Conglomerate or pressed fuel. Briquettes.
3.—Management of fires.—Priming the furnaces. Lighting. Care of the fires while the engine is in motion. Increasing and moderating fires. Banking fires. Forcing fires. Hauling fires.
4.—Feeding and blowing.—Filling the boilers. Keeping up a constant level. Feeding during a stop. Salts in solution in sea-water. Action of the salt in the boiler.

Deposit. Measure of density; salinometer; testing salinometer. Effect of deposits; means of prevention; blowing off. Losses occasioned by it. Regulation of the blowing off. Emptying the boilers.

5.—Pressure; production of steam.—Keeping up the pressure. Increasing and diminishing pressure. Coal consumed, (1) in terms of the evaporation per given weight of coal; (2) per unit area of grate in a given time.

6.—Accidents.—Foaming: causes; effects; remedies; means of prevention. Priming: causes; effects; remedies; means of prevention. Leaks: their consequences. Plugging a leaking tube. Dangerous lowering of the water level in the boilers; measures to be taken.

7.—Cold testing of boilers: object and methods.

SECOND YEAR.

I.

Principal types of ordinary engines in use, of Woolf's or compound engines, and of boat engines.

Oscillating paddle engine. Compound inverted-cylinder engine. Woolf's three-cylinder back-acting engine. Inverted-cylinder non-condensing launch engine. Silent launch engines.

II.

Description and details of various parts of the machinery.

1.—The cylinder and its appendages.—Cylinder; valve-seat; ports; bottom; cover. Clearance. Ratio between the stroke of the piston and the diameter of the cylinder. Steam-jacket; lagging.

Cylinders and steam-passages of Woolf engines. Continuous-expansion engine. Compound engines, with cranks at 90°. Three-cylinder engines, with cranks at 90° and 135°. Piston-rod stuffing-boxes for horizontal and vertical cylinders. Laying up the packing. Stuffing-box capable of adjustment while under way. Cylinder relief-valves. Blow-through cock. Ordinary balance-pressure lubricator. Thibant's lubricator. Roscoët's lubricator.

2.—Steam piston, packing, rods.—Spider of piston; piston-rings; follower. Anti-friction piston-rings. Piston-rods; fitting of the rod in the piston and in the cross-head. Trunks; their adjustment with the piston.

3.—Slide-valves.—Review of the classification. Ordinary three-ported slide-valve. Object of equilibrium packings. Box-valve on Mazeline's system. Double-ported slide-valve. Dupuy de Lôme's long D-valve. Rectangular long D-valve of Indret. Relative merits of different methods of distribution.

4.—Starting and reversing gear.—Review of the classification. Mazeline's starting gear. Stephenson's link; explanation of the action of the link.

5.—Valves: regulators and apparatus of variable expansion.—Throttle-valve. Stop-valve. Throttle-valve in boat engines.

Governors: two classes. Adaptability for marine engines. Conditions that must be fulfilled by a governor used in a marine engine. Farcot's governor: description, conditions of equilibrium, and mode of action. *Servo-moteurs :* geometric description of a *servo-moteur* applied to an engine.

Apparatus for variable cut-off; their use. Review of the classification. Expansion by means of gridiron-valve and adjustable eccentric. Meyer's expansion-valve. Expansion by means of a butterfly-valve moved by a fixed eccentric of variable throw. Expansion-valve regulated by a link.

6.—The condenser and its parts.—Analysis of the processes of condensation by injection and of surface condensation. Weight of the injection water; temperature of jet condensation. Weight of cold water in surface-condensers; temperature *i*

area of condensing surface. Advantages and disadvantages of surface condensers and reasons for employing them.
Form and volume of jet-condensers. Sea-cock and injection-pipe. Injection-valve. Bilge-injection.
Air-pump; description; volume for maintaining a proper vacuum in a given condenser. Description of lifting air-pump. India-rubber valves. The Indret jet-condenser for a screw-engine. Double-acting air-pump. Hot-well; relief-valve; outboard-delivery valve. Discharge-pipe.
Complete description of a surface-condenser for an engine of the Indret type. Method of action of the surface-condenser. Circulation of the steam and of the cold water. Methods of fixing the tube in the tube-sheet. Distilling-apparatus for fresh water.
7.—Transmission of movement.—Mazeline's cross-head. Cross-heads of the Indret type.
Connecting-rod, type Indret, with club ends. Mazeline's connecting-rod, stub end. Cross-head guides, cross-head, and connecting-rod for an inverted cylinder engine.
Mazeline's crank-shaft bearings. Bearings of the "Forges et Chantiers" type. Crank-shaft; angle between the cranks. Valve-gear counter-shaft.
Siphon lubricator. Lubricator with valve. Crank-pin and wrist-pin lubricators.
8.—Apparatus for feeding and for pumping out bilge.—Feed-pumps. Donkey engine; its valve chamber.
Giffard's injector; its principle. Description of the Giffard feeding apparatus.
Behrens's rotary engine.
Bilge-pumps. Jet-bilge pump. Friedmann's ejector. Centrifugal pump.

III.

Propellers.

1.—Paddle-wheels.—Geometrical description; mode of action. Causes of loss of power; slip. Elements of paddle-wheels. Position and number of wheels. Advance and slip. Rolling circle. Different systems of paddle-wheels; wheels with feathering paddles.
2.—The screw.—Definitions. Mode of action. Causes of loss of power; slip. Elements. Position and number of screws. Advance and slip. Fixed and variable pitch. Classification and description of various screws; type Mangin. Screw with spherical hub and bent-back blades. New model of screw.
Line of shafting. Cardan's coupling and hand-turning gear. Movable coupling and break. Stern-pipe and screw-shaft. Thrust-bearing. Stern stuffing-box. Outboard bearing of screw-shaft.

IV.

Regulation, work, and employment of marine engines.

1.—Curves representing the motions of the slide-valve.—Representation of the simultaneous movement of the piston and slide-valve; curves of piston-speeds. Motion curve of a 3-ported slide. Variation of the angle of lead. Relative positions of piston and slide-valve. Motion curve of a D slide-valve. Determination of the action of a slide-valve by means of the motion of any given point in the valve.
Determination of the elements necessary for a graphic representation of the motion of a slide-valve. Determination of the dead centers of the piston and slide. Determination of the lap. Operations necessary for determining the relative position of the piston and slide-valve.
Drawing of a sketch of the arrangement and position of the ports. Analysis of the operations of the slide-valve. Drawing of a D-valve. Adjustment of the

valve-stem and angle of eccentric by means of a motion-diagram. Case of a link. Observations relative to the valve-diagram for a backing motion.

2.—Work of the steam in the engine.—Theory of the indicator. Primitive indicator of Watt. Drawing of a diagram from the analysis of the working of the slide-valve. Measurement of the pressure by means of the indicator-diagram. The surface of the diagram represents the effective work of the steam upon a square centimeter of the surface of the piston.

Garnier's indicator, new model. Richards's indicator.

Fitting the indicator on an engine. Adjustment of the indicator. Disarrangement of the indicator. Representation of an indicator-curve. Tracing of the atmospheric line.

Calculation of the mean effort on the pistons. Mean vacuum in the cylinder; correction by means of the barometric pressure. Calculation of the power of an engine by the mean effort found by the indicator-curves. Nominal, indicated, and actual horse-power. Work on the shaft. Old formula for nominal horse-power.

Calculation of the consumption of steam by the indicator-curve; common method. Labrousse's method.

3.—Regulation and working of compound engines.—Definitions and classification. Continuous-expansion engine. Method of producing expansion. Effective admission and expansion. Actual expansion; indicator curves.

Compound engine with cranks at 90°. Method of producing expansion. Curves of volumes, and indicator-curves.

Three-cylinder compound engine, with cranks at 90° and 135°. Expansion apparatus. Curves of volumes and indicator-curves. Adjustment of the slide-valves according to the angle of the cranks.

Advantages and disadvantages of Woolf engines. Calculation of the effective horse-power of Woolf engines by means of indicator curves. Actual mean effort; fictitious mean effort. Calculation of the consumption of steam.

4.—Efficiency of engines.—Analysis of indicator-curves under various conditions. Reduction of pressure between the boiler and the cylinder, and between the cylinder and the condenser. Contracted passages and their employment with wet steam. Undulations of curves. Leakage past the slide-valve. Leakage past the piston. Insufficient steam-lead. Working with low pressure and late cut-off. Working at low pressure with early cut-off. Three-ported slide-valve, with valve-stem too short. Angular advance of eccentric too great and valve-stem two long. Length of indicator-cord badly adjusted.

Determination of average horse-power and speed at trials of engines. Relation between the speed of the ship, the indicated power of the pistons, and the slip of the propelling instrument. Coefficient of speed. Consumption of coal per indicated horse-power per hour. Consumption of coal per square meter of grate-surface per hour. Efficiency of combustible. Loss of heat in engines.

V.

Manipulation peculiar to each of the principal types of engines in actual use, including boat engines.

1.—Maneuvers.—Preparations for starting. Heating the engine; blowing through and turning over. Starting; stopping; backing. Regulating the speed; increasing and moderating speed.

2.—Care of the engine while in use.—Cylinders, slide-valves, and distribution of steam. Condensation. Movement of the machinery. Feed-pumps and bilge-pumps.

3.—Accidents to the engine.—Heating of bearings. Dangerous pounding. Leaks. Heating of the condenser. Filling up of the condenser. Air-leaks. Leaks in the condenser-tube packing.

4.—Management of boat engines.

VI.

Injuries of engines and boilers.

1.—Boiler explosions.—Two kinds, by rupture and by bursting; causes and means of prevention. Escape of water by the safety-valve. Explosive combinations in the flues.

2.—Spontaneous combustion of coal. Causes. Indications of the heating of coal; precautions to be taken. Fire in the coal-bunkers.

3.—Injuries to the cylinder; measures to be taken for working the engine with the parts left intact; engine with simple cut-off, Multiple-cylinder continuous-expansion engine. Woolf engine, with cranks at 90°. Three-cylinder engine, with cranks at 90° and 135°.

NOTE M.

BILL OF FARE.

Menu des repas des élèves pour la journée du 18 juin 1878 et pour le lendemain matin.

DÎNER.

Potage au vermicelli.
Bœuf aux choux.
Petits pois au beurre.
Fraises.

SOUPER.

Mouton rôti.
Pommes de terre (maître d'hôtel).
Compote de prunes.

DÉJEUNER.

Café au lait.
Beurre.

NOTE N.

PROGRAMME OF EXAMINATIONS.

SCHOOL OF MACHINISTS, FRENCH NAVY.

CANDIDATES FOR FIRST MASTERS.

CHAPTER I.

Arithmetic, algebra, geometry, plane trigonometry.

Square and square root of whole numbers.
Square and square root of fractions.
Cube of whole numbers and fractions.
System of weights and measures.

Relation between the metric system and the English system of weights and measures. Ratio and proportion.

Arithmetical and geometrical progression.

Theory of logarithms. Use of tables.

Elementary processes of algebra.

Solution of equations of the first degree with one or more unknown quantities.

Solution of equations of the second degree with one unknown quantity.

Properties of right lines and of angles.

Equality of triangles.

Properties of parallel lines, rectangles, and polygons; of the circle; properties of chords, secants, and tangents.

Measurement of angles. Relative position of two circumferences.

Proportional lines. Similar polygons.

Problems of plane geometry.

Regular polygons. Measurement of plane surfaces; of the circumference and of the area of a circle.

Parallel and perpendicular planes and right lines. Dihedral angles.

Properties of prisms and pyramids. Superficial area and volume. Volume of the frustum of a prism, parallelopiped, and pyramid. Superficial area and volume of a cylinder, of a cone, of the frustum of a cone, and of a sphere.

Solid contents of a coal-bunker.

Definitions of the principal trigonometric lines.

Solution of right triangles.

CHAPTER II.

Mechanics and physics.

Elementary notions of matter; inertia, motion, velocity.

Forces; weight and density of bodies. Measurement of force. Mass and its determination.

Composition of concurrent forces. Decomposition of a force into two others acting in any given direction. Composition of parallel forces.

Centers of gravity. Practical determination of the center of gravity of any body whatever. Determination of the centers of gravity of geometrical surfaces and the principal solids.

Work of forces; its graphical representation and measure. The kilogrammeter and the horse-power.

Principle of the transmission of work in the case of a uniform motion: Application to the equilibrium of simple mechanics. Passive resistances. Necessity of regulating the motion of machines. Fly-wheels. The efficiency of a machine.

Action and equilibrium of the lever; of pulleys. The differential pulley. Action of the connecting-rod, of the crank, of the eccentric, and of cams. Action, equilibrium, and drawing of parallel and conical toothed wheels. Action and equilibrium of the screw; of the endless screw. Watt's parallelogram. Equilibrium of the inclined plane and of the wedge, taking into account the effect of friction; of the winch, and of the windlass or capstan. The differential windlass.

Strength of materials.

Equality of pressure of fluids. Calculation of the pressure exerted upon a given surface.

Air, atmospheric pressure. Different methods of determining pressure in engines. Vacuum, and methods of determining it. Construction and use of the barometer.

Effects produced upon bodies by the increase or diminution of heat. Construction and use of the thermometer. Expansion and contraction of metals. Precautions to be taken in consequence of the expansion and contraction of metals in the construction, erection, repair, and management of engines. Shrinkage, tempering, annealing. Expansion of fluids. Particular effects of the action of heat on water.

Measurement of heat. Calorific capacity or specific heat of bodies, latent heat. Propagation of heat. Effects of surfaces of various colors and of polished surfaces. Good and bad conductors. Heating of liquids by circulation. Means of preventing loss of heat, and the overheating of furnace doors and chimneys. Principle of transformation of heat into work, and *vice versa*. Mechanical equivalent of heat. Carnot's theorem.

Fundamental principles of boilers, combustion, furnaces.

Gases: expansion, tension. Mariotte's and Gay-Lussac's laws.

Theory of the siphon.

Theory and description of the suction-pump; of the force-pump, and the double-acting pump; of centrifugal and rotary pumps.

Hydraulic press.

Evaporation, vaporization, boiling, conversion into steam. General properties of steam. Saturated steam, heated and superheated steam. Spheroidal condition of water. Bourdon's manometer. Compressed-air manometers.

Condensation of steam. Problems relating to jet and surface condensers. Principle of the condenser. Action of single and double acting air pumps. Hot-well. Measurement of condensation; vacuum gauge. Different forms of condenser-barometers.

CHAPTER III.

Theory and description of engines.

Fixed and variable cut-offs; their use, advantages, and inconveniences.

Theory and action of compound engines.

Theory of slide-valves; lap; angles of lap and lead; relations between the movement of the slide-valve and that of piston; steam and exhaust lead; fixed cut-off; compression. Theory of variable cut-offs. Reversement of motion; different methods of producing it.

Classification of marine engines, according to the method of using the steam, the mode of transmission of movement of piston to shaft, and according to the kind of propeller. Advantages and disadvantages of the different kinds of engines.

Necessity of a number of cylinders; equalization of movement.

Description and working of an oscillating engine; of a back-acting engine, ordinary and compound; of a trunk-engine. Brotherhood's engine. Principal types of hoisting machines.

Properties and use of metals and other substances employed in the construction of engines.

Description of slide-valves; locomotive and double-ported slide-valves; box-valves; D and piston valves, long and short. Fitting slide-valves in oscillating engines.

Principal starting-gear; Stephenson's link; system of Creusot; systems of Mazeline and of Dupuy de Lôme; systems of oscillating engines.

Description of the principal systems of variable cut-off.

Principal systems of throttle-valves of marine engines.

Action and description of the different kinds of surface-condenser and of distilling apparatus.

Bed-frame, foundation-plates, framing and flooring of engines. Lubricators.

The screw and its elements. Advance and slip. Different types of screws. Various arrangements for carrying the screw and screw-shaft. Fitting the screw upon the shaft. Description of a line of shafting. Stern stuffing-box and thrust-bearing. Paddle-wheels and their parts; different systems. Advance and slip.

Classification of marine boilers in relation to their pressure, their interior arrangement, and their shape. Complete description of a rectangular tubular boiler. Description of a cylindical tubular boiler of the authorized pattern. Detailed description of a Belleville generator. Complete system of pipes of a system of boilers.

Plates used in the construction of boilers; their assemblage. Furnaces, ash-pits, fire-bars, bridge walls, combustion-chambers, and smoke-boxes. Metal fortubes. Fixing of stationary or movable tubes in the tube-plate. Bracing, diagonal stays; their necessity; their disposition. Bridge-bracing.

Different systems of chimneys and their jackets. Steam communication; stop-valves· Dryers and superheaters. Safety-valves, and their working and weights. Escape-pipes; water-gauges, gauge-cocks; man-holes, mud-holes, and other accessories of boilers. Cocks. Pipes in general. Testing of marine boilers. Feeding of boilers; feed pumps and valves. Bilge-pumps, principal types. Giffard's injector. Ejectors. Auxiliary feeding-engines or donkey-engines. Behrens's system.

Principle of *servo-moteurs* : various types.

CHAPTER IV.

Management of engines.

Properties of combustibles. Quantity of combustible necessary to evaporate a certain weight of water. Combustibles used in the navy. Wood and different kinds of coal. Occasions for wetting the coal. Arrangement of fuel on the fire-grate. Thickness of the layer.

Lighting. Treatment of fires while under way. Manner of stoking. Forcing and easing the fires. Banking fires. Arrangements for heating with wood. Care to be given to chimneys and their stays. Fire in the chimney.

Cleaning furnaces and grate-bars. Removing ashes and clinkers. Sweeping the tubes while under way. Hauling fires. Preventing the entry of cold air.

Filling up the boilers. Keeping up a constant water level. Precautions to be taken relative to feeding, while in motion ; before and during a stoppage. Details of Belleville's generator. Dangerous lowering of the level of water in the boilers, and the measures to be taken.

Salts in solution in sea-water. Concentration and saturation. Salinometer; its construction and graduation.

Saline deposits in boilers; means of prevention. Blowing out the boiler; precautions. Continuous blowing off. Estimating the quantity to be blown off. Heat lost by blowing off.

Fatty deposits in the boilers. Apparatus for the removal of fatty substances from feed-water. Saponification of fats.

Various causes of the augmentation and diminution of pressure in boilers. Keeping up, increasing, and reducing the pressure. Depression below atmospheric pressure.

Disuse of a boiler at sea. Precautions to prevent its collapsing. Starting fires in a fresh boiler. Case when the two preceding operations are done simultaneously. Care to be given to boilers after the fires are out. Manner of emptying them. Modification of the number of boilers used in passing from one speed to another.

Causes of foaming and of priming. Means of prevention.

Leaks in the boiler and piping. Consequence of leaks in connection with feeding and blowing out.

Blowing through and turning over. Starting: precautions to be taken. Different cases where the engine does not work.

Accelerating or slackening speed; case where the partial closing of the steam-valve is preferable to varying the cut-off. Stopping. Reversing.

Adjustment of moving parts; lubrication. Various noises. Thumping. Heating. Binding.

Leaks in the engine; means of discovery and remedy.

Choking up of condensers and obstructions to injection water. Particular care in the management of surface condensers.

General precautions to be taken in regard to the apparatus while in motion, before and after arrival, before, during, and after engagement. Distribution of the personnel for getting under way, mooring, while under way, and during action.

Method of utilizing the machinery in case of fire or springing a leak.

CHAPTER V.

Care and repair of engines.

Care to be taken for the maintenance and preservation of boilers and tubing; removal of saline deposits; different processes in use.
Preservation of the machinery and propeller during long periods of disuse.
Description and use of the diving apparatus.
Measures to be taken in case of an accident to the engine or boilers. Approximate calculation of the number of cubic meters of steam, at atmospheric pressure, that can escape from a given boiler in case of explosion or rupture. Injury to cylinders, their heads, and stuffing-boxes. Provision for continuing work when a cylinder is disabled, particularly in compound engines. Injuries to the steam-piston and piston-rod; to valves; to valve-motion and eccentrics; to pipes and steam-valves; to the condensers and injection apparatus, particularly in surface condensers; to air pumps, their bonnets, piston-rods, valves and guards; to the hot-well and discharge pipes; to the cross-heads and keys, connecting-rods, link-work, and gearing; to the shaft, propeller, and cranks; to the foundation plates and framing, their straining in heavy weather; injuries to bearings, their caps and brasses; to the screw, the bearings, and the stern stuffing-box; to the paddle-wheels and their floats; to the boilers—burnings, cracks, collapsing. Replacing of a rivet, a stay-bolt, or a plate. Bursting of tubes; plugging or replacing them. Broken gauge-glasses. Damaged pressure-gauges. Accidents in the smoke-stacks. Accidents peculiar to the Belleville generator.
Accidents to the pipes, cocks, and valves of the boiler; to the blowing-out and feeding apparatus; to the bilge and donkey pumps. Rupture and explosion of boilers; their immediate causes; precautions to be taken. Inflammable mixtures in the flues and coal-bunkers. Spontaneous combustion of coal in the bunkers; measures to be taken.

CHAPTER VI.

Erection of engines.

Putting in place all the stationary parts of a screw-engine with direct-acting connect-ing-rod, a back-acting engine, or a trunk engine (at the option of the candidate), and fixing the engine in the ship. Putting in place all the movable parts of one of the above engines; complete verification of the erection of one of the above engines, and specially of the line of shafting. Putting in place the fixed parts of an oscil-lating engine, and the fixing of this engine in the ship. Putting in place all the movable parts of an oscillating engine. Complete verification of the erection of this engine. Lining up the intermediate and paddle shafts. Lining up crank and pro-peller.
Putting the boilers on board. Their erection upon the keelsons, with or without floor-ing. *Servo-moteurs.*

CHAPTER VII.

Regulation of the work of engines.

Regulation in general of the valves and variable expansion gear; relative adjustment of the valves of a compound engine. Rectification of the point of attachment of the valve-stem and of the variable cut-off; also that of the position of the eccentrics.
Theory of distribution and expansion valve diagrams. Showing by means of curves the motion of the slide-valve. Use of this diagram. Erection and verification of distribution and expansion valves by means of the regulation diagrams.
Description and use of the indicator. Indicator connections of the different types of engines. Atmospheric line; precautions to take to trace it. Tracing and analysis

of an indicator diagram in an ordinary or a compound engine. Calculation of the work upon the piston by means of the indicator diagram; mean effort determined by this measurement; case of an ordinary or a compound engine. Determination of the effective force on the shaft. Determination of the nominal force of ordinary and compound engines at high and at low pressure. Value of nominal horse-power in kilogrammeters. Calculation of the expenditure of steam. Determination by the indicator diagram of the necessary elements for this calculation. Comparison between the pressures as shown by the indicator and the real pressures of the boiler and condenser. Comparison between the pressures shown by the indicator in the small and large cylinders of a compound engine during their communication. Forms of indicator diagram in the case of running with variable cut-off, with an almost complete closing of the throttle, in case of high-pressure engine, of leaks, of priming, of bad arrangements of the indicator, &c. Comparative forms of indicator diagrams, of small and large cylinders in compound engines.

Approximate measurement of the work of air-pumps. Calculation of the thrust of the propelling instrument when the work upon the piston and the speed of the vessel are known.